GeoPlanet: Earth and Planetary Sciences

The GeoPlanet series is a forum for presenting the latest achievements in the Earth and space sciences. It is published by the GeoPlanet consortium (Earth and Planetary Research Centre) formed by five institutes affiliated with the Polish Academy of Sciences: Institute of Geophysics, Space Research Centre, Institute of Geological Sciences, and Institute of Oceanology, and Nicolaus Copernicus Astronomical Centre. Its main objective is a multidisciplinary approach to link scientific activities in various Earth-related fields (geophysics, geology, oceanology) with Solar System research. Our publications encompass topical monographs and selected conference proceedings, authored or edited by leading experts of international repute as well as by promising young scientists. The GeoPlanet series aims to provide the stimulus for new ideas and discoveries by reporting on the state of the art and laying the foundations for the future development of the Geosciences.

More information about this series at http://www.springer.com/series/8821

Monika B. Kalinowska ·
Magdalena M. Mrokowska ·
Paweł M. Rowiński
Editors

Recent Trends in Environmental Hydraulics

38th International School of Hydraulics

Springer

Editors
Monika B. Kalinowska
Institute of Geophysics
Polish Academy of Sciences
Warsaw, Poland

Magdalena M. Mrokowska
Institute of Geophysics
Polish Academy of Sciences
Warsaw, Poland

Paweł M. Rowiński
Institute of Geophysics
Polish Academy of Sciences
Warsaw, Poland

The GeoPlanet: Earth and Planetary Sciences Book Series is in part a continuation of Monographic Volumes of Publications of the Institute of Geophysics, Polish Academy of Sciences, the journal published since 1962 (http://pub.igf.edu.pl/index.php).

ISSN 2190-5193 ISSN 2190-5207 (electronic)
GeoPlanet: Earth and Planetary Sciences
ISBN 978-3-030-37104-3 ISBN 978-3-030-37105-0 (eBook)
https://doi.org/10.1007/978-3-030-37105-0

This Springer imprint is published by the registered company Springer Nature Switzerland AG
The registered company address is: Gewerbestrasse 11, 6330 Cham, Switzerland

Series Editors

Managing Editor

Anna Dziembowska
Institute of Geophysics, Polish Academy of Sciences

Advisory Board

Kon-Kee Liu
Institute of Hydrological
and Oceanic Sciences
National Central University Jhongli
Jhongli, Taiwan

Teresa Madeyska
Research Centre in Warsaw
Institute of Geological Sciences
Warszawa, Poland

Antonio Meloni
Instituto Nazionale di Geofisica
Rome, Italy

Evangelos Papathanassiou
Hellenic Centre for Marine Research
Anavissos, Greece

Kaja Pietsch
AGH University of Science
and Technology
Kraków, Poland

Dušan Plašienka
Prírodovedecká fakulta, UK
Univerzita Komenského
Bratislava, Slovakia

Barbara Popielawska
Space Research Centre
Polish Academy of Sciences
Warszawa, Poland

Tilman Spohn
Deutsches Zentrum für Luftund
Raumfahrt in der Helmholtz
Gemeinschaft
Institut für Planetenforschung
Berlin, Germany

Krzysztof Stasiewicz
Swedish Institute of Space Physics
Uppsala, Sweden

Ewa Szuszkiewicz
Department of Astronomy
and Astrophysics
University of Szczecin
Szczecin, Poland

Roman Teisseyre
Department of Theoretical Geophysics
Institute of Geophysics
Polish Academy of Sciences
Warszawa, Poland

Jacek Tronczynski
Laboratory of Biogeochemistry
of Organic Contaminants
IFREMER DCN_BE
Nantes, France

Steve Wallis
School of the Built Environment
Heriot-Watt University
Riccarton, Edinburgh
Scotland, UK

Wacław M. Zuberek
Department of Applied Geology
University of Silesia
Sosnowiec, Poland

Piotr Życki
Nicolaus Copernicus Astronomical
Centre
Polish Academy of Sciences
Warszawa, Poland

Preface

Sustainable development, even more—survival of human beings, strongly depends on aquatic ecosystems, which are exposed to human impacts as they, for example, ultimately receive all kinds of waste from agriculture, industrial and technological production. There is also much evidence of poor management of river systems, being the result of poor understanding of the processes that occur in river streams, particularly when flow–biota, flow–vegetation and geomorphic settings are taken into account. Despite huge progress in the field of environmental hydraulics in recent years, investigations on freshwater ecosystem functioning still require further research investments. The issues discussed in the present volume analyse some emerging trends in environmental hydraulics, hoping to contribute towards an improved understanding of the functioning of rivers and other aquatic systems. This book embraces studies carried out in the field, in laboratory flumes as well as computational investigations, often representing truly interdisciplinary research methodologies.

Together with the authors who contributed to these articles, we sincerely hope our readers will enjoy these papers, which are the result of the debate led during the 38th International School of Hydraulics. This school, a biannual affair, is traditionally held in Poland, this time in 21–24 May 2019 at the Dębowa Góra Hotel situated near Płock, at the picturesque lake surrounded by the forest greenery. This school has been attended by scholars from all over the world, from thirteen countries. All papers were presented during the school, thoroughly discussed there, and all of them were peer-reviewed prior to the school and if needed, also after the event. The invited speakers at the meeting were **Aronne Armanini** from University of Trento, Italy; **Alessandra Crosato** from IHE Delft Institute for Water Education, The Netherlands; **Subhasish Dey** Indian Institute of Technology Kharagpur, India; **Yafei Jia** from National Center for Computational Hydroscience and Engineering, The University of Mississippi, USA; **Juha Järvelä** from Aalto University, Finland;

Artur Magnuszewski from University of Warsaw, Poland; and **Wojciech Majewski** from Institute of Hydroengineering of Polish Academy of Sciences, Poland; world-class experts in the field and that guaranteed very high level of all the undertaking.

Warsaw, Poland

Monika B. Kalinowska
Paweł M. Rowiński
School Chairs

Acknowledgements

The editors are extremely grateful to all people who contributed to the organisation of the 38th International School of Hydraulics and the production of this volume.

We would like to express our thanks to all the authors for their contributions, and all the participants and speakers during the 38th International School of Hydraulics for sharing their knowledge and passion with others and engagement in all conference events.

The school and this book would not be possible without financial support from the Institute of Geophysics Polish Academy of Sciences statutory funds and Committee of Water Resources of the Polish Academy of Sciences. We want to express our thanks to the Institute of Geophysics Polish Academy of Sciences, CIFAL Płock, City Mayor of Płock and BMsonic company for being a sponsor of the prizes in the competition: "Best Young Researchers Presentation Award". We were also glad to be able to run the conference under the auspices of International Association of Hydro-Environment Engineering and Research, as well as under the Honorary Patronage of the City Mayor of Płock. Very special thanks go to Artur Magnuszewski for his invaluable help in organising a study trip and also to the BMsonic—a Polish company for being with us during the conference and for organising an extraordinary event—a hot air tethered balloon flights during the conference.

We would like to express our cordial thanks to all the people from the local organising committee who helped make this conference a success. They are Hanna Baczyńska, Emilia Karamuz, Michael Nones and people from Mazurkas Congress and Conference Management company. We should also name Anna Dziembowska responsible for the high quality of English of the papers, and Karolina Branicka for all help while preparing the papers for this volume.

Finally, we wish to thank and acknowledge all the reviewers, who have donated their time and expertise to assist in improving the quality of the submitted papers:

- **Jochen Aberle**, Technical University Braunschweig, Germany
- **Mário J. Franca**, IHE Delft Institute for Water Education, The Netherlands
- **Ian Guymer**, The University of Sheffield, UK

- **Monika B. Kalinowska**, Institute of Geophysics, Polish Academy of Sciences, Poland
- **Emilia Karamuz**, Institute of Geophysics, Polish Academy of Sciences, Poland
- **Leszek Książek**, The University of Agriculture in Krakow, Poland
- **Bartłomiej Luks**, Institute of Geophysics, Polish Academy of Sciences, Poland
- **J. Russell Manson**, The Richard Stockton College of New Jersey, USA
- **Andrea Marion**, University of Padova, Italy
- **Magdalena M. Mrokowska**, Institute of Geophysics, Polish Academy of Sciences, Poland
- **Jarosław Napiórkowski**, Institute of Geophysics, Polish Academy of Sciences, Poland
- **Michael Nones**, Institute of Geophysics, Polish Academy of Sciences, Poland
- **Tomasz Okruszko**, Warsaw University of Life Sciences, Poland
- **Marzena Osuch**, Institute of Geophysics, Polish Academy of Sciences, Poland
- **Artur Radecki-Pawlik**, Cracow University of Technology, Poland
- **Renata Romanowicz**, Institute of Geophysics, Polish Academy of Sciences, Poland
- **Paweł M. Rowiński**, Institute of Geophysics, Polish Academy of Sciences, Poland
- **Michał Szydłowski**, Gdansk University of Technology, Poland
- **Steve Wallis**, Heriot-Watt University, UK
- **Kaisa Västilä**, Aalto University School of Engineering, Finland

Monika B. Kalinowska
Magdalena M. Mrokowska
Paweł M. Rowiński

Contents

Turbulent Length Scales and Reynolds Stress Anisotropy in Wall-Wake Flow Downstream of an Isolated Dunal Bedform

Subhasish Dey and Sankar Sarkar

Abstract This experimental study brings the turbulent length scales and the Reynolds stress anisotropy into focus in wall-wake flow downstream of an isolated dunal bedform. The results reveal that wall-wake flow downstream of an isolated dune possesses a high-turbulence level having its peak value at the dune crest level. This level lasts up to the vertical distance until the effects of the dune on the flow disappears. Regarding the turbulent length scales, downstream of the dune, the Prandtl's mixing length in wall-wake flow is greater than that in undisturbed upstream flow, while the Taylor microscale and the Kolmogorov length scale are smaller. In Reynolds stress anisotropy analysis, the anisotropy invariant maps demonstrate the data plots form a looping trend in wall-wake flow. Below the dune crest, the turbulence is characterized with an affinity to show a two-dimensional isotropy, while above the crest, the anisotropy has a tendency to reduce to a quasi-three-dimensional isotropy.

Keywords Dune · Turbulent flow · Open-channel flow · Turbulent length scales · Reynolds stress anisotropy

1 Introduction

Natural turbulent streams often come across bed-mounted solid obstacles of different geometries that play an important role in modifying flow characteristics. These obstacles are often called *bluff-bodies* in applied hydrodynamics. The presence of a bed-mounted bluff-body in a streamflow generates wakes at its downstream, called the *wall-wake flow*. Depending on the geometry of a bluff-body and the magnitude of streamflow velocity, the wake flow downstream of the bluff-body sustains up to a

S. Dey (✉)
Department of Civil Engineering, Indian Institute of Technology Kharagpur, Kharagpur, West Bengal 721302, India
e-mail: sdey@iitkgp.ac.in

S. Sarkar
Physics and Applied Mathematics Unit, Indian Statistical Institute, Kolkata, West Bengal 700108, India

M. B. Kalinowska et al. (eds.), *Recent Trends in Environmental Hydraulics*, GeoPlanet: Earth and Planetary Sciences,
https://doi.org/10.1007/978-3-030-37105-0_1

certain distance until the undisturbed upstream velocity profile is eventually recovered. Owing to the importance of bluff-bodies in practical purposes, the wake flow downstream of a bluff-body remains a topic of immense importance in applied hydrodynamics (Bose and Dey 2016, 2018). Schlichting (1979) was the pioneer to study the wall-wake flow downstream of a circular cylinder developed by the approach shear-free flow, providing a similarity theory for the velocity defect profile in the free wake flow. However, the wall-wake flow downstream of a bed-mounted obstacle under an approach wall-shear flow is more complex being different from the free wake flow. Limited studies have been carried out to understand the turbulence characteristics downstream of bed-mounted bluff-bodies in both near- and far-wake under a shear flow. To be specific, Balachandar et al. (2000) and Tachie and Balachandar (2001) argued that the similarity for velocity profiles in the wall-wake downstream of a bed-mounted plate could be preserved even in the near-wake flow zone by using the appropriate velocity and length scales. Shamloo et al. (2001) studied the effects of hemispherical objects in characterizing the streamwise velocity and Reynolds shear stress in an open-channel flow. Downstream of a hemisphere, they observed a recirculation zone, which disappeared after a downstream distance of about twice the hemisphere diameter. Kahraman et al. (2002) observed that the effects of localized wall-roughness on shallow wake flow downstream of a bed-mounted circular cylinder are not only confined to the near-wall flow zone, but also govern the entire turbulence in a shallow wake flow covering up to the half of the flow depth. Some researchers were also interested to study the vortex shedding in wall-wake flows. Akilli and Rockwell (2002), Ozturk et al. (2008) and Ozgoren (2006) studied the near-wake flow downstream of a bed-mounted vertical circular cylinder by particle image velocimetry (PIV). Akilli and Rockwell (2002) observed that a large-scale Kármán vortex that contains upward ejections through its core is developed and ultimately becomes a streamwise vortex, inducing a distortion to the free surface. Ozturk et al. (2008) studied the flow field upstream and downstream of a bed-mounted vertical circular cylinder within the boundary layer involving the interaction among the primary and the trail of the horseshoe vortices developed at the upstream cylinder base. On the other hand, Ozgoren (2006) studied the generation of vortical structures and the turbulence quantities downstream of a bed-mounted vertical circular cylinder under a uniform flow. They reported that the length of vortex formation decreases as the Reynolds number increases. Lacey and Roy (2008) studied the turbulent flow downstream of a submerged pebble cluster and observed a dominance of longitudinal-vertical vortex shedding in the wake of the cluster. Sadeque et al. (2009) considered the bed-mounted cylinders as bluff-bodies and found that the flow far from the bed is approximately similar and can be described by the plane wake flow equation. Further progresses in understanding the wake flow characteristics downstream of bluff-bodies have been recently made by Dey et al. (2011, 2018a, b, c), Sarkar and Dey (2015a, b).

Considering an isolated dunal bedform as a bluff-body and studying the flow characteristics downstream of it, inadequate attention has so far been paid to the best of the authors' knowledge. Importantly, the existing knowledge of flow over an isolated dune is not sufficient in predicting the flow over it, because most of the cases were studied for the flow over a series of dunes. For instance, McLean and Smith (1986) and Nelson and Smith (1989) studied the turbulent flow field over the dunes and observed the existence of separation zone, reattachment point, wake flow zone, internal boundary layer, etc. Maddux et al. (2003a, b), who experimentally studied the turbulent flow over a series of two- and three-dimensional dunes observed that the three-dimensional dunes possess different turbulence characteristics from the two-dimensional ones. Based on the laboratory experiments, Best (2005) showed that the ejections dominate the instantaneous flow field over the crests of the dunes and the highest instantaneous Reynolds stresses are associated with the ejections and sweeps.

It is evident from the recent studies by Sarkar and Dey (2015a, b), and Dey et al. (2018b) that the bluff-bodies have significant impacts on the eddy sizes and the turbulence anisotropy. To be specific, the turbulent flow contains large number of eddies, which can be described by their traversing distances by using Prandtl's mixing length and their sizes by using the Taylor microscale and the Kolmogorov length scale. In addition, the characterization of turbulent flow requires an understanding of anisotropy as it provides information on the extent of departure from the isotropy (Lumley and Newman 1977; Smalley et al. 2002; Keirsbulck et al. 2002; Dey et al. 2019). However, a few studies have been reported to understand the turbulent length scales and Reynolds shear stress anisotropy for the case of an isolated dune. This study is therefore dedicated to characterize the turbulent length scales and the Reynolds stress anisotropy in the flow downstream of an isolated dunal bedform mounted on a rough bed.

2 Experimental Setup

Experiments were carried out in a re-circulatory rectangular flume of 20 m length, 0.5 m width and 0.5 m depth at the Fluvial Mechanics Laboratory in the Indian Statistical Institute, Kolkata, India. The flume was facilitated with visual observation of flow through its transparent sidewalls. A centrifugal pump was used to supply the inflow discharge, which was measured by an electromagnetic device.

The flume bed that had a bed slope of 0.03% was created rough by gluing gravels. Uniform gravels of a median size of $d_{50} = 2.49$ mm were used for this purpose. Isolated two-dimensional dunal (triangular) bedforms for Runs 1 and 2, as shown in Fig. 1, were mounted on the bed at a distance of 7 m from the flume inlet. A dune is of scalene triangular shape with its length l as the horizontal length of the triangular base summing the lengths of stoss-side l_S and leeside l_L, given by $l = l_S + l_L$ (Fig. 2). The dune height h_d is the vertical distance of the dune crest from the bed. The dune was also made rough by using the same gravel sample. Two dunes were used with height $h_d = 0.09$ m and length l $(= l_S + l_L = 0.24 + 0.16) = 0.4$ m for Run 1 and $h_d = 0.03$ m and l $(= l_S + l_L = 0.24 + 0.06) = 0.3$ m for Run 2. Both the runs were performed under the same approach uniform flow condition with a flow depth of $h \approx 0.3$ m and the depth-averaged approach velocity of $\bar{U} \approx$

Fig. 1 Photographs of bed-mounted isolated dunal bedforms in the experimental flume

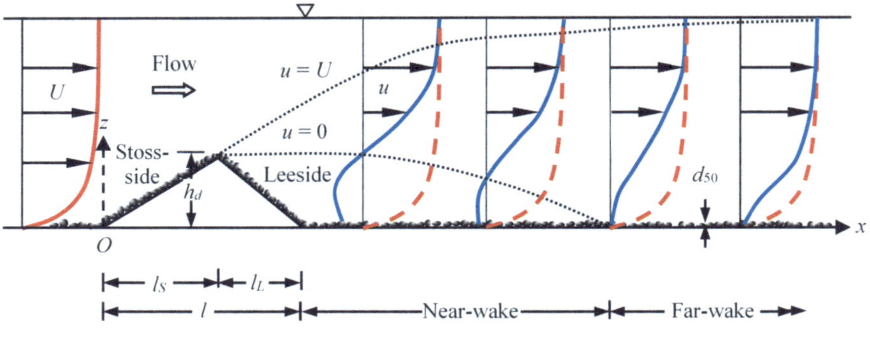

Fig. 2 Schematic of an isolated dunal bedform and velocity profiles

0.44 m s^{-1}. In both Runs, 1 and 2, the flow depth and approach velocity were slightly different (about 1.5%) owing to the experimental rearrangements. A Vernier point gauge with a precision of ±0.1 mm was used to measure the flow depth and the free surface profile. The approach shear velocity u_*, calculated from the bed slope, was 0.030 (m s^{-1}), while the values of u_* for Runs 1 and 2, obtained from the Reynolds shear stress profiles, were 0.027 and 0.025 m s^{-1}, respectively. The flow Reynolds number and Froude number were 528,000 and 0.256, respectively, for both the runs. Besides, the shear-particle Reynolds number R_* ($= d_{50}u_*/\nu$, where ν is the coefficient of kinematic viscosity of water) was 74.7 (>70) for both the runs, ensuring the flow to be turbulent-rough and subcritical. The instantaneous velocities were measured along the centerline of the flume at different nondimensional streamwise locations (according to the coordinate system shown in Fig. 2) $x/l = -0.5, -0.25, 0, 0.1, 0.2, 0.3, 0.4, 0.5, 0.6, 0.7, 0.8, 0.9, 1, 1.1, 1.3, 1.7, 2.1, 2.5$, and 3.3 by using a 5 cm downlooking *Vecrtino velocimeter*. Here, x is the streamwise distance. The Vecrtino system had an adjustable sampling volume of 6 mm diameter and 1–4 mm height. It was operated with an acoustic frequency of 10 MHz and at a sampling rate of 100 Hz. The velocity components u, v, and w were considered in the streamwise (x), spanwise (y), and vertical (z) directions, respectively. Up to the dune crest, the lowest sampling height of 1 mm was used, while above the crest, it was set as 2.5 mm. A sampling duration of 300 s was considered to be sufficient to get the time-invariant velocity and turbulence quantities. The minimum signal-to-noise ratio (SNR) and signal correlation were maintained as 18 and 70%, respectively. After the measurements, the data were filtered whenever necessary using the *acceleration thresholding method* (Goring and Nikora 2002).

3 Time-Averaged Velocity, Reynolds Shear Stress and Average Turbulence Intensity

Figure 2 presents a schematic of an isolated dunal bedform and the velocity profiles upstream and downstream of it. The flow is directed from the stoss-side to the leeside. Therefore, the approach flow before starting the stoss-side of the dune is designated herein as upstream flow and the flow past the dune as wall-wake flow. Immediate downstream vicinity of the dune ($x = \xi l$, where ξ is a factor being less than unity), the flow is characterized by the reversed flow, called the *near-wake flow*. Thereafter, the flow is called the *far-wake flow*. In Fig. 2, the upper dotted line represents the boundary layer ($u = U$) in wall-wake flow and the lower dotted line signifies the locus of $u = 0$, where $u(z)$ is the streamwise velocity in wake flow, $U(z)$ is the streamwise velocity in undisturbed upstream flow and z is the vertical distance measured from the bed. The far-wake flow having a defect velocity profiles starts recovering, as the flow travels downstream. Far downstream of the dune, the flow fully recovers the undisturbed upstream flow, signifying the fully recovered (or fully developed) open-channel flow.

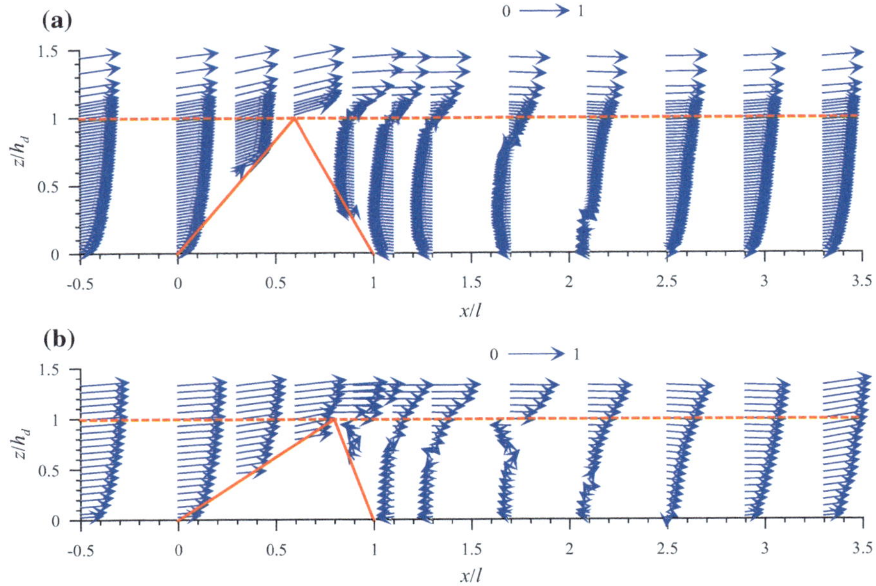

Fig. 3 Velocity vectors in flows upstream and downstream of an isolated dune in **a** Run 1 and **b** Run 2 (dune geometries not to in same horizontal and vertical scales)

The time-averaged velocity vectors for Runs 1 and 2 are depicted in Fig. 3a, b, respectively, where the abscissa and the ordinate represent the nondimensional streamwise distance x/l and vertical distance z/h_d, respectively. The magnitude and the direction of the velocity vectors are $(u^2 + w^2)^{0.5}$ and $\tan^{-1}(w/u)$, respectively, where u and w are the time-averaged streamwise and vertical velocities, respectively. Upstream of the dune ($x/l = -0.5$), the velocity vectors do not change significantly in magnitude and direction while approaching from $x/l = -0.5$ to 0 (starting point of the stoss-side of dune). However, downstream of the dune, the velocity vectors change significantly both in magnitude and direction, particularly below the dune crest. Immediate downstream vicinity of the dune ($1 \leq x/l \leq 1.75$), the velocity vectors change their directions, signifying a circulatory motion in the near-wake flow zone. The near-bed velocity vectors in the reverse direction continue up to a certain distance ($x/l \approx 1.75$) in the near-wake flow zone. However, the wall-wake flows in both the runs start recovering their respective undisturbed upstream flows with an increase in horizontal distance x/l, attaining approximately the undisturbed velocity vectors at $x/l = 3.3$ (or $x/h_d = 15$ and 33 in Runs 1 and 2, respectively).

Figure 4 shows the nondimensional streamwise velocity u/u_* profiles in flows upstream and downstream of an isolated dune at different streamwise distances for Runs 1 and 2. It is evident that the approach shear flow passing the isolated dune gets separated to form a reversed flow, in the near-wake flow zone, at the immediate downstream vicinity of the dune ($1 \leq x/l \leq 1.75$). As the flow moves further

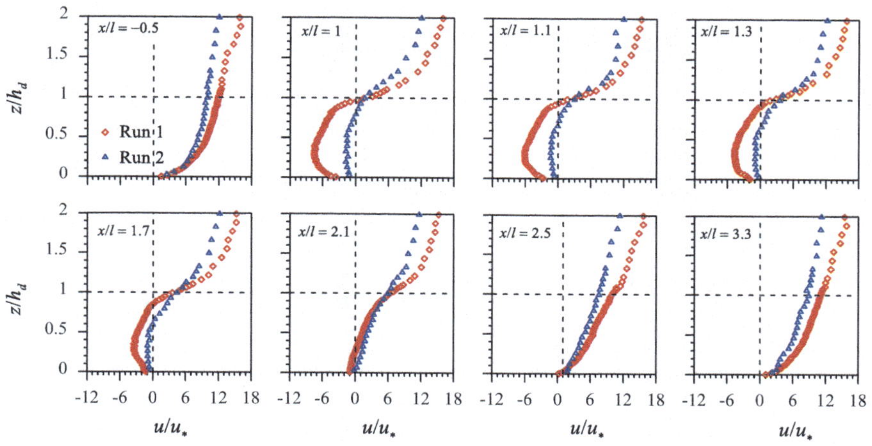

Fig. 4 Vertical profiles of nondimensional streamwise velocity u/u_* at different streamwise distances for Runs 1 and 2

downstream, the reversed flow disappears establishing the far-wake flow ($1.75 < x/l < 3.3$). Far downstream of the dune ($x/l \geq 3.3$), the velocity profiles in Runs 1 and 2 recover their respective undisturbed upstream profiles, called the *no-wake flow* or *fully recovered flow* (Sforza and Mons 1970; Sadeque et al. 2009; Dey et al. 2011). The main feature of u/u_* profiles in wall-wake flow is that the velocity gradient, which is milder and concave below the crest level, changes (forming an inflection point) at the crest level to become steeper and convex above it. However, the length of wall-wake flow zone and the distance of recovery of streamwise velocity profile for an isolated dune are different as compared to those for bed-mounted spherical obstacles (Dey et al. 2011) and bed-mounted horizontal and vertical cylinders (Dey et al. 2018a, b, c). The recovery of the streamwise velocity took place at a distance of 8.5 times the sphere diameter in Dey et al. (2011), while it was at a distance of 20 times the horizontal cylinder diameter in Dey et al. (2018a) and 16 times the vertical cylinder diameter in Dey et al. (2018b, c). In the present case, the recovery of the velocity profiles occurred approximately at 3.3 times the dune length or 15 and 33 times the dune height in Runs 1 and 2, respectively. The differences of the downstream locations to recover the undisturbed upstream streamwise velocity profiles in Runs 1 and 2 are attributed to the different geometries of the dunes.

Figure 5 displays the vertical profiles of nondimensional Reynolds shear stress (RSS) τ/u_*^2 in flows upstream and downstream of an isolated dune at different streamwise distances for Runs 1 and 2. Here, τ is the Reynolds shear stress relative to mass density ρ of water, defined by $-\overline{u'w'}$. Upstream of the dune ($x/l = -0.5$), the τ/u_*^2 profiles in both the runs follow approximately a linear law, where the peak values in τ/u_*^2 profiles appear in near the bed, decreasing linearly with an increase in vertical distance and approaching to zero at the free surface. However, downstream of the dune, the τ/u_*^2 profiles start with a finite value from the bed and increase with an

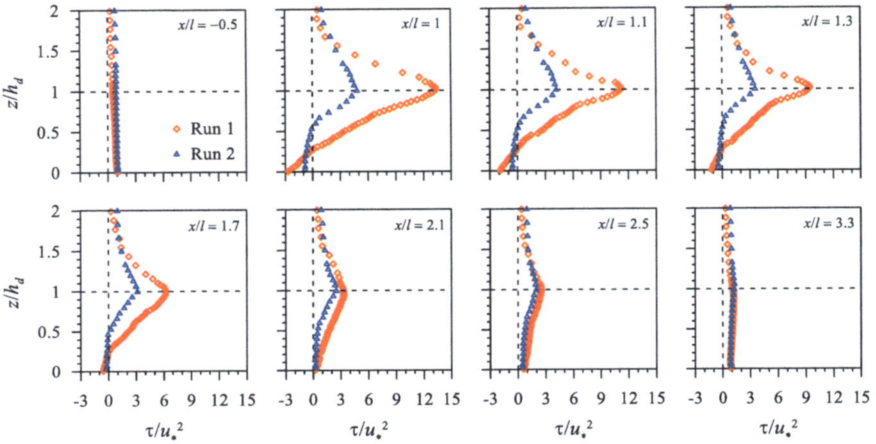

Fig. 5 Vertical profiles of nondimensional Reynolds shear stress τ/u_*^2 at different streamwise distances for Runs 1 and 2

increase in vertical distance until they attain their peaks at the dune crest. Then, they decrease, as one goes toward the free surface. The peaks in τ/u_*^2 profiles progressively diminish with an increase in downstream distance. However, far downstream of the dune ($x/l = 3.3$), the τ/u_*^2 profiles in both the runs follow the undisturbed upstream profiles of the respective runs, confirming a stress recovery. It can be concluded that at a given vertical distance, the RSS in wall-wake flow is greater than that in undisturbed upstream flow owing to the fluid mixing in wall-wake flow zone.

The vertical profiles of average turbulence intensity I_{av} at different streamwise distances for Runs 1 and 2 are shown in Fig. 6. The average turbulence intensity I_{av} is used to assess the overall turbulence level. It is expressed as $I_{av} = (2q/3)^{0.5}/U_1$; where q is the turbulent kinetic energy (TKE) [$= 0.5(\overline{u'u'} + \overline{v'v'} + \overline{w'w'})$], $\overline{u'u'}$, $\overline{v'v'}$, and $\overline{w'w'}$ are the Reynolds normal stresses relative to ρ in streamwise, spanwise and vertical directions, respectively, and U_1 is the local time-averaged resultant velocity [$=(u^2 + v^2 + w^2)^{0.5}$]. In case of a pipe flow, the ranges $I_{av} = 0.05$–0.2, 0.01–0.05 and <0.01 represent the high-turbulence, medium-turbulence and low turbulence, respectively (Russo and Basse 2016; Basse 2017). However, these ranges are deemed to be applicable to the turbulent flow in an open channel as well.

The vertical profiles of I_{av} at different streamwise distances show that upstream of the dune ($x/l = -0.5$), the I_{av} starts with a bed induced high-turbulence level from the near bed, but as the vertical distance increases, the I_{av} profiles gradually reduces to medium and then to low turbulence range. However, downstream of the dune, the I_{av} profiles start with a high-turbulence level close to the bed, but after a brief decrease, the I_{av} profiles increase sharply with vertical distance, attaining their peak values at the dune crest. It indicates the maximum value in high-turbulence level owing to an intense fluid mixing. Above the crest, the I_{av} profiles decrease with an increase in vertical distance and above $z/h_d = 1.5$, the I_{av} profiles fall sharply to

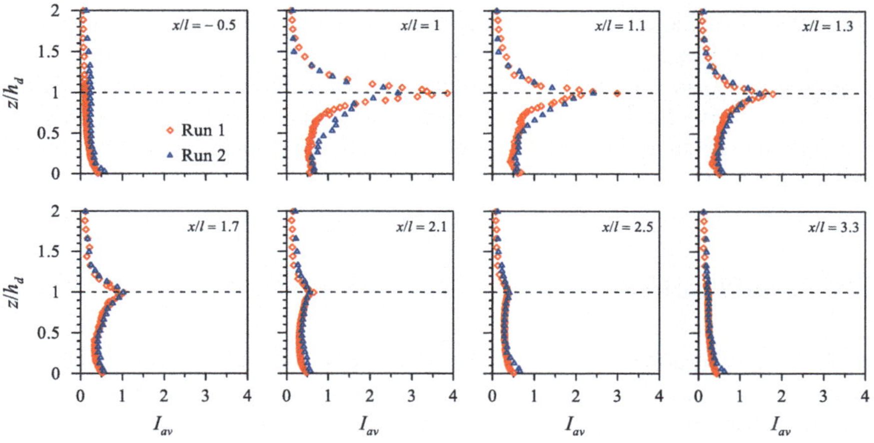

Fig. 6 Vertical profiles of average turbulence intensity I_{av} at different streamwise distances for Runs 1 and 2

follow the upstream profiles of respective runs. Importantly, the peaks of I_{av} taking place at the dune crest give an indication of the occurrence of the maximum TKE at the crest. The present findings are in conformity with the observations of Sarkar et al. (2016), who also found that the maximum TKE occurs at the crests of the protruding array of large gravels owing to the fluid mixing at the crest. In essence, the wall-wake flow possesses a high-turbulence level having a peak at the crest level indicating an intense fluid mixing. The high-turbulence level continues up to the vertical distance of 1.5 times the dune height, and beyond that the effects of the dune vanish.

4 Turbulent Length Scales

The turbulent fluid flow comprises the eddies that are dynamically self-perpetuate, as the fluid flows. The eddies collapse unremittingly from the largest scale to the smallest one, initiating from the intrinsic flow instabilities (Pope 2000). While the size and the travelling space of individual eddies are hard to determine uniquely, it is however feasible to delineate the length scales by characterizing the behavior of eddies. Remarkably, Prandtl (1925) was the innovator of the mixing-length concept that elucidates the mixing phenomenon by the turbulent eddies in a fluid flow stemming from the fluid momentum interchange in a macro-scale. The Prandtl's mixing-length thus indicates the travelling space of eddies. On the other hand, the turbulent length scale is a physical quantity to describe the behavior of the large energy-containing eddies in a turbulent flow. The large eddies in the flow account for most of the transport of momentum and energy. If a larger eddy collapses, its kinetic energy is transmitted to the smaller eddies in an energy cascade process. By contrast, when small eddies merge, as they often do in a mixing process, the TKE is transmitted to the larger eddies. The length scale of larger eddies is controlled by the physical boundaries of a flow, while the length scale of smaller eddies can be defined by the TKE dissipation rate and the kinematic viscosity of fluid in terms of the *Taylor microscale* and the *Kolmogorov length scale*. The former refers to the length scale in the inertial subrange and the latter to the dissipation range (Pope 2000; Jovanović 2004; Dey 2014).

4.1 Prandtl's Mixing Length

Prandtl (1925) argued that in a turbulent shear flow, the eddies forming the fluid parcels generate and then degenerate to exchange their momentum after travelling an average distance, termed the *mixing length* (also well-known as *Prandtl's mixing length*). The process of generation and degeneration of eddies perpetuates as long as the fluid flows at a high Reynolds number. According to the Prandtl's mixing-length theory, the mixing-length l_p is expressed as

$$l_p(z) = \frac{(-\overline{u'w'})^{0.5}}{d\bar{u}/dz} \tag{1}$$

The nondimensional Prandtl's mixing-length l_p/d_{50} as a function of nondimensional vertical distance z/h_d at different nondimensional streamwise distances x/l for Runs 1 and 2 is plotted in Fig. 7. Upstream of the dune ($x/l = -0.5$), the l_p/d_{50} profiles increase almost linearly with an increase in vertical distance obeying $l_p = \kappa z$, where κ is the von Kármán constant (=0.4). Downstream of the dune ($x/l \geq 1$), the l_p/d_{50}

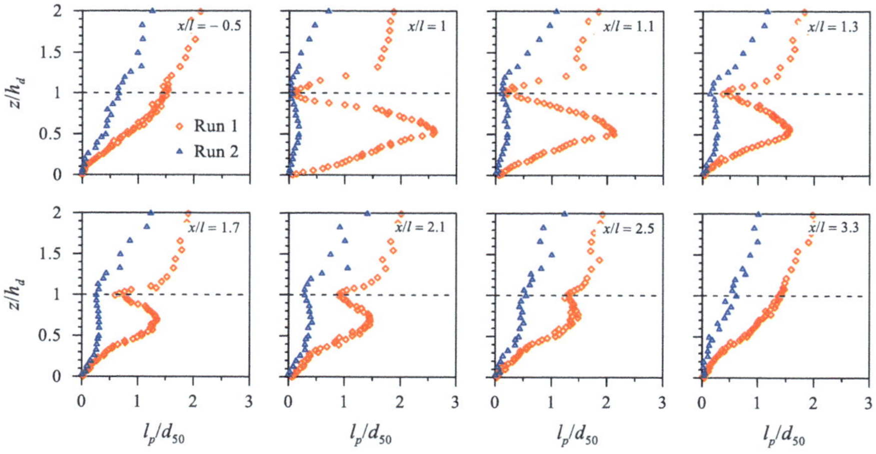

Fig. 7 Vertical profiles of nondimensional Prandtl's mixing length l_p/d_{50} at different streamwise distances for Runs 1 and 2

profiles increase sharply with an increase in vertical distance, attaining their peak values at $z/h_d \approx 0.5$–0.6. Then, they decrease to a second lowest critical value at the dune crest ($z/h_d \approx 1$) and, thereafter, gradually increase, as one goes toward the free surface. It is clearly noticeable that at a given z, the mixing-length l_p in wall-wake flow is greater than that in undisturbed upstream flow. The increased values of l_p/d_{50} gradually disappear with an increase in downstream distance. Far downstream of the dune ($x/l = 3.3$), the l_p/d_{50} profiles in both the runs become almost similar to their corresponding undisturbed upstream profiles displaying a recovery of mixing-length. It can be concluded that at a given vertical distance, the average travelling distance of turbulent eddies in wall-wake flow is greater than that in the undisturbed upstream flow. The reason is accredited to an increased RSS and a reduced velocity gradient in wall-wake flow.

4.2 Taylor Microscale

The Taylor microscale λ represents the eddy size in the inertial subrange of the Kolmogorov spectrum of velocity fluctuations being the pertinent length scale of the pure turbulence. It is expressed as

$$\lambda = \left(\frac{15\nu\sigma_u}{\varepsilon}\right)^{0.5} \tag{2}$$

where σ_u is the streamwise Reynolds normal stress relative to ρ, given by $\overline{u'u'}$, and ε is the TKE dissipation rate. In this study, the ε was estimated from the Kolmogorov second hypothesis, as was done by Dey and Das (2012).

The vertical profiles of nondimensional Taylor microscale λ/d_{50} at different nondimensional streamwise distances x/l are presented in Fig. 8. Upstream of the dune, the λ/d_{50} profiles gradually increase with an increase in vertical distance up to $z/h_d \approx 1.25$ and thereafter become almost invariant to the vertical distance, as one goes toward free surface. Downstream of the dune, the λ/d_{50} profiles increase gradually with vertical distance up to $z/h_d \approx 0.75$, and then decrease sharply up to the dune crest. Above the crest, they again start increasing and becoming almost invariant to vertical distance as one goes toward the free surface. It is however evident that at a given z, the Taylor microscale λ in wall-wake flow is smaller than that in undisturbed upstream flow. As the downstream distance increases, the effects of the dune weaken; and far downstream of the dune ($x/l = 3.3$), the λ/d_{50} profiles in both the runs regain their corresponding undisturbed upstream profiles. In conclusion, at a given vertical distance, the length scale of turbulent eddies in the inertial subrange in wall-wake flow is smaller than that in undisturbed upstream flow. The reason is ascribed to an enhanced TKE dissipation rate in wall wake flow.

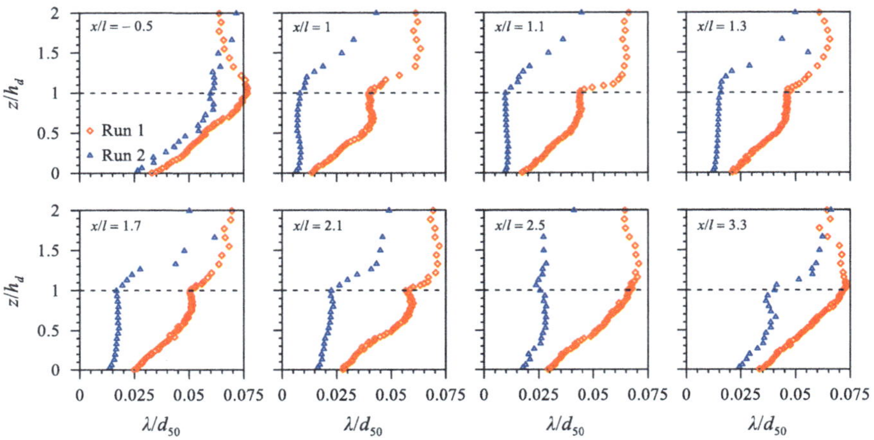

Fig. 8 Vertical profiles of nondimensional Taylor microscale λ/d_{50} at different streamwise distances for Runs 1 and 2

4.3 Kolmogorov Length Scale

In the dissipation range, the viscosity dominates and the TKE is dissipated into heat. The dissipation of TKE takes place at a length scale of the order of the Kolmogorov length scale η, which is expressed as

$$\eta = \left(\frac{\nu^3}{\varepsilon}\right)^{0.25} \tag{3}$$

Figure 9 shows the vertical profiles of nondimensional Kolmogorov length scale η/d_{50} at different nondimensional streamwise distances x/l for Runs 1 and 2. In general, the characteristics of η/d_{50} profiles are almost similar to λ/d_{50} profiles with difference in magnitudes. At a given vertical distance, the Kolmogorov length scale in wall-wake flow is approximately 1/10 times the Taylor microscale.

5 Reynolds Stress Anisotropy

In an *isotropic turbulence*, the velocity fluctuations in a turbulent fluid flow are independent of the axis of reference and invariant to the rotation of axis. It implies the Reynolds normal stresses to be identical, e.g. $\sigma_x = \sigma_y = \sigma_z$, where $(\sigma_x, \sigma_y, \sigma_z) = (\overline{u'u'}, \overline{v'v'}, \overline{w'w'})$. Therefore, the statistics of instantaneous fluid motion at a point correspond to a pure chaotic motion having no preferential direction. On the other hand, in an *anisotropic turbulence*, the velocity fluctuations $u'_i[= (u', v', w')$ for $i =$

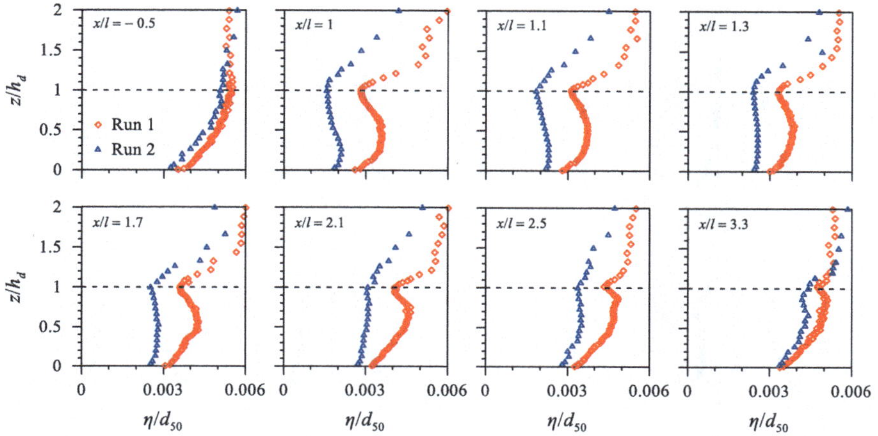

Fig. 9 Vertical profiles of nondimensional Kolmogorov length scale η/d_{50} at different streamwise distances for Runs 1 and 2

(1, 2, 3)] in a turbulent fluid flow are directionally preferred; and hence, the Reynolds normal stresses are not identical.

The Reynolds stress anisotropy remains an important feature in the realm of many natural and industrial flow characteristics, where the notion of isotropic turbulence breaks down. The *Reynolds stress anisotropy tensor* a_{ik} is defined as $a_{ik} = \overline{u'_i u'_k} - (2/3)q\delta_{ik}$ (Rotta 1951), where $q = \overline{u'_i u'_i}/2$ and δ_{ik} is the Kronecker delta function, defined as $\delta_{ik}(i \neq k) = 0$ and $\delta_{ik}(i = k) = 1$. In nondimensional form, the Reynolds stress anisotropy tensor a_{ik} is expressed as $b_{ik}[= a_{ik}/(2q)] = \overline{u'_i u'_k}/(2q) - \delta_{ik}/3$. In essence, the b_{ik} remains a symmetric and traceless tensor ranging from −1/3 to 2/3 and vanishes for an isotropic turbulence ($b_{ik} = 0$). The contribution from the Reynolds stress relative to the TKE corresponds to the sign of each diagonal component in b_{ik}. Figure 10a, b illustrate the vertical profiles of Reynolds stress anisotropy tensor components b_{ik} at different streamwise distances for Runs 1 and 2, respectively. Downstream of the dune, the b_{11} and b_{22} components suggest the wall-wake flow yields less anisotropic turbulence in streamwise direction and more in spanwise direction below the crest. On the other hand, the b_{33} component produces more anisotropic turbulence in the lower half and less in the upper half of the dune, as compared to their undisturbed upstream values. However, the b_{13} component, which signifies the ratio of RSS to TKE, exhibits little change in wall-wake flow. For all b_{ik} components in both the runs, they display a kink at the crest level. However, a recovery is evident, as downstream distance increases.

To portrait the degree and the nature of Reynolds stress anisotropy, it is quite convenient to introduce the second and third principal invariants, $I_2(=-b_{ik}b_{ik}/2)$ and $I_3(=b_{ij}b_{jk}b_{ki}/3)$, respectively. Since the b_{ik} is a traceless tensor, the first invariant, $I_1(=b_{ii})$, becomes zero. A quantitative description of total Reynolds stress anisotropy is ascertained by plotting $-I_2$ against I_3, termed *anisotropy invariant map* (AIM). In an AIM, the feasible turbulence states are confined to the space bounded by the

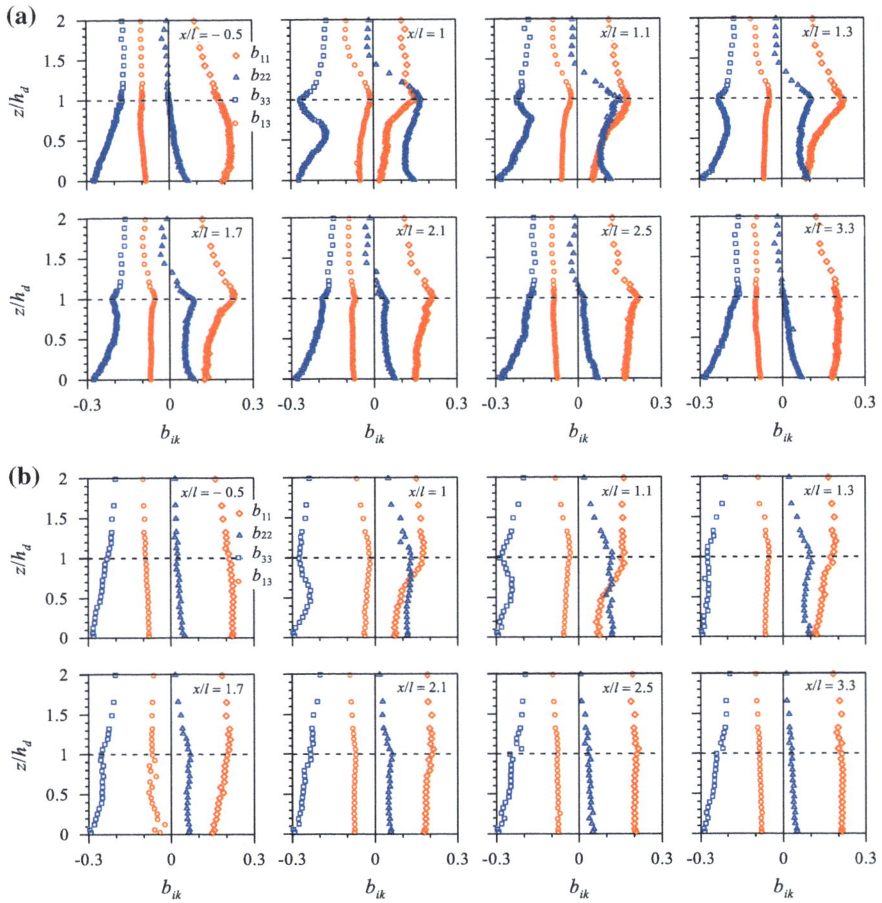

Fig. 10 Vertical profiles of Reynolds stress anisotropy tensor b_{ik} at different streamwise distances for **a** Run 1 and **b** Run 2

left-curved, the right-curved and the top-linear boundaries, forming a triangle, called the *Lumley triangle* (Fig. 11). The left-curved and the right-curved boundaries, being symmetric about the *plane-strain limit* ($I_3 = 0$), evolve from the isotropic limit ($I_2 = I_3 = 0$), obeying a generic relationship as $I_3 = \pm 2(-I_2/3)^{3/2}$. On the other hand, the top-linear boundary follows $I_3 = -(9I_2 + 1)/27$. The current understanding of the states of Reynolds stress anisotropy has so far been achieved on the basis of two different perspectives—the shape of turbulent eddies (Reynolds and Kassinos 1995) and the shape of ellipsoid that the Reynolds stress tensor forms (Simonsen and Krogstad 2005). To be specific, Simonsen and Krogstad (2005) provided a clarification concerning the traditionally used terminologies to describe the states of anisotropy founded on the concept of the shape of turbulent eddies. They concluded

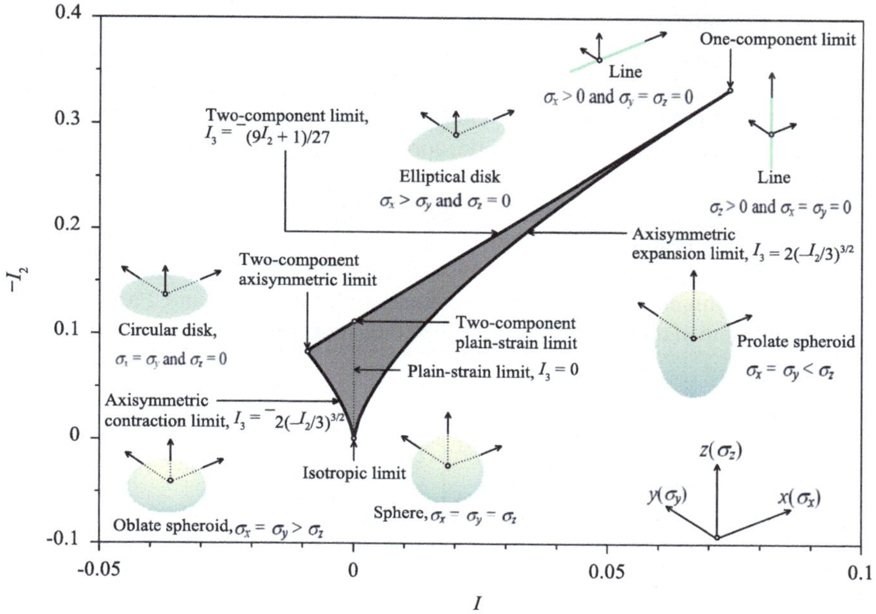

Fig. 11 Definition sketch of the Lumley triangle after Dey et al. (2019)

that a better way of representation of the Reynolds stress anisotropy is to introduce the states of anisotropy that underpins the concept of the shape of ellipsoid created by the Reynolds stress tensor. This is attributed to the former concept that could create ambivalences towards the terminologies that are commonly used in describing the states of anisotropy.

In a recent attempt, Dey et al. (2019) put forward a subtle description of the Reynolds stress anisotropy to elicit an enriched understanding of the subject, envisioning the Reynolds stress anisotropy from the viewpoint of the shape of ellipsoid created by the Reynolds normal stress components (σ_x, σ_y, σ_z) in (x, y, z). For an idealized scenario preserving the notion of an isotropic turbulence ($I_2 = I_3 = 0$ or $\sigma_x = \sigma_y = \sigma_z$), the stress ellipsoid is a sphere (Fig. 11). The left-curved boundary corresponds to the *axisymmetric contraction limit*, where one component of Reynolds normal stress is smaller than the remaining two components that are equal ($\sigma_x = \sigma_y > \sigma_z$). It turns out that the stress ellipsoid takes the form of an oblate spheroid. However, at the left vertex of the Lumley triangle, corresponding to the *two-component axisymmetric limit*, one component of Reynolds normal stress vanishes leaving the other two equal components ($\sigma_x = \sigma_y$ and $\sigma_z = 0$). Therefore, the stress ellipsoid becomes a circular disc (Fig. 11). On the other hand, the right-curved boundary corresponds to the *axisymmetric expansion limit*, where one component of Reynolds normal stress is larger than the remaining two components that are equal ($\sigma_x = \sigma_y < \sigma_z$). It follows that the stress ellipsoid is a prolate spheroid (Fig. 11). In addition, the top-linear boundary corresponds to the *two-component limit*, because one component of Reynolds normal stress is larger than the other component in conjunction with a third component that disappears ($\sigma_x > \sigma_y$ and $\sigma_z = 0$). This makes the stress ellipsoid an elliptical disk. Note that the intersecting point of the two-component limit and the plain-strain limit is called the *two-component plain-strain limit* (Fig. 11). Lastly, at the right vertex of the Lumley triangle, corresponding to the *one-component limit* [($\sigma_x > 0$, $\sigma_y = \sigma_z = 0$) or ($\sigma_x = \sigma_y = 0$, $\sigma_z > 0$)], one component of Reynolds normal stress prevails along with the other two vanishing components. It results a stress ellipsoid to become a line (Fig. 11).

The AIMs constructed for the data plots $-I_2$ against I_3 within the AIM boundaries, at different nondimensional streamwise distances x/l for Runs 1 and 2 are shown in Fig. 12. Upstream of the dune ($x/l = -0.5$), the data plots start from the near left vertex

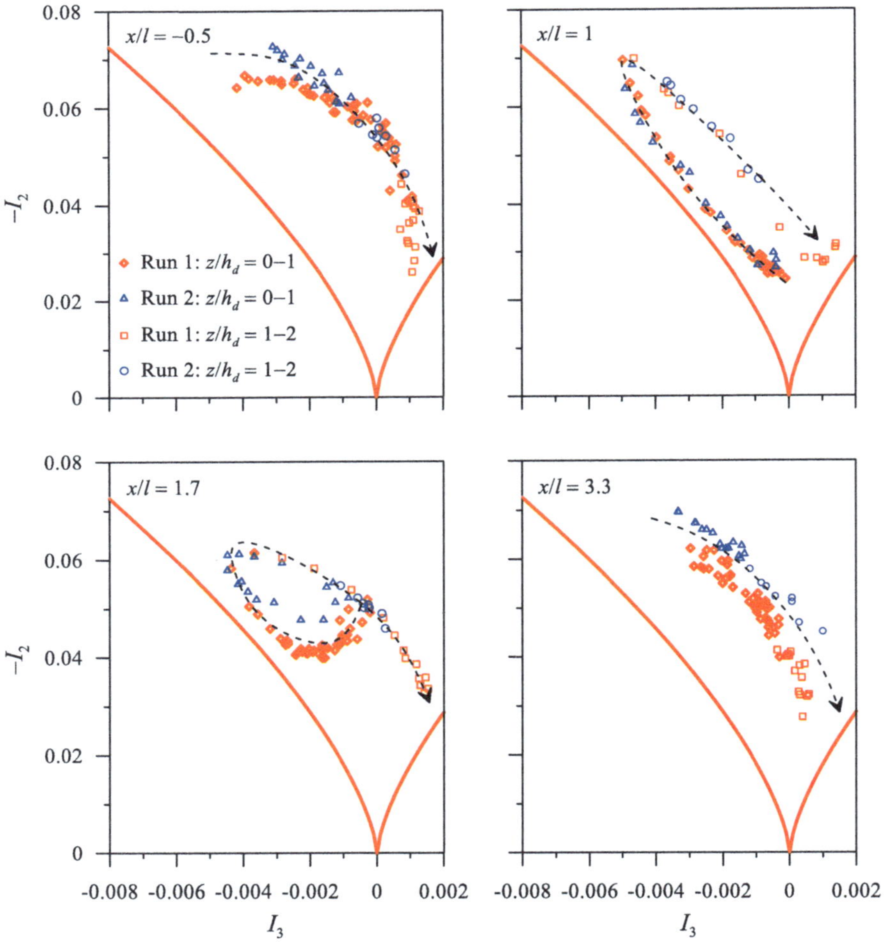

Fig. 12 AIMs at different streamwise distances for Runs 1 and 2

of the Lumley triangle, moving initially toward the bottom cusp, and cross the plain-strain limit ($I_3 = 0$), shifting toward the right-curved boundary, with an increase in vertical distance. The trends of these data plots for both the runs are fairly monotonic. The AIM of the upstream indicates that the turbulence anisotropy has a tendency to reduce toward a quasi-three-dimensional isotropy, as the vertical distance increases. On the other hand, immediate downstream of the dune, the data plots have a tendency to form a stretched loop inclined to the left-curved boundary. Below the crest (z/h_d < 1), the near-bed data plots start from the plain-strain limit and shift toward the left vertex closely along the left-curved boundary with an increase in vertical distance up to the crest, suggesting the turbulence anisotropy to have an affinity toward a two-dimensional isotropy. The reason is attributed to the fact that the vertical velocity fluctuations are less than both the streamwise and the spanwise velocity fluctuations that are approximately equal. Above the crest, the data plots take a sharp right turn and move toward the bottom cusp almost parallel to the left-curved boundary, as the vertical distance increases further, signifying the turbulence anisotropy tend to reduce toward a quasi-three-dimensional isotropy. Further downstream ($x/l = 1.7$), the loop formed by the data plots reduces its size forming a tail and finally disappears at $x/l =$ 3.3, corroborating a recovery of the undisturbed upstream trend. It can therefore be concluded that below the dune crest, the turbulence is characterized with an affinity to show a two-dimensional isotropy, while above the crest, the anisotropy has a tendency to reduce to a quasi-three-dimensional isotropy.

In the framework of the ellipsoid shapes formed by the Reynolds stress tensor, as shown in Fig. 11, the AIMs presented in Fig. 12 can be further discussed. Figure 12 shows that in the wall-wake flow, an oblate spheroid axisymmetric turbulence is prevalent below the crest. The curve through the data plots touches the line of plain-strain limit ($I_3 = 0$) in the vicinity of bed. It suggests that the axisymmetric contraction to the oblate spheroid increases with an increase in vertical distance up to the crest. However, above the crest, the axisymmetric contraction to oblate spheroid relaxes with a further increase in vertical distance.

Further, the use of the anisotropy invariant function **F** is an easier method to demonstrate the overall state of turbulence to be two- or three-dimensional isotropy. The **F** is calculated as $\mathbf{F} = 1 + 9I_2 + 27I_3$. Importantly, $\mathbf{F} = 0$ corresponds to the two-component limit, while $\mathbf{F} = 1$ refers to the three-dimensional isotropic state. Figure 13 demonstrates the vertical profiles of anisotropy invariant function **F** at different nondimensional streamwise distances x/l for Runs 1 and 2. Upstream of the dune ($x/l = -0.5$), the data plots have a tendency to start with a two-dimensional isotropy from the near-bed flow and move monotonically toward a quasi-three-dimensional isotropy, as the vertical distance increases. On the other hand, downstream of the dune, the data plots in wall-wake flow have an affinity to a quasi-three-dimensional isotropy in the inner flow zone. Then, they sharply turn toward a two-dimensional isotropy with an increase in vertical distance until they attain their individual maximum two-dimensional isotropy state at the dune crest. Above the crest, as the vertical distance increases, the data plots sharply move toward a quasi-three-dimensional isotropy almost following the outer zone of the undisturbed upstream profile of an individual run. It is interesting to note that the cusps formed by the $\mathbf{F}(z/h_d)$ curves at the crest

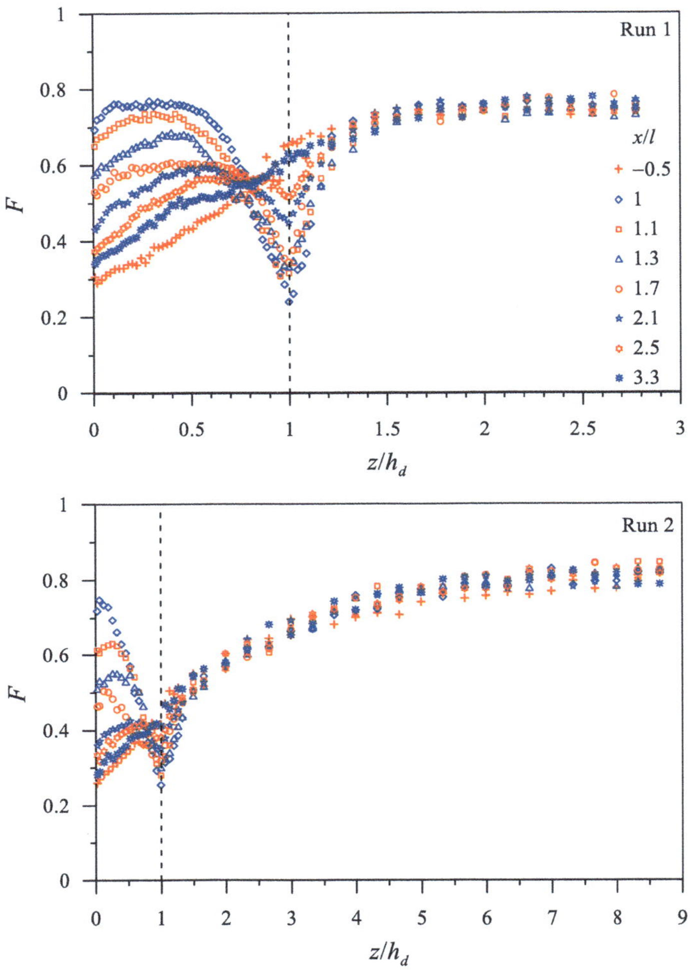

Fig. 13 Vertical profiles of anisotropy invariant function **F** at different streamwise distances for Runs 1 and 2

level in the wall-wake flow gradually disappear with an increase in downstream distance becoming a monotonic trend to follow the undisturbed upstream profile at far downstream ($x/l = 3.3$).

6 Conclusions

The turbulent length scales and the Reynolds stress anisotropy in wall-wake flow downstream of an isolated dunal bedform are studied experimentally measuring instantaneous velocity by a Vecrtino velocimeter. The main findings are as follows:

The velocity in wall-wake flow downstream of an isolated dune suffers from a defect, where the Reynolds shear stress enhances owing to the fluid mixing. The wall-wake flow has a high-turbulence level with a peak at the dune crest. This level continues up to the vertical distance until the effects of the dune vanish.

The average traveling distance of turbulent eddies in wall-wake flow is greater than that in undisturbed upstream flow, while the length scales of turbulent eddies in the inertial subrange and dissipation range are smaller.

In Reynolds stress anisotropy analysis, the anisotropy invariant maps illustrate that the data plots form a looping trend in wall-wake flow. They start from the plain-strain limit near the bed and shift toward the left vertex of the Lumley triangle along the left-curved boundary up to the dune crest, showing an affinity to a two-dimensional isotropy. Above the crest, the data plots take a sharp turn and move toward the bottom cusp, showing an affinity to a quasi-three-dimensional isotropy. Further, it implies that an oblate spheroid axisymmetric turbulence is predominant below the crest with an axisymmetric contraction to the oblate spheroid to increase with vertical distance up to the crest. Above the crest, the axisymmetric contraction to oblate spheroid reduces, as one goes toward the free surface.

References

Akilli H, Rockwell D (2002) Vortex formation from a cylinder in shallow water. Phys Fluids 14(9):2957–2967

Balachandar R, Ramachandran S, Tachie MF (2000) Characteristics of shallow turbulent near wakes at low Reynolds numbers. J Fluids Eng 122(2):302–308

Basse NT (2017) Turbulence intensity and the friction factor for smooth- and rough-wall pipe flow. Fluids 2(30):1–13

Best JL (2005) The kinematics, topology and significance of dune-related macroturbulence: some observations from the laboratory and field. In: Blum MD, Marriott SB, Leclair S (eds) Fluvial sedimentology VII. Special Publication of International Association of Sedimentologists, vol 35, pp 41–60

Bose SK, Dey S (2016) Circular far-wake flow behind a sphere: solutions to the second-order. J Eng Mech 142(1):06015005

Bose SK, Dey S (2018) Far-wake flows downstream of cylinders: a novel generalized similarity method. Eur J Mech/B Fluids 67(January-February):65–69

Dey S (2014) Fluvial hydrodynamics: hydrodynamic and sediment transport phenomena. Geoplanet Book Series. Springer, Berlin

Dey S, Das R (2012) Gravel-bed hydrodynamics: a double-averaging approach. J Hydraul Eng 138(8):707–725

Dey S, Sarkar S, Bose SK, Tait S, Castro-Orgaz O (2011) Wall-wake flows downstream of a sphere placed on a plane rough wall. J Hydraul Eng 137(10):1173–1189

Dey S, Lodh R, Sarkar S (2018a) Turbulence characteristics in wall-wake flows downstream of wall-mounted and near-wall horizontal cylinders. Environ Fluid Mech 18(4):891–921
Dey S, Swargiary D, Sarkar S, Fang H, Gaudio R (2018b) Turbulence features in a wall-wake flow downstream of a wall-mounted vertical cylinder. Eur J Mech-B/Fluids 69(May-June):46–61
Dey S, Swargiary D, Sarkar S, Fang H, Gaudio R (2018c) Self-similarity in turbulent wall-wake flow downstream of a wall-mounted vertical cylinder. J Hydraul Eng 144(6):04018023
Dey S, Ravi Kishore G, Castro-Orgaz O, Ali SZ (2019) Turbulence length scales and anisotropy in submerged plane offset jets. J Hydraul Eng 145(2):04018085
Goring DG, Nikora VI (2002) Despiking acoustic Doppler velocimeter data. J Hydraul Eng 128(1):117–126
Jovanović J (2004) The statistical dynamics of turbulence. Springer, Berlin
Kahraman A, Sahin B, Rockwell D (2002) Control of vortex formation from a vertical cylinder in shallow water: effect of localized roughness elements. Exp Fluids 33(1):54–65
Keirsbulck L, Labraga L, Mazouz A, Tournier C (2002) Influence of surface roughness on anisotropy in a turbulent boundary layer flow. Exp Fluids 33(3):497–499
Lacey RWJ, Roy AG (2008) Fine-scale characterization of the turbulent shear layer of an instream pebble cluster. J Hydraul Eng 134(7):925–936
Lumley JL, Newman GR (1977) The return to isotropy of homogeneous turbulence. J Fluid Mech 82:161–178
Maddux TB, Nelson JM, McLean SR (2003a) Turbulent flow over three-dimensional dunes: I. free surface and flow response. J Geophys Res 108(F1):6009
Maddux TB, Nelson JM, McLean SR (2003b) Turbulent flow over three-dimensional dunes: II. Fluid and bed stresses. J Geophys Res 108(F1):6010
McLean SR, Smith JD (1986) A model for flow over two-dimensional bed forms. J Hydraul Eng 112(4):300–317
Nelson JM, Smith JD (1989) Mechanics of flow over ripples and dunes. J Geophys Res 94(C6):8146–8162
Ozgoren M (2006) Flow structure in the downstream of square and circular cylinders. Flow Meas Instrum 17(4):225–235
Ozturk NA, Akkoca A, Sahin B (2008) Flow details of a circular cylinder mounted on a flat plate. J Hydraul Res 46(3):344–355
Pope SB (2000) Turbulent flows. Cambridge University Press, Cambridge
Prandtl L (1925) Bericht über untersuchungen zur ausgebildeten turbulenz. Z Angew Math Mech 5(2):136–139
Reynolds WC, Kassinos SC (1995) One-point modelling of rapidly deformed homogeneous turbulence. Proc R Soc Lond A 451(1941):87–104
Rotta JC (1951) Statistische theorie nichthomogener turbulenz. Z Phys 129(6):547–572
Russo F, Basse NT (2016) Scaling of turbulence intensity for low-speed flow in smooth pipes. Flow Meas Instrum 52:101–114
Sadeque MAF, Rajaratnam N, Loewen MR (2009) Shallow turbulent wakes behind bed-mounted cylinders in open channels. J Hydraul Res 47(6):727–743
Sarkar S, Dey S (2015a) Turbulent length scales and anisotropy downstream of a wall mounted sphere. J Hydraul Res 53(5):649–658
Sarkar S, Dey S (2015b) Turbulence anisotropy in flow at an entrainment threshold of sediment. J Hydraul Eng 141(7):06015007
Sarkar S, Papanicolaou AN, Dey S (2016) Turbulence in a gravel-bed stream with an array of large gravel obstacles. J Hydraul Eng 142(11):04016052
Schlichting H (1979) Boundary layer theory. McGraw-Hill, New York
Sforza PM, Mons RF (1970) Wall-wake: flow behind a leading edge obstacle. AIAA J 8(2):2162–2167
Shamloo H, Rajaratnam N, Katopodis C (2001) Hydraulics of simple habitat structures. J Hydraul Res 39(4):351–366

Simonsen AJ, Krogstad PÅ (2005) Turbulent stress invariant analysis: clarification of existing terminology. Phys Fluids 17(8):088103
Smalley RJ, Leonardi S, Antonia RA, Djenidi L, Orlandi P (2002) Reynolds stress anisotropy of turbulent rough wall layers. Exp Fluids 33(1):31–37
Tachie MF, Balachandar R (2001) Shallow wakes generated on smooth and rough surfaces. Exp Fluids 30(4):467–474

A Free-Surface Immersed-Boundary Lattice Boltzmann Method for Flows in Porous Media

Ayurzana Badarch, John D. Fenton and Hosoyamada Tokuzo

Abstract The immersed-boundary Lattice Boltzmann method directly coupled with a free-surface flow treatment is introduced as an alternative numerical model for flows in porous media at the representative elementary volume scale. The liquid fraction values representing the presence of void in porous media are used as the modelled parameter which explicitly relates to the modelled permeability. The exact relation of the liquid fraction and the modelled permeability is examined by the analytical and numerical computations of flows through a U-tube with a porous zone. The proposed model with the modelled permeability relation and scaling of the permeability to the modelled permeability obtained provides good agreement with the analytical results convincing the extended applicability to flows in porous media. The model retains the inherent advantages for parallelization and the smooth treatment of a moving boundary.

Keywords Seepage flow · Immersed boundary · Free surface flow · Lattice Boltzmann method

1 Introduction

Flows through porous media, which can be referred to as seepage flow, are complex physical problems often encountered in engineering fields, and also in nature. Examples of flows in porous media in hydraulics can be flows under or past hydraulic structures, groundwater flows and part of the river flows percolated into the bed, to remark a few. Generally, a flow through any type of solid scattered structure is considered as a flow in porous media (Scheidegger 1958) and its flow definition can

A. Badarch (✉) · J. D. Fenton
Institute of Hydraulic Engineering and Water Resources Management, Vienna University of Technology, Karlsplatz 13/222, 1040 Vienna, Austria
e-mail: ayur_426@yahoo.com

H. Tokuzo
Department of Civil and Environmental Engineering, Nagaoka University of Technology, 1603-1 Kamitimioka, Nagaoka 940-2188, Japan

M. B. Kalinowska et al. (eds.), *Recent Trends in Environmental Hydraulics*, GeoPlanet: Earth and Planetary Sciences,
https://doi.org/10.1007/978-3-030-37105-0_2

be classified into pre-Darcy, Darcy and Forchheimer zones, depending on the magnitude of flow velocity through the porous medium. In environmental hydraulics, a flow in porous media is often described by Darcy's law (Darcy 1856), in which the velocity is proportional to the pressure gradient of the flow. Using Darcy's law and its extended version with Brinkman and Forchheimer, the numbers of analytical and numerical solutions have been applied for various problems to address static and time dependent dynamic behavior of the flow through porous media (Philip 1970). The governing equations for many models are the generalized Navier-Stokes equations (NSEs) with resistance term of the Brinkman-Forchheimer extended Darcy model (Gray and O'Neill 1976). Numerical methods both in Eulerian (Lacy and Prevost 1987) and Lagrangian approaches (Huang et al. 2013) were successfully applied to solve the generalized equations with different conditions such as with a free surface (Neuman and Witherspoon 1970). However, difficulties and limitations of those models due to the model formulation and complexity of porous medium and flows have still remained (Losada et al. 2016).

In the last two decades, the lattice Boltzmann method (LBM) has been extensively applied to fluid flows including porous-media flows (Chen and Doolen 1998) and free surface flows (Körner et al. 2005). The LBM has been dealing with porous-media flows at the pore-scale and the representative elementary volume scale (REV) (Guo and Zhao 2002). It can directly be applied to the pore-scale porous-media flow, where detailed interactions between fluid flow and solid are attained (Kutay et al. 2006). While there are two types of LBM at REV scale, one based on the generalized NSEs (Guo and Zhao 2002) and the other taking advantages of partial bounce-back rules in the LBM (Dardis and McCloskey 1998), have been developed.

Inspired by the simple nature of the immersed-boundary LBM (IB-LBM) proposed by Noble and Torczynski (1998), we introduce a novel REV scale IB-LBM to model porous-media flows. The proposed model allows, with the free-surface formulations of LBM (Körner et al. 2005), to be applied to free surface flows through unconfined porous media.

2 Numerical Models

2.1 Free-Surface Immersed-Boundary Lattice Boltzmann Method

Without losing generality, an additional collision term formulated from mass and momentum conservation perspectives and accounting for the interactions with each solid obstacle representing the porous media is introduced by Noble and Torczynski (1998) to the lattice Boltzmann equation. The evolution equation in the immersed-boundary lattice Boltzmann method is:

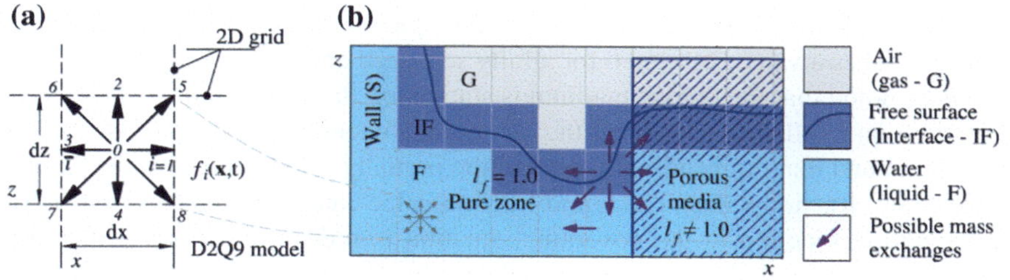

Fig. 1 Schematic description of a numerical model: **a** D2Q9 lattice arrangement on a cell, **b** free surface representation and zones for porous media and pure fluid according to the liquid fraction

$$f_i(\mathbf{x} + \mathbf{c}_i\delta t, t + \delta t) - f_i(\mathbf{x}, t) = -\frac{\delta t(1-\beta)}{\tau}\left(f_i(\mathbf{x}, t) - f_i^{eq}(\mathbf{x}, t)\right) + \beta f_i^m(\mathbf{x}, t) + \delta t(1-\beta)A_i, \tag{1}$$

where $f_i(\mathbf{x}, t)$—density distribution function (DF) at time t (as shown in Fig. 1a) in a two dimensional space, $\mathbf{x} = (x, z)$—position vector (m), $\mathbf{c}_i = (cx_i, cz_i)$—discrete lattice velocity, δt—lattice time step, τ—dimensionless relaxation time, which can be the total relaxation time, if a turbulent model is employed, $\beta = \left(1 - l_f(\mathbf{x}, t)\right)$—immersed boundary weighting function, $l_f(\mathbf{x}, t)$—lattice liquid fraction value, $f_i^{eq}(\mathbf{x}, t)$—equilibrium distribution function, $f_i^m(\mathbf{x}, t)$—additional collision term, and A_i—force term. At each lattice time step, the evolution equation is solved by two steps, namely collision and streaming, on fluid and interface cells, as shown in Fig. 1b. After the streaming step, the boundary conditions realized in forms of distribution functions should be applied. The macroscopic physical variables, pressure, p, and velocity vector, $\mathbf{u}$, can be defined by the evaluated distribution functions as

$$p = \frac{\rho}{3} = \frac{1}{3}\sum_{i=0}^{8} f_i \quad \text{and} \quad \frac{\rho\mathbf{u}}{l_f(\mathbf{x}, t)} = \sum_{i=0}^{8} \mathbf{c}_i f_i + \frac{\mathbf{F}\delta t}{2}, \tag{2}$$

where $\mathbf{F}(= \rho\mathbf{g}_L)$—force due to the gravitational acceleration, $\mathbf{g}_L\left(= \mathbf{g}_R\Delta t^2/\Delta x\right)$, L and R subscript stands for lattice and real physical values, and Δt and Δx are physical time step (s) and grid spacing (m), respectively.

In the present model, the equilibrium distribution for the fluid flow can be calculated by the second order of the expansion of the Maxwell distribution function (Chen and Doolen 1998) and the force term by Guo et al. (2002) is used. Among the several forms (Noble and Torczynski 1998), the additional collision term in Eq. (1) can be expressed as

$$f_i^m(\mathbf{x}, t) = f_i^{eq}(\rho, \mathbf{u}_s) - f_i(\mathbf{x}, t) + \left(1 - \frac{\delta t}{\tau}\right)\left[f_i(\mathbf{x}, t) - f_i^{eq}(\rho, \mathbf{u})\right], \tag{3}$$

where $\mathbf{u}_s$—velocity of the moving solid at new time on the cell. The liquid fraction of the cell takes a value between 0 for a fully solid cell and 1 for a fully fluid cell. It should be noted that the porous medium is stationary, the velocity, $\mathbf{u}_s$, will be zero. The additional collision term with the weighting function allows some degree of momentum and mass transmission through the partially filled solid cells depending on the liquid fraction value. The liquid fraction value can be used solely instead of the weighting function in the governing equation to reduce unnecessary memory consumption, only if the relationship between the liquid fraction and weighting function is linear and a problem is just a flow through porous media.

As shown in Fig. 1b, the four types of cells in the computational domain indicate materials (Badarch and Tokuzo 2018a, b), such as a motionless wall (W), air (G) and water (IF and F), while the cells are also differentiated according to pure fluid and porous medium regions. Zoning for the latter can be done by the liquid fraction value which can be a space function to be extended to the heterogeneous porous media. Coupling of the free surface model (Körner et al. 2005) with the immersed boundary lattice Boltzmann method is straightforward, so that one should ensure that the computation process occur only on the cells other than the G and W cells. The detailed explanation of the free surface model can be found in Körner et al. (2005), and the application to the free surface flows in hydraulics are given in Diao et al. (2018), Badarch et al. (2016), and many others. The free-surface-immersed boundary-LBM (FS-IB-LBM) has been applied for various problems, such as fluid-structure interactions (Diao et al. 2018; Badarch and Tokuzo 2018a, b), liquid-solid phase transitions (Badarch and Tokuzo 2017), and moving boundary problems (Strack and Cook 2007), but never for the porous-media flow. In following sections, we will discuss the applicability of the FS-IB-LBM for porous-media flows.

2.2 *The FS-IB-LBM for Flows in Porous Media*

From Darcy's law, the permeability of a porous medium can be defined as

$$K_p = \mathrm{u}\nu\Delta L/g\Delta H, \tag{4}$$

where u—seepage flow velocity (m s^{-1}), ν—kinematic viscosity of fluid (m^2 s^{-1}), g—magnitude of the gravitational acceleration (m s^{-2}), $\Delta H/\Delta L$—hydraulic gradient. By defining the variables in Eq. (4) in terms of the lattice Boltzmann method, the several types of permeability relation derived by different authors, such as Dardis and McCloskey (1998), Ginzburg (2016) and Walsh et al. (2009). By comparing the IB-LBM to models described in those works, one can simply check the differences. The difference between the IB-LBM with the model described in Walsh et al. (2009) or in Dardis and McCloskey (1998) is $f_i^{eq}(\mathbf{u}_s, \rho) - f_i^{eq}(\mathbf{u}, \rho)$, which can vanish if the moving velocity of solid is equal or very close to the seepage flow. Therefore, the relation by Walsh et al. (2009),

$$k = \frac{l_f \nu_L}{2(1 - l_f)}, \tag{5}$$

is chosen to be used for the simulation of porous media flows with the IB-LBM. In the relation, $k\ (\sim K_{pL})$ can be called the modelled permeability, not necessarily the same as lattice permeability, K_{pL}, which is scaled from the real permeability, ν_L—lattice fluid viscosity and $l_f(\mathbf{x}, t)$ is considered here instead of the solid fractions. The relation in Eq. (5) not only enables the IB-LBM to simulate heterogeneous porous media, but also helps to define correct connections for lattice parameters to their physical parameters.

3 Numerical Experiments: Flows in U-Tube Porous Media

To define the connections between parameters in the FS-IB-LBM to their physical counterparts, we have conducted the numerical experiments with different cases. The procedure for this is done by accurately matching the results obtained by the analytical and numerical simulations.

Simple gravity driven seepage flow through porous media packed in a U-tube was solved as a benchmark problem for numerical experiments. This problem has been considered for various numerical validations with different sizes (Huang et al. 2013). In this study, the problem has the same size configuration (see Fig. 2a) with Wang et al. (2016), however the simulation cases are different. Parameters of the problem were: length of the porous media $L = 1.0$ m, kinematic viscosity of water $\upsilon_k = 1.05 \times 10^{-6}$ m s^{-2}, simulation time 200 s.

The analytical solution of the head difference and seepage velocity in terms of elapsed time, t, can be computed as (Peng et al. 2017)

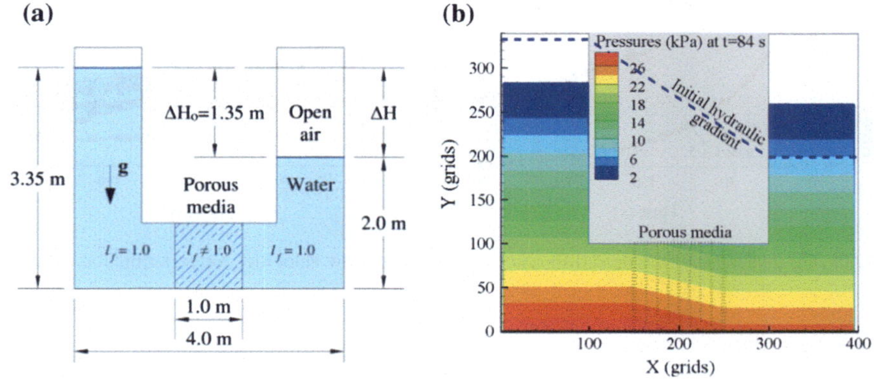

Fig. 2 **a** The geometry and initial condition of the flow through U-tube, **b** temporal hydrostatic pressure evaluation by the FS-IB-LBM at $t = 84$ s in Case 1

$$\Delta H = \Delta H_o \exp(-2K_h t/L) \quad \text{and}$$
$$u = \Delta H_o K_h \exp(-2K_h t/L)/L, \tag{6}$$

respectively. In the formulae, the permeability, $K_p(\mathrm{m}^2)$, can be inserted instead of the hydraulic conductivity using the relation $K_h = K_p g/\upsilon_k$.

The problem geometry is discretized into 400×340 grids spacing as $\Delta x = \Delta z = 0.01$ m for the simulation cases with the FS-IB-LBM. For all cases, the following parameters except the liquid fraction values were used: lattice gravity $\mathbf{g}_L = 1.28 \times 10^{-4}$, which gives the relaxation time $\tau = 0.5315$ and the physical time step $\Delta t = 3.61 \times 10^{-4}$ s. The initial condition is given by hydrostatic condition in the fluid region (Badarch and Tokuzo 2018a, b). For the solid wall, the slip boundary condition (Chen and Doolen 1998) is assigned.

The simulated results of the cases agree well with those of analytical solutions, as shown in Fig. 3. The corresponding case parameters for the analytical and numerical simulations are given in Table 1.

The FS-IB-LBM properly predicts the flow and pressure gradients (an example shown in Fig. 2b), which agree with Darcy's law. However, with the very low permeability, the model shows some degree of discrepancies in head differences and seepage flow velocities, as seen in Fig. 3, due to the low liquid fraction values. It is seen that the model under/overestimate the head difference and repeatedly overestimate the seepage flow for lower permeability. In the simulations, the seepage velocity was measured as average on the horizontal axis of the porous zone.

The permeability and its error were again calculated from the results shown in Fig. 3 by using Eq. (4). Maximum errors of the simulation results reached 37% in Case 1, which is slightly difficult to accept. However the global average error under 17% in the all calculated permeability in Case 1, errors drastically reduced in other cases as

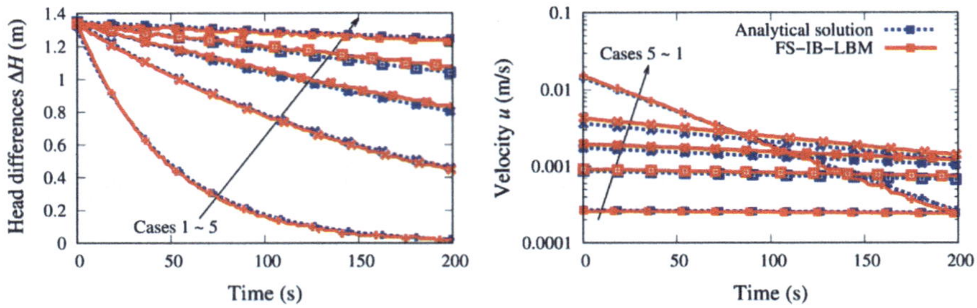

Fig. 3 Comparisons of the head differences and seepage flow velocities for experimental cases

Table 1 Parameters for experimental cases

Cases	1	2	3	4	5
Analytical solution K_h (m s^{-1})	0.01	0.0027	0.0013	0.00065	0.00020
Values of l_f in FS-IB-LBM	0.9	0.7	0.5	0.3	0.1

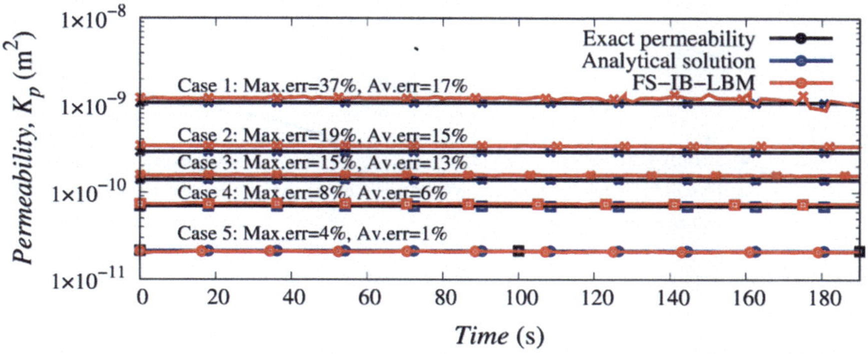

Fig. 4 Time series of recalculated permeability by the analytical solution and the FS-IB-LBM

seen in Fig. 4, gives much closer agreement with the exact permeability. Interestingly, the recalculated permeability error decreases with decreasing given permeability, while the accuracy of results improves with increasing given permeability. Thus, the model might be more sensitive in cases with higher permeability.

4 Discussion

The experiments show that with the correct relation of the permeability the original formulation of the IB-LBM with an appropriate additional collision term can be applied to the porous media flow directly. It can be seen that if l_f becomes lower approaching its limit, unlimited modelled permeability will be provided. In turn, the additional operator for the IB-LBM correctly reflects the distribution functions from the impermeable solid. In this sense, the model is also applicability for a flow through pore scale porous media, since the collision step functions as the simple bounce back boundary condition on the current cell.

In other case of $l_f = 1$, the additional operator is vanished and the standard LBM for a fluid flow will recover. It is worthwhile noting that viscosity in Eq. (5) should be regarded as the lattice fluid viscosity, which relates to the relaxation time, rather than the effective viscosity that is influenced by the eddy viscosity term if the sub-grid scale turbulent model is present. Also the liquid fraction value should be considered as the internal modelled parameter which directly relates to the modelled or lattice permeability. From the numerical experiments, we derived the scaling of the modelled permeability to the lattice permeability, as shown in Fig. 5a. The correct scaling of the lattice permeability to the modelled permeability or vice versa is essential for the accuracy of the model. The computed relations (Fig. 5b) between the modelled permeability and the liquid fraction values were similar to the Kozeny-Carman equation (Scheidegger 1958) and other relations derived in the lattice Boltzmann framework (Walsh et al. 2009).

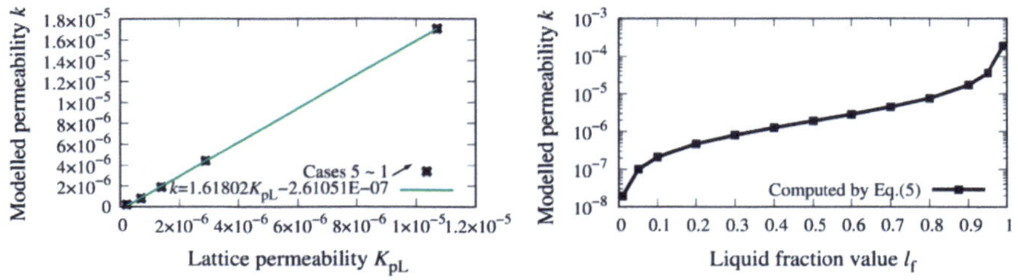

Fig. 5 Relations of the modelled permeability to the lattice permeability and liquid fractions

5 Conclusions

The paper has presented the applicability of the free-surface immersed-boundary lattice Boltzmann method to porous media flow which includes heterogeneous and unconfined media. The model employs the liquid fraction value as a modelled parameter of which a relation to the permeability has been defined and examined by the numerical experiments. The scaling of the physical permeability to the modelled permeability is found as $k \cong 1.618K_p/\Delta x^2$ based on the numerical experiments. The model can provide accurate solutions with low to high permeable media. The parallel nature of the model remains. The model can also be applied to the force estimation on porous objects and application can be found in coastal engineering and river engineering as well as other engineering fields involving porous and flexible media.

Acknowledgements The corresponding author would like to acknowledge the financial support by the Ernst Mach Grant—Eurasia-Pacific Uninet together with OeAD provided by the Austrian Federal Ministry of Education, Science and Research.

References

Badarch A, Tokuzo H (2017) Application of the lattice Boltzmann method to liquid-solid phase change in free surface flow: an example of Mongolian small hydropower plant. J Jpn Soc Civ Eng Ser B1 73(4):I_607–I_612

Badarch A, Tokuzo H (2018a) Development of numerical wave channel by LBM with immersed boundary method and its application to estimation of wave forces acting on a caisson breakwater. J JSCE Ser B2 (Coast Eng) 74(2):I_49–I_54

Badarch A, Tokuzo H (2018b) Lattice Boltzmann method for the numerical simulations of the melting and floating of ice. In: Kalinowska MB, Mrokowska MM, Rowiński PM (eds) Free surface flows and transport processes. GeoPlanet: earth and planetary sciences. Springer, Cham, pp 143–154

Badarch A, Tokuzo H, Narantsogt N (2016) Hydraulics application of the free-surface lattice boltzmann method. In: 11th international forum on strategic technology (IFOST). IEEE, pp 195–199

Chen S, Doolen GD (1998) Lattice Boltzmann method for fluid flows. Annu Rev Fluid Mech 30(1):329–364
Darcy H (1856) Les fontaines publiques de la ville de Dijon. Victor Dalmont, Paris
Dardis O, McCloskey J (1998) Lattice Boltzmann scheme with real numbered solid density for the simulation of flow in porous media. Phys Rev E 57(4):4834
Diao W, Cheng Y, M Xu, Wu J, Zhang C, Zhou W (2018) Simulation of hydraulic characteristics of an inclined overflow gate by the free-surface lattice Boltzmann-immersed boundary coupling scheme. Eng Appl Comput Fluid Mech 12(1):250–260
Ginzburg I (2016) Comment on "An improved gray Lattice Boltzmann model for simulating fluid flow in multi-scale porous media": intrinsic links between LBE Brinkman schemes. Adv Water Resour 88:241–249
Gray WG, O'Neill K (1976) On the general equations for flow in porous media and their reduction to Darcy's law. Water Resour Res 12(2):148–154
Guo Z, Zhao TS (2002) Lattice Boltzmann model for incompressible flows through porous media. Phys Rev E 66(3):036304
Guo Z, Zheng C, Shi B (2002) Discrete lattice effects on the forcing term in the lattice Boltzmann method. Phys Rev E 65(4):046308
Huang Y, Zhang W, Dai Z, Xu Q (2013) Numerical simulation of flow processes in liquefied soils using a soil–water-coupled smoothed particle hydrodynamics method. Nat Hazards 69(1):809–827
Körner C, Thies M, Hofmann T, Thürey N, Rüde U (2005) Lattice Boltzmann model for free surface flow for modeling foaming. J Stat Phys 121(1–2):179–196
Kutay ME, Aydilek AH, Masad E (2006) Laboratory validation of lattice Boltzmann method for modeling pore-scale flow in granular materials. Comput Geotech 33(8):381–395
Lacy SJ, Prevost JH (1987) Flow through porous media: a procedure for locating the free surface. Int J Numer Anal Methods Geomech 11(6):585–601
Losada IJ, Lara JL, del Jesus M (2016) Modeling the interaction of water waves with porous coastal structures. J Water Way Port Coast Ocean Eng 142(6):03116003
Neuman SP, Witherspoon PA (1970) Finite element method of analyzing steady seepage with a free surface. Water Resour Res 6(3):889–897
Noble DR, Torczynski JR (1998) A lattice-Boltzmann method for partially saturated computational cells. Int J Mod Phys C 9(08):1189–1201
Peng C, Guofang X, Wei W, Hai-sui Y, Chun W (2017) Multiphase SPH modeling of free surface flow in porous media with variable porosity. Comput Geotech 81:239–248
Philip JR (1970) Flow in Porous Media. Ann Rev Fluid Mech 2:177–204
Scheidegger A (1958) The physics of flow through porous media. University Of Toronto Press, London
Strack OE, Cook BK (2007) Three-dimensional immersed boundary conditions for moving solids in the lattice-Boltzmann method. Int J Numer Methods Fluids 55(2):103–125
Walsh SD, Burwinkle H, Saar MO (2009) A new partial-bounceback lattice-Boltzmann method for fluid flow through heterogeneous media. Comput Geosci 35(6):1186–1193
Wang C, Wang Y, Peng C, Meng X (2016) Smoothed particle hydrodynamics simulation of water-soil mixture flows. J Hydraul Eng 142(10):04016032

Modelling River Flow Through In-Stream Natural Vegetation for a Gravel-Bed River Reach

Simon D. A. Clark, James R. Cooper, Ponnambalam Rameshwaran, Pamela Naden, Ming Li and Janet Hooke

Abstract Macrophytes (aquatic vegetation) are known to modify river flow by reducing local velocities, increasing turbulence generation, and reducing channel conveyance capacity resulting in increased flow depth. Understanding flow response to vegetation change is imperative to inform flood mitigation strategies, however the field is nascent with much uncertainty surrounding the estimation of flow conveyance in vegetated channels and, subsequently, the best vegetation management practices and possible flood risk. Therefore, in order to develop an understanding of how macrophytes modify flow conveyance modelling techniques must be developed which can effectively represent the three-dimensional effect of natural vegetation within an open channel flow environment. This study simulated vegetation-flow interaction using a 3D finite-element modelling method to investigate the feasibility to which a vegetated natural gravel-bed river can be represented using the drag-force approach. Two methods of parameterising vegetation-induced drag were explored to investigate the effect of flow behaviour within a vegetated gravel-bed river reach.

Keywords Computational fluid dynamics (CFD) · Macrophytes · Plant patches · Fluvial · Finite element

1 Introduction

Previous research has implicated macrophytes (aquatic vegetation) as a key factor controlling flow conveyance. Macrophytes increase local momentum absorption and turbulence generation whilst decreasing channel conveyance capacity (Green 2006). This effect is amplified by the increased turbulence-inducing shear zones resulting from a greater vegetated surface area provided by a larger cross-sectional biomass (Rameshwaran et al 2014). This results in lower velocity, exaggerating water depths

S. D. A. Clark (✉) · J. R. Cooper · M. Li · J. Hooke
The University of Liverpool, Liverpool, UK
e-mail: s.d.a.clark@liverpool.ac.uk

P. Rameshwaran · P. Naden
UK Centre for Ecology and Hydrology, Wallingford, UK

M. B. Kalinowska et al. (eds.), *Recent Trends in Environmental Hydraulics*, GeoPlanet: Earth and Planetary Sciences,
https://doi.org/10.1007/978-3-030-37105-0_3

(Marjoribanks et al. 2014). This has ramifications for river management strategies concerned with flood mitigation, where the influence of macrophytes may exaggerate flood depths. Despite this, macrophytes have been a neglected component within flood estimation models, resulting in a high degree of uncertainty surrounding weed management in flood risk mitigation (Marjoribanks et al. 2014). Macrophytes act as a source of hydraulic resistance, thus their impact can be determined by introducing additional drag (Nepf 2011). Previous research has described how vegetation alters the three dimensional (3D) flow structure (Folkard 2011), suggesting that 3D modelling approaches are required to fully capture the impacts of vegetation. Current numerical approaches investigating reach-scale influences of vegetation generally favour Reynolds-averaged Navier-Stokes equations (RANS) methods and their derivatives which are used to resolve local flow and turbulence induced features of the time-averaged flow field. RANS approaches can model flow through vegetation by introducing additional source terms to the RANS and turbulence transport equations to account for vegetative drag effects (Wang et al. 2015). The drag force approach has increasingly been adopted to model the influence of stem drag by using a drag term in the momentum equations (Nepf 2011). This study attempts to assess the feasibility to which a natural, vegetated gravel-bed river can be represented using a 3D free surface flow application with a drag force approach.

The paper focuses on a vegetated reach of the River Blackwater, Hampshire, UK, a site of historical research which provided necessary information regarding local hydraulics and seasonal macrophyte distribution (Fig. 1). This study used a 140 m reach with an average width of 7.22 m. Flow data was measured at five cross-sections using a 3-D Acoustic Doppler Velocimeter (ADV) during May 2001 and at a single cross-section at the apex of the channel in February 2003. Measurements of the water surface profiles, local topography, and vegetation patches were taken using the Zeiss Elta R55 Total Station. Measurements are detailed in Naden et al. (2006).

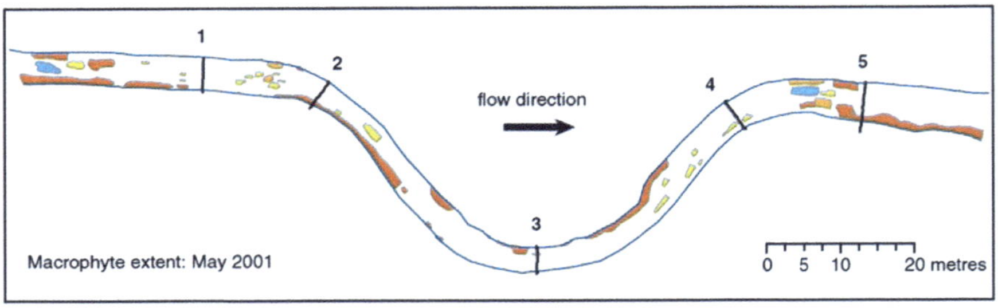

Fig. 1 The River Blackwater, Hampshire, UK, during the May (2001) vegetated period showing the five sampled cross-sections and the dominant macrophyte species: *Sparganium erectum* (orange), *Sparganium emersum* (yellow), *Potamogeton natans* (blue) (Naden et al. 2006)

2 The Three-Dimensional Flow Model

This study used Telemac 3D, an open-source finite-element method modelling suite used for free surface flows (Hervouet 2007). The flow variables were solved using the RANS equations and a k-ε turbulence closure model. The boundary conditions for the model were prescribed as the mean discharge at the inlet, and mean water depth at the outlet.

Gravel-bed drag: The gravel-bed roughness was parameterized using the spatial-averaging approach detailed in Rameshwaran et al. (2011). The drag term for the gravel-bed was represented by prescribing four layers with a vertical spacing of 0.01 m at the bottom of the channel using the spatially averaged properties of the gravel bed (Table 1) and estimated as (Rameshwaran et al. 2014):

$$F_{gi} \approx -\frac{1}{2} C_{dg} S_{fg} A_p |\langle \overline{u_i} \rangle| \langle \overline{u_i} \rangle \tag{1}$$

where C_{dg} represents a single roughness element on the bed, S_{fg} is the sheltering factor due to the proximity of other roughness elements, A_p is the average projected area per unit volume, and $|\langle \overline{u_i} \rangle|$ is the time-space averaged velocity.

Method 1: Representing vegetation using bulk drag: The vegetation-induced drag force F_{vi} can be approximated by the bulk drag coefficient C_{dv} (Rameshwaran et al. 2014):

$$F_{vi} \approx -\frac{1}{2} C_{dv} S_{fv} A_s |\langle \overline{u_i} \rangle| \langle \overline{u_i} \rangle \tag{2}$$

where C_{dv} represents vegetation roughness, S_{fv} is the sheltering factor due to the proximity of other vegetation, and A_s is the averaged surface area of the vegetation element per unit volume.

Method 2: Representing vegetation using patch shape: At patch scale, the drag force is significantly influenced by biomass volume and the patch geometry, thus describing the patch in terms of aspect ratio is appropriate. Using this approach the drag force term for a vegetation patch can also be approximated as (Rameshwaran and Naden 2012):

Table 1 Averaged areal values for gravel-bed elements

Layer	Projected area A_{px} (m^2/m^3)	Projected area A_{py} (m^2/m^3)	Surface area A_s (m^2/m^3)
1 (0–10 mm)	1.6546	1.9889	13.8340
2 (10–20 mm)	12.0348	15.0840	77.5191
3 (20–30 mm)	12.3361	14.8194	76.6161
4 (30–40 mm)	2.1330	2.4954	15.3330

Table 2 Averaged areal values for represented vegetation species

Species	Surface area A_s (m^2/m^3)
Spargnium erectum	24.381
Spargium emersum	77.666
Potamogeton natans	27.000

$$F_i \approx -\frac{1}{2} C_{d0} R_f A_s |\langle \overline{u_i} \rangle| \langle \overline{u_i} \rangle \qquad (3)$$

where the new drag coefficient C_{d0} represents the vegetation patch with an aspect ratio almost equal to zero, and the R_f factor describes the aspect ratio of the individual patch as $R_f (= C_d / C_{d0})$ estimated using the relationship described by Blevins (2003). The vegetative drag term was represented by prescribing head losses at all finite element mesh nodes on the vertical within the patch boundaries and applying the areal values (Table 2) to Eqs. (2) or (3).

Quiescent initial conditions with appropriate boundary conditions were used, and the model was run until convergence was reached. The model was calibrated for both vegetated and non-vegetated cases by successively adjusting the respective drag coefficient until discrepancies between the observed and simulated free surface data were at a minimum.

2.1 Flow Over the Non-vegetated Gravel Bed River

For the non-vegetated case, the gravel bed bulk drag coefficient $C_{dg} S_{fg}$ calibrated at 0.54 which is within the same order of magnitude as similar studies and falls within a similar range for non-vegetated gravel-bed rivers such as, for example, 0.45 (Carney, et al. 2006; Rameshwaran et al. 2014), 0.27–0.85 (Rameshwaran et al. 2011), 0.69 (Rameshwaran and Naden 2012), and 0.23–0.87 (Zeng and Li 2012). Figure 2 shows that the flow profiles for the modelled streamwise velocity and turbulent kinetic energy (TKE) matched with the observed profiles reasonably well, however discrepancies between the velocity profiles and measured data occurred at the banks. TKE is underpredicted at the near-surface zone for both profiles towards the left bank, and overpredicted near the bed at the right bank (D). These discrepancies may be the result of local heterogeneity not captured by the averaging approach, or environmental factors not considered in the model, such as dead or overhanging riparian vegetation. Generally, the model profiles exhibit a close match with those of the measured flow quantities and indicates that the spatial-averaging method effectively represents hydraulics of a natural non-vegetated river. The cross-sectional flow profile shows the velocity vectors for secondary flow circulation, a rotation pattern that has been described by other measurements investigating secondary flow patterns at meander apex (Rameshwaran et al. 2013) suggesting that the flow pattern is predicted reasonably well.

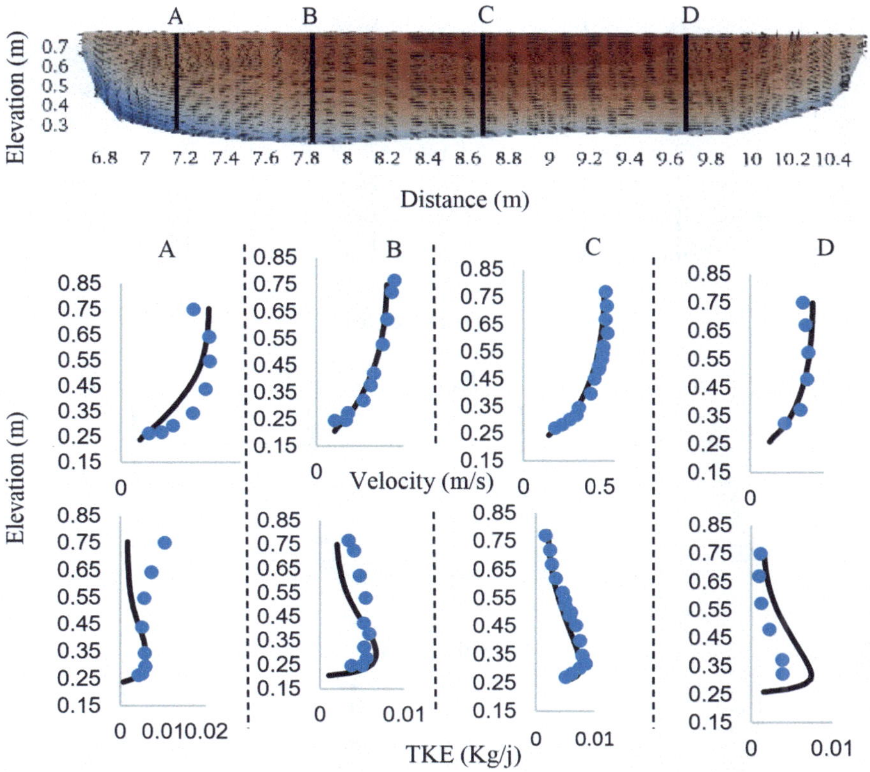

Fig. 2 Calibrated velocity and TKE profiles for the non-vegetated February period against the measured (circles) and modelled (solid line) data. Secondary flow circulation is shown with the location of the measured profiles (black lines)

2.2 *Flow Through a Vegetated Gravel Bed River*

For the May (2001) vegetated case, method one calibrated at $C_{dv}S_{fv}$ 0.04, whilst method two calibrated at C_{d0} 0.34. Both methods predicted the surface energy slope well with little differences between them. The calibrated value of the first approach is significantly lower than most other values discovered in the literature which reported values of, for example, 0.2–1.1 (Nepf 1999), 0.01–0.1 (Sand-Jensen 2003) 0.15 (Rameshwaran and Naden 2012), and 0.65–1.06 (Mulahasan and Stoesser 2017). For the second approach the coefficient is of the same order of magnitude and falls within the range of the values found in similar studies. Figure 3 compares the simulated streamwise velocity and TKE profiles with the measured data at Sect. 3 of the reach. For both methods the streamwise velocity is approximated reasonably well, capturing the characteristic logarithmic curve, and differences between the methods are small. The most explicit difference is the turbulent profile at the right bank, D, where

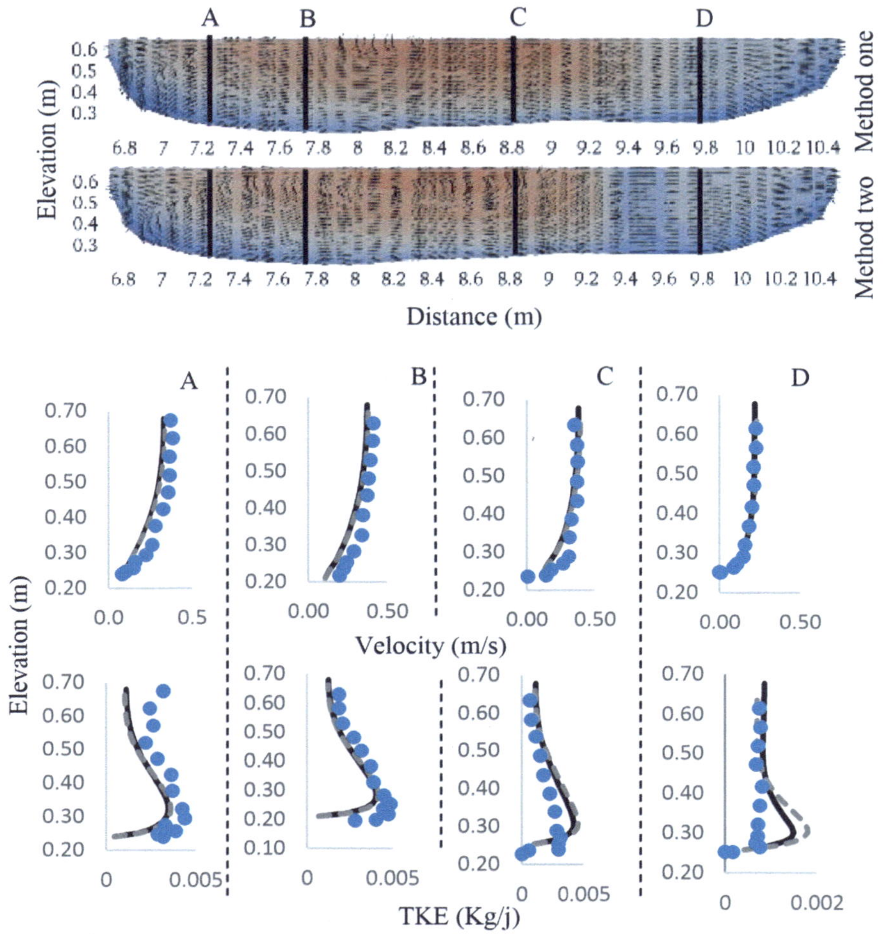

Fig. 3 Calibrated profiles for the vegetated May period for both method one (dashed line) and method two (solid line) compared against the measured data (circles) for the apex section of the channel

the lower half of the profile is overpredicted for method one. The velocity profiles match the measured data well at locations C and D, but are slightly underpredicted at A. The TKE profiles are generally underpredicted left of the channel centre, and overpredicted towards the right bank (D). The greatest deviation in TKE is at the banks where the turbulent energy is underpredicted for the near-free surface zone (A) and underpredicted at the shear zone for D. Discrepancies are apparent in the description of secondary flow between the two methods, with method two presenting a similar flow pattern to the non-vegetated case. However, method one describes a less distinct circulation pattern across the channel, with rotation occurring at the near-surface region near the channel centre.

2.3 Modification of Flow Conveyance by Patch

Figure 4 illustrates the simulated streamwise velocity and TKE quantities affecting the free surface at the leading limb of the meander for the vegetated May case. The influence of the vegetation patches is obvious in the pattern of both the velocity and TKE fields. Velocity is clearly shown to lose momentum both within the patch and immediately downstream where a wake is produced. An increase in velocity occurs at the patch boundary due to flow bifurcation. An increase in TKE is evident at the patch boundaries as the accelerated flow induces a strong shearing effect. Cross-sections of an affected region show how, upstream of a patch (A), the velocity field is largely affected by roughness at the channel bottom and acceleration due to the meander bend. Following bifurcation at the patch (B) velocities accelerate throughout the flow column in the non-vegetated channel whilst a significant deceleration is observed within the patch. Local variations in flow around and within vegetation patches can be attributed largely to the geometry of the patch profile and planform.

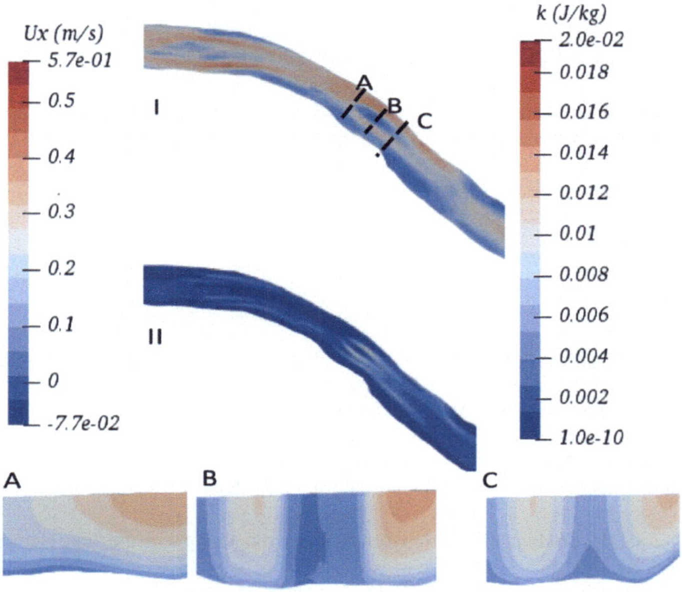

Fig. 4 Spatial distribution of streamwise velocities (I) and TKE (II) for the initial limb of the meander, including cross-sections of in-stream vegetation patch

3 Conclusion

This paper has discussed a numerical method for parameterising vegetation-induced drag at the patch scale, an analysis of model results for non-vegetated (February 2003) and vegetated (May 2001) cases, and assessed the impact of vegetation patches upon 3D flow. Both drag approaches perform reasonably well, simulating the heterogenic flow field resulting from vegetation-induced hydrodynamic flow behaviour. Flow profiles for streamwise velocity and turbulent kinetic energy (TKE) generally match the observed data. Deviations within both profiles could result from difficulties in capturing the natural heterogeneity of channel geometry during surveying and the presence of other drag sources in the channel, such as dead vegetation. Vegetated patches have been shown to be important flow modifiers within natural rivers. The production of high values of TKE and flow velocity around the edges of patches correspond to flow acceleration and subsequent shearing at the patch boundary. Further research focusing on the impact of vegetation on flow in natural channels is required. This study will provide a basis for developing scenario-based experiments investigating the impact of climate change on variation in patch size and river flow. This has profound implications for river management strategies, such as those concerned with natural-based solutions to flood risk mitigation.

Acknowledgements This study has been financed by the Engineering and Physical Sciences Research Council as part of their Centre for Doctoral Training at the Institute for Risk & Uncertainty, Liverpool, UK.

References

Blevins RD (2003) Applied fluid dynamics handbook. Krieger Publishing Company, USA, 558 pp

Carney SK, Bledsoe BP, Gessler D (2006) Representing the bed roughness of coarse-grained streams in computational fluid dynamics. Earth Surf Proc Land 31(6):736–749

Folkard AM (2011) Flow regimes in gaps within stands of flexible vegetation: laboratory flume simulations. Environ Fluid Mech 11(3):289–306

Green JC (2006) Effect of macrophyte spatial variability on channel resistance. Adv Water Resour 29(3):426–438

Hervouet JM (2007) Hydrodynamics of free surface flows: modelling with the finite element method. Wiley, New York

Marjoribanks TI, Hardy RJ, Lane SN (2014) The hydraulic description of vegetated river channels: the weaknesses of existing formulations and emerging alternatives. Wiley Interdiscip Rev: Water 1(6):549–560

Mulahasan S, Stoesser T (2017) Flow resistance of in-line vegetation in open channel flow. Int J River Basin Manag 15(3):329–334

Naden P, Rameshwaran P, Mountford O, Robertson C (2006) The influence of macrophyte growth, typical of eutrophic conditions, on river production. Hydrol Process 20(18):3915–3938

Nepf HM (1999) Drag, turbulence, and diffusion in flow through emergent vegetation. Water Resour Res 35(2):479–489

Nepf HM (2011) Flow over and through biota. In: Wolanski E, McLusky D (eds) Treatise on estuarine and coastal science. Academic Press, Waltham, pp 267–288

Rameshwaran P, Naden PS (2012) Modelling the influence of macrophyte patches on river flow. In: Murillo Muñoz RE (ed) River flow 2012. CRC Press, Boca Raton, pp 163–170
Rameshwaran P, Naden PS, Lawless M (2011) Flow modelling in gravel-bed rivers: rethinking the bottom boundary condition. Earth Surf Proc Land 36(10):1350–1366
Rameshwaran P, Naden P, Wilson CAME, Malki R, Shukla DR, Shiono K (2013) Inter-comparison and validation of computational fluid dynamics codes in two-stage meandering channel flows. Appl Math Model 37(20):8652–8672
Rameshwaran P, Sutcliffe A, Naden P, Wharton G (2014) In: Schleiss AJ, de Cesare G, Franca MJ (eds) Modelling river flow responses to weed management, pp 467–474
Sand-Jensen K (2003) Drag and reconfiguration of freshwater macrophytes. Freshw Biol 48(2):271–283
Wang C, Zheng SS, Wang PF, Hou J (2015) Interactions between vegetation, water flow and sediment transport: A review. J Hydrodyn 27(1):24–37
Zeng C, Li CW (2012) Modeling flows over gravel beds by a drag force method and a modified S-A turbulence closure. Adv Water Resour 46:84–95

On the Use of Surface PIV for the Characterization of Wake Area in Flows Through Emergent Vegetation

J. Leonardo Corredor-Garcia, Alexandre Delalande, Virginia Stovin and Ian Guymer

Abstract New results from surface PIV (Particle Image Velocimetry) measurements are presented. Surface PIV can potentially provide researchers with a cheap and versatile method for mapping 2D flow fields. This technique was evaluated in a laboratory flume with a random distribution of rigid plastic straws, to simulate flows through emergent vegetation. Velocities were computed via an open-source tool for conventional PIV, and a sensitivity analysis conducted, in which the factors seeding particle size, particle image density, size of interrogation window, number of passes and contrast were evaluated. Results show that, with the appropriate settings, 98.7% of data points were considered to be reliable. It was found that the best quality velocity maps were obtained with small seeding particles and intermediate window resolutions (16 × 16 pixels). The practical use of this technique is illustrated by using the data to identify the portion of flow through vegetation occupied by wakes. For this, a straightforward criterion, related to the incident flow conditions and generated vorticity, is proposed. Further refinements of this research can lead to applications in several branches of fluid mechanics, such as in situ measurements of the flow field and analysis of scalar dispersion processes in ecohydraulics.

Keywords Surface PIV · Vegetated flow · Vorticity · Wake area · Ecohydraulics · PIV · Large scale PIV

1 Introduction

In the study of ecohydraulics, it is well known that hydrodynamic properties play an important role in determining flow and scalar transport processes (Nepf 2012), bed erosion and siltation, and the condition of vegetation species (Jadhav and Buchberger

J. L. Corredor-Garcia (✉) · V. Stovin · I. Guymer
University of Sheffield, Sheffield, UK
e-mail: jlcorredorgarcia1@sheffield.ac.uk

A. Delalande
INSA (Institut National des Sciences Appliquées) Lyon, Lyon, France

M. B. Kalinowska et al. (eds.), *Recent Trends in Environmental Hydraulics*, GeoPlanet: Earth and Planetary Sciences,
https://doi.org/10.1007/978-3-030-37105-0_4

1995). The specific nature of this dependency is not well known. This problem is exacerbated in flows through emergent vegetation, where complexities in the distribution and morphology of macrophytes (i.e. plants) induce largely non-deterministic turbulence patterns (Ricardo et al. 2016a) and additional sources of resistance (Nepf 1999). Furthermore, experimental techniques currently available for the quantification of such phenomena are limited to small study areas to provide temporally and spatially accurate results. A trade-off must be made between the scope and reliability of information and the amount of resources available for the tests.

Two types of measurement techniques have been used to retrieve flow information from vegetated flow experiments. First are the point measurements of velocities via Acoustic Doppler Velocimetry (ADV) or Laser Doppler Anemometry (LDA) (Lohrmann et al. 1994), which, although can achieve high sampling frequencies, are limited to single locations within the flow field. These are useful for profiling and bulk quantities, but not for extracting information regarding coherent flow structures. The measurements of the second type are non-intrusive, optical and taken over planar sections of the flow field, which generate velocity maps from illuminated planes via PIV (Particle Image Velocimetry) or PTV (Particle Tracking Velocimetry) (Raffel et al. 2018); they compute the relative movement of particles through image correlation.

PIV is widely used for spatial analysis of the velocity field; however, its usage is limited due to the necessity of high-specification optical, computational and illumination equipment. Alternatively, a simpler variation of the technique, called Large Scale PIV (Fujita et al. 1998), has been proposed to measure velocities from surface images. LSPIV has been used to obtain discharges and mean velocities from surface measurements in laboratory flumes (Osorio-Cano et al. 2013; Gharahjeh and Aydin 2016) and rivers (Dobson et al. 2014), but no evidence of its use for a detailed measurement of small scale flow structures, in vegetated flow, has been reported.

This experimental investigation has, therefore, a twofold purpose: first, to present a sensitivity analysis of the relevant parameters for a successful application of the technique, for small scale mapping of flows through vegetation; and second, to relate the velocity maps obtained to the problem of quantifying the proportion of a flow field occupied by stem wakes, a key descriptor of longitudinal dispersion in flow through emergent vegetation (White and Nepf 2003). The term Surface PIV is used instead of LSPIV, to indicate that, while a corpus of methods has been developed for LSPIV (Osorio-Cano et al. 2013; Gharahjeh and Aydin 2016; Patalano et al. 2017), particularly for large scale applications (e.g. rivers, estuaries), this research uses tools devised for conventional PIV. The experimental configuration is described in Sect. 2, the sensitivity analysis and results are reported in Sect. 3 and the conclusions presented in Sect. 4.

2 Experimental Configuration

The experiments were conducted in a 300 mm wide flume, the specific details of which are covered in Sonnenwald et al. (2018), with the camera being located above

Table 1 Flow conditions for the tests under study. U_∞ is the free stream velocity, Re_d is the stem Reynolds number and S_0 is the bed slope of the flume

Test	U_∞ (mm s^{-1})	Re_d –	S_0 –
Test 1	13	47.27	0.0
Test 2	20	72.73	7.0×10^{-6}
Test 3	40	145.45	1.1×10^{-4}
Test 4	69	250.91	4.6×10^{-4}

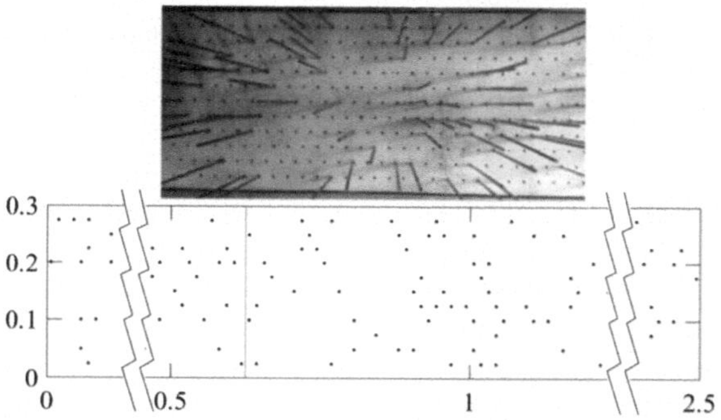

Fig. 1 Random array of synthetic emergent vegetation. Measurement section: 550 mm × 300 mm; depth of water in the flume: 150 mm

the flume so that the focus was perpendicular to the bed, at a small enough distance to avoid projection errors, thereby avoiding the need for orthogonal corrections (Fujita et al. 1998). Surface PIV measurements were made with random synthetic emergent vegetation, composed of stems (plastic straws) with the following morphological parameters: diameter (d), 4 mm; density (n), 398 stems m^{-2}; frontal facing area of stems per unit volume (Nepf 1999) (a), 1.29 m^{-1}; and solid volume fraction (ϕ), 0.005. The velocity measurements had uniform flow and a constant depth of 150 mm for all tests; with the slope of the flume changed to maintain uniformity. The specific flow conditions are presented in Table 1.

Velocity measurements were taken in a section of the flume covering a plan area of 550 mm x 300 mm (the latter being the entire width of the flume), and the emergent vegetation was placed randomly, as shown in Fig. 1. The surface PIV illumination and image acquisition devices were as follows.

- **Camera**: Panasonic LUMIX DMC-GX80 + polarizing filter.
- **Lens**: Panasonic H-FS12032/G VARIO 12–32 mm F3.3-5.6 ASPH OIS.
- **Illumination**: 500 LED Ultra Bright, 5500 K Panel (35.50 × 20.30 × 10.0 cm).
- **Image resolution**: 3840 × 2160 pixels - **Image frequency**: 25 fps.

In addition to the camera and illumination settings, the seeding particles also require particular attention. To date, experimental analyses have focused on the use of large plastic seeding particles for PTV (Osorio-Cano et al. 2013; Patalano et al. 2017) and PIV (Novak et al. 2017), natural surface foam (Dobson et al. 2014), and fine grains

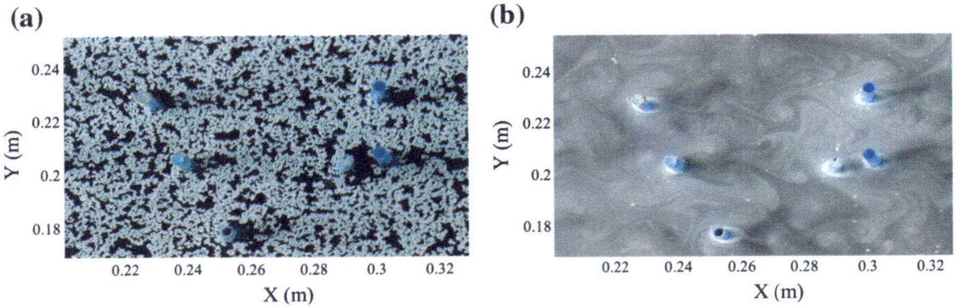

Fig. 2 Influence of particle image density on image quality for surface PIV analysis. **a** Higher local contrasts for a better computation of velocities through correlation. **b** The ideal case of vortices being visible. However, uniform colour and reflectivity of the smaller particles make it difficult to compute accurate velocity maps in any PIV analysis software. Flow moves from left to right

(Muste et al. 2014; Gharahjeh and Aydin 2016). However, the discussions revolve around its use for surface PIV in rivers, disregarding any application for visualization of turbulence structures. To look at the effects associated with the selection of seeding particles for flow through vegetation, two types were used in this study, which had similar physical properties (density and reflection) but different diameters, leading to differences in particle image density. One important feature of surface PIV is that it requires comparatively larger particle image densities than conventional PIV to avoid transparency effects. The extent to which excessively high, or low, particle image densities can affect the velocity measurements depends on the particle size. For smaller particles, excessively high densities can lead to an undesirable uniformity in the pixel intensity, whereas very low densities can lead to interference from the channel bottom. Figure 2 shows frames for large and small particles, respectively, where translucent spots can be identified behind stems (Fig. 2a) and some uniform zones are also visible (Fig. 2b).

Coarser particles can also lead to undesired effects. First, for extremely high particle image densities, the particles can travel sufficiently packed together as to be insensitive to the effects of the flow, particularly at the smallest scales. Second, for low particle densities the effects of light going through the surface and inducing transparency in the images persist. While larger particles (Fig. 2a) offer higher local intensity contrast, which allows PIV analysis tools to compute velocities more reliably, smaller particles allow for a better visual identification of turbulent structures which may be overlooked by larger particles.

3 Calculation of the Wake Area from Velocity Maps

3.1 Computation of Velocity Maps and Sensitivity Analysis

The image pre-processing, evaluation and post-processing settings for surface PIV are different to those for conventional PIV and LSPIV, and hence require further discussion. To get reliable information, it is necessary to know the limitations of the experiment and adjust the computational parameters accordingly. To account for the particle image density, it is necessary to select an appropriate size of interrogation window. Also, depending on the range of velocities expected and extent of the flow field, the correlation method to compute velocities must be chosen.

Flows through emergent vegetation comprise a wide range of velocities, which, if computed with a single interrogation window, will include significant background noise due to relative displacement of particles, caused by velocity variations at different scales (Thielicke and Stamhuis 2014; Raffel et al. 2018). To capture the entire spectrum of velocities, correlations are computed sequentially with interrogation windows of different sizes (Stanislas et al. 2008). Calculating multiple correlations in the spatial domain involves a larger computation time than in the frequency domain, with no difference in accuracy; therefore, correlations were calculated via the Discrete Fourier Transform (DFT) method, which guaranteed the same accuracy with a lower computational cost. Each correlation computed with an interrogation window size is called a pass (Thielicke and Stamhuis 2014) and the passes used to calculate velocities for the experiments are given in Table 2.

To process the information, the free-access, open-source PIV analysis tool, PIVlab (Thielicke and Stamhuis 2014), was used. Table 2 shows the size of interrogation areas, for different velocity resolutions and particle sizes. For the large particle size, the velocity fields were calculated using the Intermediate and Fine Resolution; and all resolutions were tried on the smaller particles. Also, to test the effect of the number of passes, all tests for the small particles size were processed using the last three interrogation areas only (columns 3, 4 and 5 in Table 2).

To quantify the effect of using a particular configuration, all tests were evaluated considering the proportion of points with valid velocities, which were defined as values that were not non-numeric errors (NaN in MATLAB), which result from poor correlations between consecutive frames (Thielicke and Stamhuis 2014), and were also inside the range of three standard deviations from the global mean. Table 3 shows

Table 2 Number of passes and sizes of interrogation windows used for each of the tests, and for both the big and small particles. Abbreviations: Res - Resolution; CR - Coarse Resolution; IR - Intermediate Resolution; FR - Fine Resolution

Res.	Pass 1	Pass 2	Pass 3	Pass 4	Large	Small
	Interrogation area px × px				Part. Size	Part. Size
CR	256 × 256	128 × 128	64 × 64	32 × 32		×
IR	128 × 128	64 × 64	32 × 32	16 × 16	×	×
FR	64 × 64	32 × 32	16 × 16	8 × 8	×	×

Table 3 Percentage of valid data per test, calculated as a ratio between the numeric points within three standard deviations from the global mean, and the total number of velocity points per test

Test	Small particles					Large particles	
	4 Passes			3 Passes		4 Passes	
	CR (%)	IR (%)	FR (%)	CR (%)	IR (%)	IR (%)	FR (%)
Test 1	99.2	99.4	97.3	99.3	99.3	98.8	98.2
Test 2	98.9	98.8	95.1	99.2	98.6	98.9	98.1
Test 3	98.6	98.2	85.8	98.8	97.0	98.6	97.8
Test 4	97.4	97.9	80.4	97.6	94.2	98.4	97.0
Mean	98.5	**98.6**	89.7	98.7	97.3	**98.7**	97.8

Note Percentage of valid data per test, calculated as a ratio between the numeric points within three standard deviations from the global mean, and the total number of velocity points per test. Numbers in bold font represent the best performing configuration for each particle size

the proportion of valid measurements for each test and each configuration. It is evident that the number of passes does not affect the computation of velocities, meaning that 3 passes are enough to cover all velocity scales for these experiments. Calculations are more sensitive to the size of the interrogation area, with the finer resolution performing consistently worse than the other ones. The tests with small particle sizes, together with intermediate resolutions (Table 2) were found to perform best.

This performance analysis allows for the selection of the settings that require less post-processing. To correct erroneous information, the PIV post-processing method PODDEM (Proper Orthogonal Decomposition Detection and Estimation) (Higham et al. 2016; Higham 2017) was used. To illustrate the post-processing analysis, Fig. 3 shows the velocity calculated for a frame of Test 2 ($U_\infty = 20$ mm s^{-1}), using the Intermediate Resolution configuration of Interrogation Areas (Table 2) for large particles (Fig. 3a), and the result of the correction process for that frame (Fig. 3b); then compared to the Intermediate Resolution for small particles (Fig. 3c), and the result of the data correction for that frame (Fig. 3d).

The presence of outliers does not allow for a clear visualization of local velocity gradients, because these extreme values unrealistically expand the range of velocities, as can be seen in the colour scales of Fig. 3a, c, and the fact that the velocity spectrum is visually more uniform compared with the reduction in the range of velocities for Fig. 3b, d, after the post-processing. Notwithstanding the initial accuracy, there will always be a degree of post-processing necessary to reveal the relevant velocity gradients. In general, small particles allow for a better identification of local velocity gradients, particularly between defect zones behind stems and high velocities between stems (Fig. 3c, d), something usually overlooked by tests with large particles due to their insensitivity to small scale movements.

In addition to detecting outliers, selecting the correct configuration parameters for a detailed description of the flow field influences the accuracy of the velocity gradients obtained. Figure 3b, d represent the same experimental conditions, and quantitatively their mean values will be similar; however, the velocity maps reveal substantial differences in the velocity gradients and the effect of vegetation on the distribution

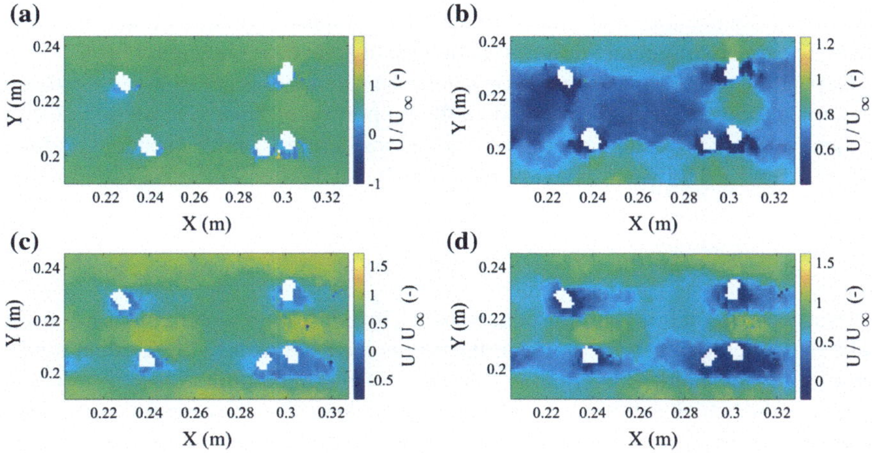

Fig. 3 Difference between results of the velocity analysis of a frame from Test 2 with: **a** Large Particles and Intermediate Resolution; **b** Post-processing of frame (**a**); **c** Small particles and Intermediate Resolution, **d** post-processing outcome of frame (**c**)

of velocities. Defining a wake requires a detailed knowledge of the flow field, which evidently will not be revealed with the methods currently used in applications of LSPIV (Patalano et al. 2017; Dobson et al. 2014; Osorio-Cano et al. 2013).

3.2 Definition of a Criterion for the Wake Area

To relate longitudinal dispersion to flow through emergent vegetation, White and Nepf (2003) defined the wake as the combination of the immediate recirculation zone behind a stem, and the defect velocity zone that spreads further downstream, and described the dispersion coefficient (D_x) in terms of the fraction of the flow field occupied by wakes. However, emergent vegetation also has an impact on shear layer detachment and reattachment (Sumner 2010), secondary vortex formation and changes in shedding frequency (Wang et al. 2010), which are highly dependent on the relative distance and orientation angle between adjacent stems and diameters (Zhou and Mahbud 2016).

Despite the limitations of surface PIV, the velocity maps produced can support the investigation of a practical metric to describe the wake area. First, the existence of a recirculation regime, as a response to stem boundary layers, suggests the existence of vorticity fluxes in close proximity to the stems. Since random cylinder arrays increase the shedding frequency and limit the longitudinal extent of the vorticity flux (Ricardo et al. 2016b) it is expected that this will be concentrated close to the stems. Secondly, velocity defect zones refer to areas of reduced streamwise velocity due to turbulence generation. Both effects need to be considered without temporal fluctuations, hence a time-averaged map of longitudinal velocity and vorticity can illustrate these two separate descriptors of a steady wake, as shown in Fig. 4.

Depending on the flow upstream of vegetation patches, it is possible to investigate thresholds for each experiment based on the proportion of both streamwise velocity (U) and vorticity (W_z) that lie within a wake. Considering **U** and **V** to be longitudinal and transverse velocity components, respectively, the vorticity map $\mathbf{W_z}$ is defined as

$$\mathbf{W_z} = \frac{\partial \mathbf{V}}{\partial x} - \frac{\partial \mathbf{U}}{\partial y} \tag{1}$$

First, as recirculation is proportional to vorticity generation, a recirculation-related wake is defined as the area in which the points have a vorticity higher than a predetermined threshold, expressed as a fraction of the maximum value of vorticity. Similarly, the velocity-defect-related wake can be defined as the proportion of points in which the streamwise velocity is lower than a predetermined level, expressed as as fraction of the free stream velocity (U_∞). Both definitions can be stated as the conditions in Eq. 2.

$$\frac{|w_{z,wake}|}{max|w_z|} \geq \epsilon \qquad \frac{U_{wake}}{U_\infty} \leq \alpha \tag{2}$$

For illustration purposes, Fig. 4 shows consecutive images of a section of test 3, and the fraction of the frame occupied by wakes, assuming ϵ and α equal to 10% and 50%, respectively. Evidently, time-averaged maps tend to smooth out turbulent fluctuations in the mixing layer of each individual wake. The values ϵ and α presented in this research, are associated with the quantification of the wake area for the study of scalar dispersion in ecohydraulics, and thus require experimental validation.

Muste et al. (2014) found that vortical structures (generated by submerged obstructions) can be revealed from instantaneous velocity maps, calculated from surface images. However, to quantify random wake fields, additional conditions are imposed when obstructions (stems) are added. First, the reliability of the velocity gradient obtained depends on the PIV computation settings, as well as on the seeding particles and image characteristics. Second, time-averaged maps, such as those shown in Fig. 4b, c, are relevant only under steady conditions, even under spatially heterogeneous flow fields. The criterion presented in Eq. 2 is only valid under the aforementioned conditions, which are, nevertheless, achievable in practical applications in the field, such as estuaries, stormwater ponds and wetlands.

4 Conclusion

Surface PIV (LSPIV) allows for the computation of velocity maps, following the same principles of Particle Image Velocimetry, but with simpler set-up requirements. The use of surface PIV for the quantification of the areal extent of the wake generated

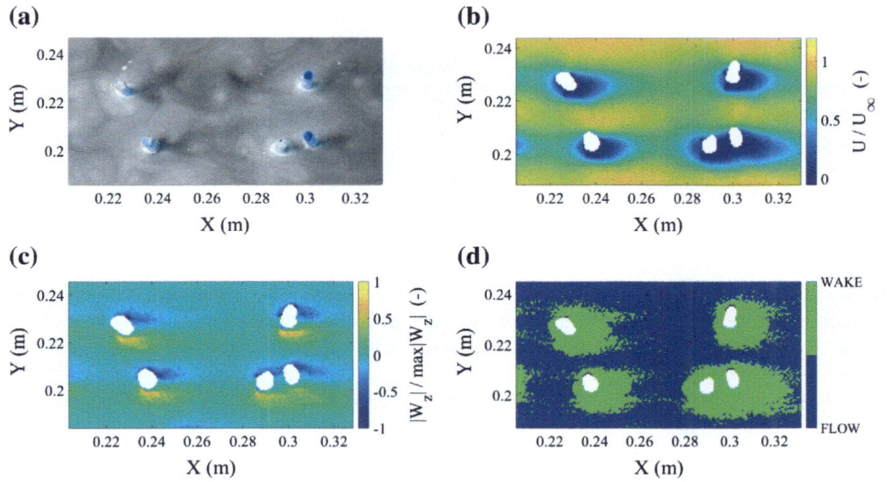

Fig. 4 Defined criteria to determine the wake area in flow through vegetation: **a** Experimental section for Test 3. **b** Time-averaged longitudinal velocity, as a fraction of the free stream velocity (U_∞). **c** Time-averaged vorticity, as a fraction of the maximum value of vorticity. **d** Wake area, shown in green for the conditions in Eq. 2 with α and ϵ equal to 50% and 10%, respectively

by an array of randomly distributed stems is presented. This approach is intended to provide insight into the role of macrophyte distribution on scalar transport.

It was found that the best velocity maps could be obtained with smaller seeding particles and intermediate resolutions. However, some errors due to uniformity in the pixel intensity persist. To improve the accuracy of these tests it is necessary to devise alternatives, such as: having particles with different colours or reflection indices to allow for a better identification of local velocity gradients.

A dual criterion for numerically defining the extent of a random wake field is proposed, from experimental data obtained via surface PIV. First, the recirculation leading to vortex shedding has been related to a proportion of the increase in vorticity induced by stems; second, the areas of decreased longitudinal velocity, characterizing mixing zones behind obstructions, are described as areas with a specific fraction of the incident velocity.

Further experimental studies are needed to relate this wake area descriptor to parameters of the flow field. Additionally, connections between the measurements of the wake fractions thus described and dispersion coefficients will be done for future contributions.

References

Dobson DW, Holland KT, Calantoni J (2014) Fast, large-scale, particle image velocimetry-based estimations of river surface velocity. Comput Geosci 70:35–43
Fujita I, Muste M, Kruger A (1998) Large-scale particle image velocimetry for flow analysis in hydraulic engineering applications. J Hydraul Res 36(3):397–414
Gharahjeh S, Aydin I (2016) Application of video imagery techniques for low cost measurement of water surface velocity in open channels. Flow Meas Intrum 51:79–994
Higham J (2017) The application of modal decomposition techniques for the analysis of environmental flows. Ph.D. thesis, University of Sheffield
Higham JE, Brevis W, Keylock CJ (2016) A rapid non-iterative proper orthogonal decomposition based outlier detection and correction for PIV data. Meas Sci Technol 27:1–16
Jadhav R, Buchberger S (1995) Effects of vegetation on flow through free water surface wetlands. Ecol Eng 4:481–496
Lohrmann A, Cabrera R, Kraus NC (1994) Acoustic doppler velocimeter (adv) for laboratory use. In: Pugh CA (ed) Proceedings of symposium on fundamentals and advancements in hydraulic measurements and experimentation. ASCE, pp 351–365
Muste M, Hauet A, Fujita I, Legout C, Ho H-C (2014) Capabilities of large-scale particle image velocimetry to characterize shallow free-surface flows. Adv Water Resour 70:160–171
Nepf H (1999) Drag, turbulence and diffusion in flow through emergent vegetation. Water Resour Res 35(2):479–489
Nepf H (2012) Flow and transport in regions with aquatic vegetation. Annu Rev Fluid Mech 44:123–142
Novak G, Rak G, Preseren T, Bajcar T (2017) Non-intrusive measurements of shallow water discharge. Flow Meas Intrum 56:14–17
Osorio-Cano JD, Osorio AF, Medina R (2013) A method for extracting surface flow velocities and discharge volumes from video images in laboratory. Flow Meas Intrum 33:188–196
Patalano A, Garcia CM, Rodriguez A (2017) Rectification of image velocity results (river): a simple and user-friendly toolbox for large scale water surface particle image velocimetry (piv) and particle tracking velocimetry (ptv). Comput Geosci 109:323–330
Raffel M, Willert CE, Scarano F, Kahler CJ, Wereley ST, Kompenhans J (2018) Particle image velocimetry: a practical guide, 3rd edn. Springer, Berlin
Ricardo AM, Franca M, Ferreira RML (2016a) Turbulent flows within random arrays of rigid and emergent cylinders with varying distributions. J Hydraul Eng 142(9):1–6
Ricardo AM, Sanches P, Ferreira RML (2016b) Vortex shedding and vorticity fluxes in the wake of cylinders within a random array. J Turbulence 17(11):999–1014
Sonnenwald F, Stovin VR, Guymer I (2018) A stem spacing-based non-dimensional model for predicting longitudinal dispersion in low-density emergent vegetation. Acta Geophys pp 1–18. https://doi.org/10.1007/s11600-018-0217-z
Stanislas M, Okamoto K, Kahler CJ, Westerweel J, Scarano F (2008) Main results of the third international PIV challenge. Exp Fluids 45:27–71
Sumner D (2010) Two circular cylinders in cross-flow: a review. J Fluid Struct 26:849–899
Thielicke W, Stamhuis EJ (2014) Pivlab - towards user-friendly, affordable and accurate digital particle image velocimetry in matlab. J Open Res Software 2(1). http://dx.doi.org/10.5334/jors.bl
Wang S, Tian F, Jia L, Lu X, Yin X (2010) Secondary vortex street in the wake of two tandem circular cylinders in cross-flow. Physics Rev E 81:1–9
White BL, Nepf HM (2003) Scalar transport in random cylinder arrays at moderate reynods number. J Fluid Mech 487:43–79
Zhou Y, Mahbud MA (2016) Wake of two interacting cylinders: a review. J Heat Fluid Flow 62:510–537

Dominant Hydraulic Conditions in the 2-D Model—Vistula River from Zawichost to Słupia Nadbrzeżna

Jacek Florek, Maciej Wyrębek and Agnieszka Woś

Abstract Complexity of the natural river system, creating diversified habitat conditions, was always a focus of attention and in the presented context is addressed by the use of the 2-D modeling results. Recognizing the importance of riverbed morphological elements and identification of hydraulic parameters suitable for its description, the use of 2-D modeling and spatial relations with channel parts was presented. The section of the Vistula River Gorge of Lesser Poland was modeled, segmented according to proposed method and the results were interrelated pointing out limited diversification of narrow parts in the channel. The measurement material was collected in the summer of 2017. Data including measurements of riverbed bottom geometry, water surface position and flow, as well as data obtained from the numerical model of the terrain was used. Calculations were made using discharges with relation to the hydrology of the river's section. The spatial joining involved data located in the individual parts of the channel and hydraulic parameters derived from the model. Channel parts were analyzed: dominating depth on wide parts of the channel were found, impact of flow was established, dynamics of water velocities and depths on mid channel bars and sandbars point them out as the most sensitive places in the whole river. Minimal width of the narrow parts of the river was found as insufficient. The advantage of proposed approach is the free access to any possible questions about hydraulic and spatial interrelationships. The results prove that spatial and contextual analysis can be used as a tool to increase usability of 2-D hydraulic model in the field of scientific purposes.

Keywords Dominating hydraulic conditions · 2D hydraulic model · Spatial analysis

---

J. Florek (✉) · M. Wyrębek · A. Woś
University of Agriculture in Kraków, Mickiewicza Ave. 24/28, 30-059 Kraków, Poland
e-mail: rmflorek@cyf-kr.edu.pl

M. B. Kalinowska et al. (eds.), *Recent Trends in Environmental Hydraulics*, GeoPlanet: Earth and Planetary Sciences,
https://doi.org/10.1007/978-3-030-37105-0_5

1 Introduction

Rivers are a complex, dynamic system creating habitat conditions for aquatic organisms. Naturally induced diversification of hydrodynamic parameters of water flow is a key element of the biodiversity and proper ecological processes of the river (Palmer et al. 1997). The nature of the flow creates an abiotic environment determined by the shape of the river bed. Diversity of the riverbed morphology can be an important parameter characterizing habitat conditions. One of methods describing hydro-morphological river's diversity is the description of cross-section: average depth, maximum depth, width, width/depth relation, area, wetted perimeter, hydraulic radius (Knithon 1981). Another tool is the identification and description of morphological units along the river. These units are temporally and spatially (Belletti et al. 2017) dynamic areas developed on the basis of interaction between bed morphology and hydrological regime (Maddock 1999). Parameters describing morphological and habitat units are related to bottom morphology: unit surface, depth, and water flow: velocity, sheer stresses (Hauer et al. 2008) velocity/depth, Froude number (Jowett 1993; Kemp et al. 2000). Modern methods of acquiring data of the riverbed geometry allow for a precise projection for the purpose of 2-D hydraulic modeling. Flow modeling provides a number of hydraulic parameters for any discharge.

Previously used methods describing river's channel can be supplemented by analyzing hydro-morphological variability in the results of the model's grid. One such characteristic is the hydro-morphological index of diversity HMID (Gostner et al. 2013), which also describes variability of depth and water velocity in the section, applied in the context of eco-hydraulic modeling (Vanzo et al. 2016; Lane et al. 2018). This paper presents another method where analysis of the variability in the hydro-morphology meets with data from the model's grid.

The aim of the study is to analyze hydraulic conditions of the Vistula River channel on the section from Zawichost (50° 48′ 19.3″ N 21° 51′ 53.9″ E) to Słupia Nadbrzeżna (50° 56′ 48.8″ N 21° 48′ 38.6″ E) calculated in 2-D hydraulic model, divided and compared among separated riverbed areas. The measurement material was collected in the summer of 2017. Data included measurements of riverbed bottom geometry, water surface position and flow, as well as data of the numerical model of the terrain obtained from the resources of the Head Office of Geodesy and Cartography in Poland.

2 Materials and Methods

The analysis uses spatial relationships between results of the numerical model and the specified areas of the riverbed (the whole model is considered). The spatial joining involved data on the location of individual parts in the channel and hydraulic parameters derived from the model (Książek et al. 2018). As a result, the following parts of the channel were distinguished at grid points: wide, widening, narrow, narrowing, mid channel bars (including occasional islands) and side bars.

This parts of the channel were distinguished based on the following rules: Calculations were made for discharges of 100, 435 and 1000 m^3s^{-1}. The values were chosen for several reasons. According to the station gauge in Zawichost, the average discharge (1951–2010) is 436 m^3s^{-1} and the value of 435 m^3s^{-1} covers all the mid channel bars of the modeled section. The low discharges are: average flow from the lowest 135 m^3s^{-1} and the lowest flow 84 m^3s^{-1} (ISOK).[1] In Zawichost, in the years 2015–2017 the lowest flow of 142 m^3s^{-1} was recorded, the flow of 1000 m^3s^{-1} was exceeded four times. The Vistula on the section is classified as a large lowland river. The valley was formed by waters flowing northward down from the Carpathians in the Pleistocene era (Maruszczak 1982). The Vistula River Gorge of Lesser Poland is located in a deep fossil channel, covered with a layer of 25-30 m sediments. The shape of the channel indicates a significant dynamics of Vistula Valley in this section. In the southern part of the Gorge, from the Zawichost to Annopol, Jurassic deposits emerge from underneath the Paleogene, Neogene and Quaternary period strata, mainly in the form of limestones and sandstones (Łoś 2006). The channel pattern is braided with mid channel bars and occasional islands, side bars appear naturally or to some degree being an effect of river training works such as groynes, cross-dikes and longitudinal dikes.

3 Numerical Modeling of Hydraulic Conditions

The numerical model of the riverbed segment consists of data from three sources: bottom configuration measured with ADCP probe (98 cross-sections), geometry of the mid channel bars and islands - GPS RTK, channel edges - DTM. The numerical model of the water surface was created on the basis of water surface measurements and points were taken along with the bottom configuration measurement. Roughness coefficients on floodplain terraces were estimated based on the terrain coverage. The measurements were taken in 2017. Numerical modeling was carried out using the CCHE2D model. The program uses Navier-Stokes equations (Zhang 2006).

The Vistula model had a length of 18 km and its range included the left and right floodplain and the main river channel. The number of computation nodes on the length of the model (2854) and on the width (107) was constant. In order to better reflect the morphology of the bottom of the main channel, the model was divided into three parts (left floodplain—23 nodes, main channel—50 nodes, right floodplain—36 nodes). The length between computational nodes was approx. 6 m, while the width was variable. This is due to the varying widths of individual parts of the model along the length of the river (Table 1).

[1]Project: ISOK Informatyczny system osłony kraju przed nadzwyczajnymi zagrożeniami. POIG.07.01.00–00–025/09.

Table 1 Basic geometric parameters of the 2-D hydraulic model

	Width of block (m)		Distance between nodes (m)		Number of nodes
	Min	Max	Min	Max	
Right bank	107.7	617	4.7	26.8	65,642
Channel	211.6	786.3	4.2	15.7	142,700
Left bank	81.5	806.3	2.3	22.4	102,744

Calibration of the model was done for steady-state conditions and measured flow of 205 m^3s^{-1}. By comparing the measured (GSP RTK) and the calculated water surface, coefficients of roughness in the main channel were estimated. After the model was calibrated, numerical modeling of three flow values was performed.

Analysis of water flow conditions in selected zones required a clear distinction of their spatial interaction. This means individual zones have been interpreted both independently and in relation to others. The zones where the bed is expanding, narrowing, wide or narrow have been taken into consideration without regulation areas. The purpose of this is to be able to distinguish between the impact of selected features without including places (in the model) above the water (or places where the water does not move). In this way it was possible to compare water flow conditions in the channel, without other unwanted elements. Interesting might be also to point out those places where water movement is not present at a lower flow and appears at the higher one. Further modeling of sediment transport and geometry changes was neglected as the presented method was already time consuming and other study of different part of the river taking into consideration this type of calculation was conducted in the same time. The basic number of possible combinations of the analyzed relationships with the following assumptions is:

- 6 number of discharges and changes between them,
- 4 parts of the channel (wide, widening, narrow, narrowing),
- 2 mid channel bars (rare islands) and side bars,
- 2 non-zero values of the depth and water velocities.

It should be understood that every additional assumption, e.g. change of surface, water velocity average and maximum value in the points of the grid and other multiplies the number of combinations. 24 combinations of channel part (4), discharge (3), depth and water velocity (2) with up to three dominating values were analyzed, 24 cases of bottom form (2), velocity and surface change (3) and channel area (4). Changes related to discharges and location of surfaces with predominant resulting computational values were also taken into consideration.

4 Spatial Analysis of Hydraulic Modeling Results

The comparison of water movement conditions is based on determination of the proportion between the spatial occurrence of areas with known depth and velocity. As an example, a comparison of the frequency distribution of dominating velocities in the part of the riverbed considered to be a wide section is presented (Fig. 2). For this channel section and flow value of 1000 m^3s^{-1} the water velocity changes in the range of 0–1.75 ms^{-1}. Analyzing velocity distribution on these surfaces, in this example, four of them can be distinguished: 0.1 ms^{-1} occurs on 4% of the surface, 0.6 and 0.9 ms^{-1} occupies respectively 14 and 15% of the surface, while the rest of the area characterized by variety of water velocities ranging from 0.25 to 1.75 ms^{-1} averaged 0.8 ms^{-1} (Fig. 1).

The allocation of a given area to the group is as follows:

- a riverbed much wider than the average width on a section of several kilometers counting up and down the river length is a wide channel and much narrower is considered narrow (average width of a wide parts 386 m and narrow parts 273 m),
- significant change in width (in practice at least 20%) is widening or narrowing,
- bottom forms not connected to the banks at a low discharge are mid channel bars or islands and connected are sandbars.

It is worth noting that the areas of larger mid channel bars, islands and side bars do not include their part above the water surface but only the submerged part is participating in the calculations of the model. This area will grow with flow. This allocation was done using a map made in the measurement season. Water velocities and depth values dominating in the certain areas were selected and presented according to the part of the channel they were found in. This method provides information about certain tendencies in diversification of local parameters, not only absolute dominant values. The remaining values represent the average (average of the rest).

The following graphs (Fig. 3) present the dominating depths and water velocities within selected parts of the riverbed for the whole modeled part of the river. This should be understood as the dominance of certain conditions of water movement during the computational flow and in a specific part of the channel.

The basic values appearing in the areas of the separated parts of the riverbed (narrow, wide, narrowing, expanding) are listed in Table 2. Three most-dominant values (and average of the rest) of velocity and depth were selected with relation to the rest of the calculated area in the model and with discharge.

In the wide part of the channel, in the areas of mid channel bars and water-covered side bars, the relationship between flow, riverbed and the water velocity was established.

A common feature of the widening parts of the channel is the lack of dominating water velocities on at least 61% of the surface (Fig. 3d, h) and the occurrence of one dominant depth, which, regardless of the flow, covers over 45% of the surface (Fig. 3j, k). As for the depth, it is not the same surface, because it's dominating value changes differently than the average of the rest (dominant 1.25, 2.02, 3.75 m and respectively

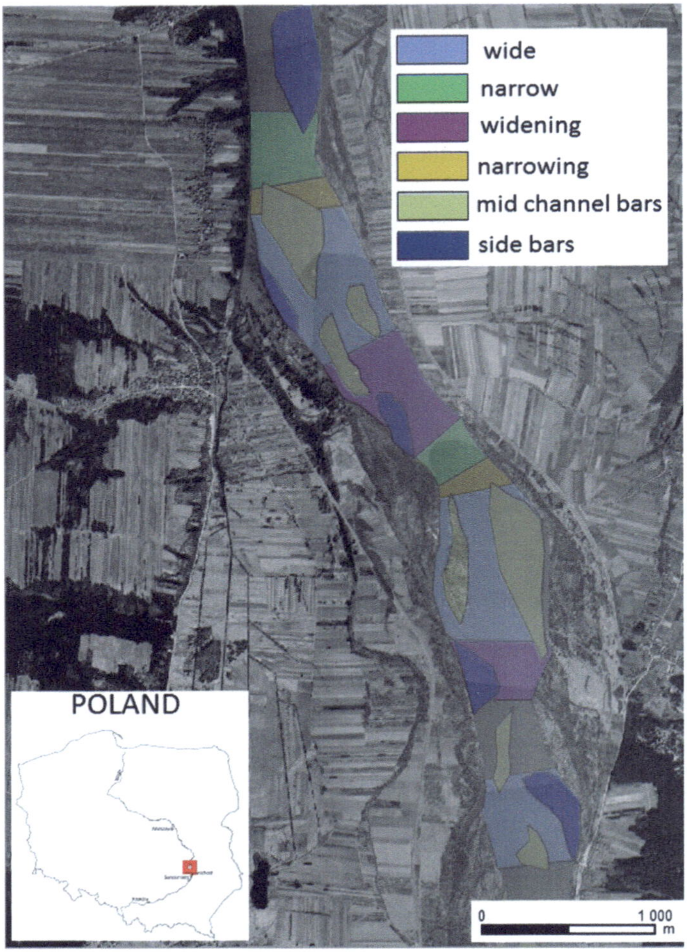

Fig. 1 Location of the sections and an example of dividing river bed area

average 1.3, 2.46, 3.72 m). In the case of the flow 435 m^3s^{-1} dominating depth appears on the area being covered by approximately 0.88 m of water at discharge of 100 m^3s^{-1} (at the 435 m^3s^{-1} dominant is 2.02 m, and the depth difference of the whole part is $2.43 - 1.29 = 1.14$ m in this way at 100 m^3s^{-1} it should be $2.02 - 1.14 = 0.88$ m).

The narrowing parts of the channel have a dominating depth greater than the average of this part of channel at medium flows 435 m^3s^{-1} and lower than the average at high and low flows. The average depth increase between discharges of 435 and 1000 m^3s^{-1} is greater by 0.24 m than the average increase of the entire channel (2.4 and 2.16 m), which means that conditions in the area of the narrowing channel change more rapidly with flow.

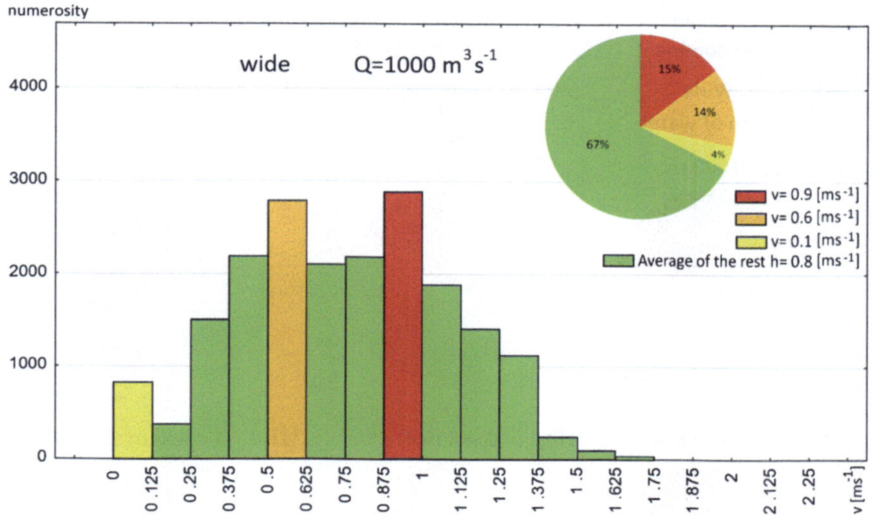

Fig. 2 The method of determining the dominating surfaces of the velocity on the example of the wide channel part, discharge 1000 m^{3}s^{-1}

In the wide channel during low discharge there is a large area, covering 64% of this part with an average depth of 0.74 m (Fig. 3c), which at higher flows decreases while still dominating (31 and 22%). The increase in discharge does not compensate for the different conditions in this area; on the contrary, it increases the share of areas with considerable depth variation (average of the rest, Fig. 3c, b, a, respectively, is 12, 65 and 77%).

The dominating depths of widening and narrowing parts are larger than those on the narrow channel parts although, as expected, the relation of average's is opposite.

There are large similarities between the widening and narrowing channel parts conditions, especially in terms of velocity and depth, while varying in the total area (Table 2). Another repeatable pattern for the depth is almost the separation by half on the expanding channel at each discharge (Fig. 3i, j, k) and narrowing at the lowest on the dominating surface and the differentiated rest (Table 2).

The area occupied by the wide channel part is 60% larger than narrow. In the area of a wide channel, all forms occupy a total of 37.78% while mid channel bars take 27.53% of it and in the narrow channel parts the values are, respectively, 1.09% and only 0.06%.

There are regulations in the narrowing parts, 60% of which are closed (water movement in negligible at low stages) and the rest is open (if the morphology is favorable water can enter at low stages).

The areas under the influence of river training structures constitute a small component (at least in currently existing riverbed) usually not exceeding a few percent. The exception are the narrow parts of the channel where up to 22.3% of the area is

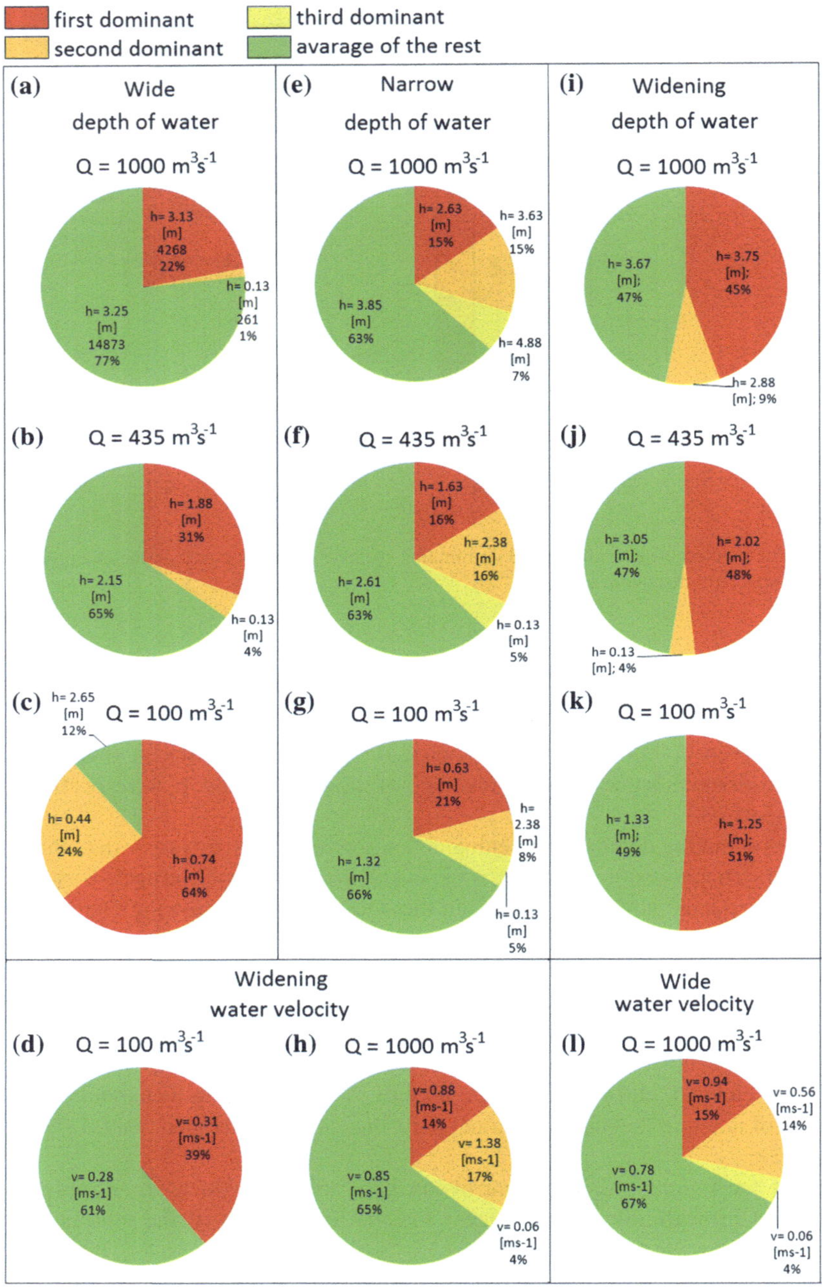

Fig. 3 Selected example of differences among dominating (calculated) values in the parts of the channel (along the whole model)

Table 2 Dominating values of depth and velocity, divided by discharge, narrowing, widening, narrow, wide part of the channel and correlated with rest of the channel part and the whole model

Average value, whole model, h (m), v (ms^{-1})	Parameter	Discharge (m^3s^{-1})	Area for the whole grid (m^2)	Average, h (m), v (ms^{-1})	Dominant 1 area (m^2)	Dominant 1 value, h (m), v (ms^{-1})	Dominant 2 area (m^2)	Dominant 2 value, h (m), v (ms^{-1})	Dominant 3 area (m^2)	Dominant 3 value, h (m), v (ms^{-1})	Average value of the rest, h (m), v (ms^{-1})
Widening part of the channel											
3.2	h	1000	37,2835	3.64	166,175	3.75	32,352	2.88			3.72
2.11	h	435	34,9950	2.43	169,254	2.02	14,521	0.13			2.46
1.04	h	100	34,6291	1.29	176,921	1.25					1.3
0.84	v	1000	353,435	0.95	64,066	1.38	52,797	0.88	13,301	0.06	0.97
0.61	v	435	341,818	0.64	75,856	0.56	25,324	0.07			0.64
0.31	v	100	343,096	0.29	133,823	0.31					0.29
Narrowing part of the channel											
3.2	h	1000	28,0657	3.58	64,066	3.12	45,363	3.88	3833	0.13	3.65
2.11	h	435	275,662	2.33	51,636	2.38	11,036	0.13			2.37
1.04	h	100	271,015	1.18	131,268	1.01	15,276	0.13			1.17
0.84	v	1000	277,695	0.91	49,603	0.94	9642	0.06			0.93
0.61	v	435	266,891	0.67	105,188	0.63	20,155	0.06			0.67
0.31	v	100	267,588	0.32	82,362	0.31					0.32
Wide part of the channel											
3.2	h	1000	1,126,926	3.18	247,898	3.13	15,160	0.13			3.25
2.11	h	435	1,107,759	1.98	340,134	1.88	45,711	0.13			2.03

(continued)

Table 2 (continued)

Average value, whole model, h (m), v (ms^{-1})	Parameter	Discharge (m^3s^{-1})	Area for the whole grid (m^2)	Average, h (m), v (ms^{-1})	Dominant 1 area (m^2)	Dominant 1 value, h (m), v (ms^{-1})	Dominant 2 area (m^2)	Dominant 2 value, h (m), v (ms^{-1})	Dominant 3 area (m^2)	Dominant 3 value, h (m), v (ms^{-1})	Average value of the rest, h (m), v (ms^{-1})
1.04	h	100	1,084,526	0.89	695,254	0.74	263,058	0.44			0.89
0.84	v	1000	1,111,767	0.77	167,395	0.94	162,516	0.56	47,628	0.06	0.79
0.61	v	435	1,072,212	0.56	350,705	0.38	180,115	0.81			0.57
0.31	v	100	1,062,280	0.32	277,927	0.19					0.32
Narrow part of the channel											
3.2	h	1000	696,299	3.61	114,830	2.63	107,744	3.63	50,765	4.88	3.68
2.11	h	435	683,695	2.4	116,805	1.63	115,237	2.38	37,289	0.13	2.43
1.04	h	100	676,725	1.26	147,299	0.63	56,515	2.38	36,244	0.13	1.25
0.84	v	1000	688,226	1.12	187,202	1.31	56,108	0.63			1.15
0.61	v	435	665,108	0.77	313,300	0.88	75,508	0.06			0.78
0.31	v	100	669,290	0.36	263,058	0.44					0.37

occupied by closed regulation zones. These are areas where the movement of water is limited completely and only appears at discharges close to 435 m^3s^{-1}.

Relations between water velocity and area taken by the mid channel bars and sandbars (Table 3) in the wide channel parts are as follows:

- during discharge of 100 m^3s^{-1} maximum velocities appear in the area of the side bars, during the 435 m^3s^{-1} in the area around the mid channel bars and at 1000 m^3s^{-1} in other parts of the riverbed,
- lowest average velocities occur around the mid channel bars at the lowest flows, and this area covers over 25% of the wide channel parts and over 7% of the entire riverbed in the studied section,
- the flow increase most strongly changes the conditions in the shallows around mid channel bars (between discharges of 100 and 435 m^3s^{-1}, the increase in average water velocity by 150%).

5 Conclusions

There are no mid channel bars or islands on the segments of the river with a narrow channel while on a wide part a quarter of the area is occupied in this way. Generally speaking, the majority of the channel areas with a dominant depth occur in the widening, but in narrowing and wide channel parts they occur only during low stages. As for the water velocities, dominating parts can be found at the low stage in a narrow channel. The shallows around channel bars and islands are the most affected by the flow changes. This section of the river was modified many times in the past (Łajczak 2006) and, at least in some places, it reached width well below which mid channel forms do not appear anymore. However there is a process of sedimentation and some of the training works does not function properly anymore due to the period of 30 years of structural deterioration in the new Polish economic situation. Because of this some of limited and narrow artificial parts slowly loss their parameters but still according to the 2-D model results, 200 m of width (in some places) is not enough for the river of this size. Only limited surfaces with dominating depth could be found in narrow parts of the river (at the average width of 273 m) and when the flow rises the total dominating surface only in narrow parts of the channel occupies still the same relative area.

The preparation of data providing opportunity to analyze the spatial relationship between the water flow conditions and the part of channel area is labor-intensive. In addition to building the hydraulic model itself, it is necessary to verify large group of variable relations in order to find relevant results. In this method, we worked on a relatively large and uncomfortable data resources. However, the advantage of this approach is an access to any possible questions about hydraulic and spatial interrelationships. In the obtained material, the total possible number of relations between computational flows, hydraulic parameters and river channel parts could

Table 3 Depth and water velocity changes in response to discharge on side bars and mid channel bars

	Area (%)			Average water velocity (ms^{-1})			Maximum water velocity (ms^{-1})		
Discharge (m^3s^{-1})	100	435	1000	100	435	1000	100	435	1000
Sandbars	9.58	9.68	10.42	0.29	0.49	0.61	1.06	1.01	1.2
Mid channel bars	25.37	25.49	26.01	0.18	0.45	0.64	0.9	1.38	1.74

potentially reach hundreds of millions. If the analysis is limited to the relationship between flow, computational value of depth, water velocity and area, the number of relations is 96. In our opinion, a deeper analysis (more combinations) using this technique should be applied to the original measurement data, taking into account changes in time. The approach was developed when it became clear the calculation results from the hydraulic model are beyond visual perception. In the river there are disproportions between length and width, the whole picture provides too much information at once. Similar, comparable conditions appear in places separated by long distances. The impact of discharge depending on location is unequal. On the other hand operation on much more limited in number, measured data will pose a challenge and the solution seems to be selective approach concentrated only on the most important relations.

The results of hydraulic models inform about spatial details of water movement within computational geometry of the model with relation to the chronology of events during floods. The nature of the processes and the amount of obtained data make the perception of all results difficult. Spatial and contextual analysis can be used as a tool to increase usability of hydraulic 2-D model in the field of scientific purposes.

The authors consider analysis of the modeling results as a potentially useful tool to utilize measured data of riverbed geometry, water stage and velocity gathered in the years 2014–2018 (with the perspective of continuation).

Acknowledgements This research has been funded by the EOG PL02 2009-2012. The authors wish to acknowledge Prof. Wojciech Bartnik, Dr. hab. Leszek Książek and Dr. Małgorzata Leja and all of those who contributed during the development of the research.

References

Belletti B, Rinaldi M, Bussettini M, Comiti F, Gurnell AM, Mao L, Nardi L, Vezza P (2017) Characterising physical habitats and fluvial hydromorphology: a new system for the survey and classification of river geomorphic units. Geomorphology 283:143–157. https://doi.org/10.1016/j.geomorph.2017.01.032

Gostner W, Alp M, Schleiss AJ, Robinson CT (2013) The hydro-morphological index of diversity: a tool for describing habitat heterogeneity in river engineering projects. Hydrobiologia 712:43–60. https://doi.org/10.1007/s10750-012-1288-5

Hauer C, Tritthart M, Habersack H (2008) Computer-aided mesohabitat evaluation, Part I: Background, model concept, calibration and validation based on hydrodynamic numerical modeling. In: Altinakar M, Kokpinar MA, Darama, et al. (eds) River flow. Kubaba Congress Department and Travel Services, Cesme-izmir, 1967–1974

Jowett IG (1993) A method for objectively identifying pool, run, and riffle habitats from physical measurements. New Zeal J Mar Freshw Res 27:241–248

Kemp JL, Harper DM, Crosa GA (2000) The habitat-scale ecohydraulics of rivers. Ecol Eng 16:17–29. https://doi.org/10.1016/S0925-8574(00)00073-2

Knighton AD (1981) Local variations of cross-sectional form in a small gravel bed stream. J Hydrol (New Zealand) 20:131–146

Książek L, Wyrębek M, Strutyński M, Woś A (2018) Numerical modeling of water flow conditions with spatial distribution of boulders in main channel. In: Kalinowska M, Mrokowska M, Rowiński

P (eds) Free surface flows and transport processes. GeoPlanet: earth and planetary sciences. https://doi.org/10.1007/978-3-319-70914-7_17
Lane BA, Pasternack GB, Sandoval Solis S (2018) Integrated analysis of flow, form, and function for river management and design testing. Ecohydrology 11(5):e1969. https://doi.org/10.1002/eco.1969
Łoś MJ (2006) Uwarunkowania użytkowania rzeki i doliny w Małopolskim Przełomie Wisły. Conditions for utilization of the river and valley in the Vistula Gorge of the Little Poland. Gospodarka Wodna, ISSN: 0017-2448, 2006/7:268–278
Maddock I (1999) The importance of physical habitat assessment for evaluating river health. Freshw Biol 41:373–391. https://doi.org/10.1046/j.1365-2427.1999.00437.x
Maruszczak H (1982) Wisła Lubelska. In: Piskozub A (ed) Wisła Monografia rzeki. Wydawnictwo Komunikacji i Łączności, pp 125–136
Palmer MA, Hakenkamp CC, Nelson-Baker K (1997) Ecological heterogeneity in streams: why variance matters. J North Am Benthol Soc 16:189–202. https://doi.org/10.2307/1468251
Vanzo D, Zolezzi G, Siviglia A (2016) Eco-hydraulic modelling of the interactions between hydropeaking and river morphology. Ecohydrology 9:421–437. https://doi.org/10.1002/eco.1647
Zhang Y (2006) CCHE-GUI—Graphical user interface for the NCCHE model. User's manual—Version 3.0 NCCHE. School of Engineering, The University of Mississippi. MS 38677

Water Level Uncertainties Due to Uncertain Bedform Dynamics in the Dutch Rhine System

Matthijs R. A. Gensen, Jord J. Warmink and Suzanne J. M. H. Hulscher

Abstract Quantitative estimations of water level uncertainties are essential for the design and assessment of flood protection systems. This work aims to quantify the water level uncertainties in the bifurcating Dutch river Rhine system as a result of main channel roughness uncertainty. An one-dimensional hydraulic model of the Rhine branches is used to estimate the water levels in the system for several roughness scenarios. Model results show that the roughness effect has a large influence on the modelled water levels. However, for the larger Waal branch, the changing discharge distribution counteracts the roughness effect, thereby decreasing the range of possible water levels. For the smaller Nederrijn and IJssel branch it is possible that the discharge in the respective branch increases even though the branch has a high roughness. Thereby, for these branches the discharge distribution effect increases the range in modelled water levels. The large and varying effects on water levels by roughness uncertainty and changing discharge distributions in a bifurcating river system indicate the importance to consider the system as a whole instead of as separate branches in the design and assessment of river engineering works.

Keywords River bifurcation · Uncertainty · Hydraulic modelling · Rhine · Bedforms

1 Introduction

In a river delta the river system consists of multiple bifurcating river branches, creating a complex and dynamically active system. These systems experience a large risk of flooding with the threat coming from both the seaward and landward side. In a bifurcating river, local changes in the system can cause changes in water levels of the entire system due to a varying discharge distribution over downstream branches. These system effects, via a changing discharge distribution, may have an important influence on the future planning of river engineering works.

M. R. A. Gensen (✉) · J. J. Warmink · S. J. M. H. Hulscher
University of Twente, 217, 7500AE Enschede, The Netherlands
e-mail: m.r.a.gensen@utwente.nl

M. B. Kalinowska et al. (eds.), *Recent Trends in Environmental Hydraulics*, GeoPlanet: Earth and Planetary Sciences,
https://doi.org/10.1007/978-3-030-37105-0_6

In several countries around the world, e.g. the UK, the USA and the Netherlands, probabilistic approaches are incorporated in flood risk practice. Under a probabilistic approach, flood protection systems are not designed and assessed on the basis of exceedance probabilities, but instead on the basis of probabilities of flooding. As a consequence, it is important to take into account the full probability distribution of all water levels. Furthermore, in a probabilistic framework, uncertainties play an important role as uncertainties influence the probability of flooding. Therefore, there is an increasing need to include the influence of uncertainties in river management.

The most important sources of uncertainty in river water levels are the upstream discharge and the main channel roughness due to large-scale river bedforms (Gensen 2018; Warmink et al. 2013). In the bifurcating river system uncertainties in water levels may strongly affect the distribution of discharge over branches. Such interaction is one of the most complex factors if performing an uncertainty analysis (Merz et al. 2015). This study aims to quantify the system effects, caused by the presence of a river bifurcation, on the uncertainty in river water levels for a range of discharges. It is investigated how uncertain river bedform characteristics in the bifurcating river affect the water levels throughout the system.

2 Study Area: Dutch River Rhine System

The study area consists of the upper reaches of the Dutch Rhine branches (Fig. 1). Near Lobith the Upper Rhine enters the Netherlands. At the Pannerdensche Kop the river splits into the river Waal and the Pannerdensch Kanaal. The division of discharge at this bifurcation point is approximately 2/3 to the river Waal and 1/3 to the Pannerdensch Kanaal. The ten-kilometer long Pannerdensch Kanaal bifurcates into the river Nederrijn (further downstream called river Lek) and the river IJssel.

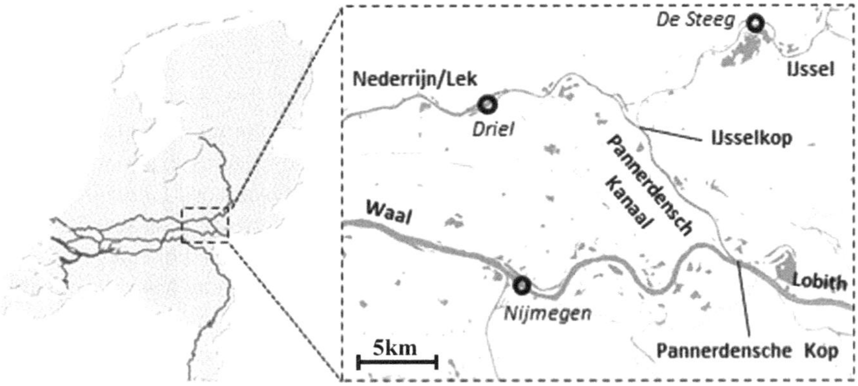

Fig. 1 The area of interest. The circles indicate the representative locations for the branches at which water levels are determined

Table 1 General characteristics of the Dutch Rhine branches

Branch	Discharge [m^3/s]	Water depth [m]	Flow velocity [m/s]	Observed dune heights [m]
Waal	500–11,000	1.5–17	0.7–2.0	0–2.5
Pannerdensch Kanaal	50–6000	1.5–17	0.3–1.5	0–1.2
IJssel	50–2700	1.5–13	0.5–2.0	0–1.5
Nederrijn	0–3400	1.5–13	0–1.5	0–1.2

The discharge distribution is approximately 2/3 and 1/3 towards the Nederrijn and IJssel, respectively. General characteristics of the upper reaches of these branches are given in Table 1. The maximum recorded discharge at Lobith occurred in 1995 and is approximately 12,000 m^3/s. The branches all have a relatively low bed gradient, which is approximately 0.1 m/km. Generally, the branches have wide floodplains, conveying approximately 1/3 of the total discharge at extremely high discharges (Warmink et al. 2013).

3 Methods: Hydrodynamic Model and Roughness Scenarios

In the Netherlands, the Directorate-General for Public Works and Water Management (Dutch: Rijkswaterstaat) uses a Sobek model (Deltares 2015) of the Rhine branches for operational purposes. Sobek is a modelling environment which numerically solves the one-dimensional Saint-Venant equations. A schematization of the Rhine branches in the Sobek environment allows the computation of water levels. In this study the most recent schematization is applied: the 2016 version (*Rijn-j16_5_v1*). The river engineering works which have been executed in the years before, are implemented in this model.

The river is schematized with cross-sections with a longitudinal spacing of approximately 500 m. The cross-sections consist of a flow profile and a storage area. The upstream boundary of the model is a static discharge boundary placed at Lobith (see Fig. 1). Downstream water level boundaries, corresponding to the constant upstream discharge, lie several tens of kilometers away from the area of interest in this study. The model is run with static discharges ranging from 5000 to 18,000 m^3/s to estimate the for varying upstream conditions.

For every branch a discharge-roughness relationship defines the main channel roughness for the entire branch (Fig. 2). To define these scenarios, data of bedform characteristics derived from bed level measurements in the Rhine branches in various studies (Frings and Kleinhans 2008; Wilbers and Ten Brinke 2003; Sieben 2008) was used. These bedform characteristics were translated into roughness predictions, expressed in Nikuradse roughness height, using the predictors of Van Rijn (1993)

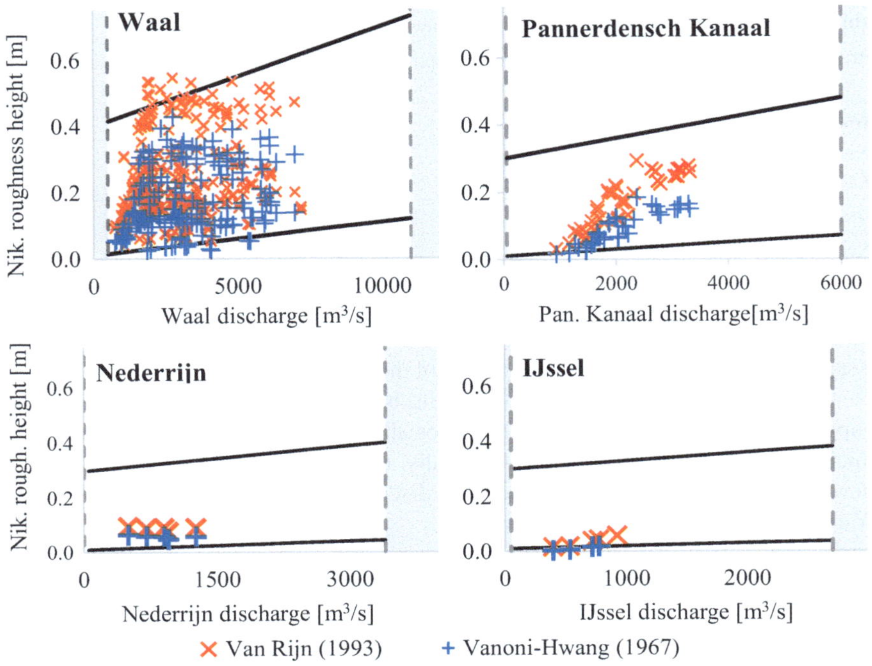

Fig. 2 Predicted roughness values using the Van Rijn (1993) and Vanoni and Hwang (1967) roughness predictors. The black lines show the defined roughness scenarios

and Vanoni and Hwang (1967). For each branch a linear higher limit and a linear lower limit of the discharge-dependent roughness was visually defined based on the roughness predictions (black lines in Fig. 2). These lines represent extreme scenarios of possible low and high roughness for each branch. Further details on the roughness limit lines are found in Gensen et al. (subm.). Finally, the combination of a higher limit or a lower limit per branch for all branches leads to 16 roughness scenarios. The defined roughness scenarios serve as input for the one-dimensional hydrodynamic Sobek model.

4 Results

The results are shown for three representative locations in the branches: Nijmegenhaven for the Waal, De Steeg for the IJssel and Driel for the Nederrijn (see Fig. 1). Figure 3 displays the modelled water levels at the three locations as a result of the 16 roughness scenarios as a function of upstream Lobith discharge. It shows the average water levels and bandwidth of water levels conditional on the roughness in

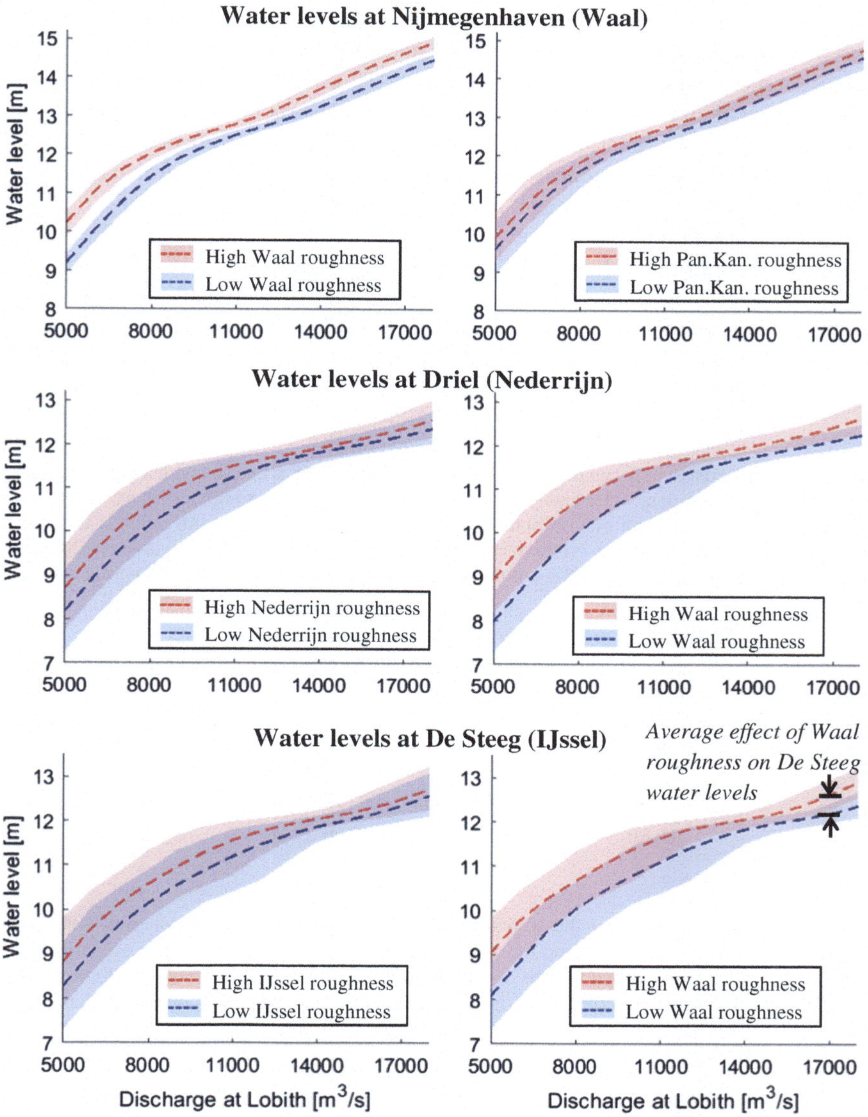

Fig. 3 Water levels at the three representative locations (see Fig. 1) as a function of upstream Lobith discharge for the 16 roughness scenarios. The dashed lines give the average of the 8 scenarios with either a high or a low roughness in the signified branch. The shaded area shows the range in possible water levels given a high or low roughness in the signified branches. The black arrows in the lower right figure illustrate the average water level effect of a change in roughness in the signified branch on the water levels at the location. Values of this average effect for all combinations of branches and locations are given in Table 2

one of the branches. Tables 2 and 3 summarize the average effects of the roughness per branch on the water levels (indicated in the lower right panel of Fig. 3) and discharges, respectively, for every location. This average effect is defined as the difference between the average water levels at the location or the average discharge in a branch for the 8 scenarios with a high roughness in the signified branch and the 8 scenarios with a low roughness in the signified branch, for an upstream discharge of 17,000 m^3/s, which is the design discharge of the Dutch Rhine system.

Figure 3 and Table 2 show that, on average, high local roughness values result in high local water levels (e.g. +0.47 m at Nijmegenhaven in the Waal). This value is affected by two mechanisms. Firstly, a high roughness will result in a high water level. The two roughness limit lines in a branch result in two possible (local) stage-discharge relationships, in which a high roughness clearly gives high water levels. Secondly, a high water level in a downstream branch of a river bifurcation, will

Table 2 The difference between the average water level over the 8 scenarios with a high roughness in a branch and the average water level over the 8 scenarios with a low roughness in a branch for an upstream discharge at Lobith of 17,000 m^3/s (design discharge). In Fig. 3 this value is illustrated for the effect of Waal roughness on the water levels at De Steeg (0.45 m). The values in this table can be interpreted as the average effect of the roughness in a branch on the water levels at a certain location

Discharge at Lobith: 17,000 m^3/s		Average effect on the water level at:		
		Nijmegenhaven (Waal)	Driel (Nederrijn)	De Steeg (IJssel)
High versus low roughness of:	Waal	+0.47 m	+0.31 m	+0.45 m
	Pan. Kanaal	+0.21 m	−0.13 m	−0.18 m
	Nederrijn	+0.07 m	+0.17 m	+0.25 m
	IJssel	+0.08 m	+0.23 m	+0.15 m

Table 3 The difference between the average discharge in a branch over the 8 scenarios with a high roughness in the signified branch and the average discharge in a branch over the 8 scenarios with a low roughness in the signified branch for an upstream discharge at Lobith of 17,000 m^3/s. The values in this table can be interpreted as the average effect of the roughness on the discharge in a certain branch. For the Pannerdensch Kanaal, the change in discharge is naturally opposite to that of the Waal branch

Discharge at Lobith: 17,000 m^3/s		Average effect on the discharge in branch:		
		Waal	Nederrijn	IJssel
High versus low roughness of:	Waal	−932 m^3/s (−8.9%)	+354 m^3/s (+9.6%)	+578 m^3/s (+20.1%)
	Pan. Kanaal	+391 m^3/s (+3.8%)	−156 m^3/s (−4.2%)	−235 m^3/s (−8.2%)
	Nederrijn	+126 m^3/s (+1.2%)	−441 m^3/s (−11.9%)	+316 m^3/s (+11.0%)
	IJssel	+148 m^3/s (+1.4%)	+261 m^3/s (+7.1%)	−409 m^3/s (−14.2%)

change the discharge distribution with a lower amount of discharge for the branch with a high water level (see Table 3, e.g. 937 m^3/s less discharge to the Waal). This in turn will result in a lower water level, thereby counteracting the effect of roughness alone. Table 2 shows that on-average the balance the two counteracting mechanism has a positive balance for every branch (+0.47 m at Nijmegenhaven, +0.17 m at Driel and +0.15 m at De Steeg).

Figure 2 and Tables 2 and 3 show that for the IJssel and Nederrijn, the influence of the Waal roughness on the water levels in these branches is very large. This is caused by the third mechanism: through changes in the discharge distribution, the water levels in a branch are affected by changes in roughness in another branch. For example, a change from low to high roughness in the Waal branch causes, on-average, an increase in discharge towards the IJssel of 578 m^3/s (Table 3), which causes an average increase in water levels of 0.45 m (Table 2). For the IJssel and Nederrijn branches, the average effect of the Waal roughness is larger than the average effect of their own roughness. Thereby, it is also possible that discharges in these branches increase due to a high Waal roughness, even though their own roughness is high as well.

The size of the bandwidths in Fig. 3 also gives an indication on the balance of the three mechanisms. It is observed that for the Waal branch, all scenarios with a high Waal roughness result in higher water levels compared to the scenarios with a low Waal roughness (the red-shaded and blue-shaded areas do not overlap). This indicates that a low Waal roughness will always result in a below-average water level and thereby an above-average discharge. The range of possible water levels is thereby reduced, compared to the situation in which the discharge distribution would not adapt. Simultaneously, as the Waal branch draws more discharge, the other branches see both a reduction in water levels and discharge. This causes a large range in possible water levels in the IJssel and Nederrijn branch, as observed in Fig. 3.

5 Discussion

In this paper the changes in the water levels resulting from uncertainty in river bedforms in a bifurcating river system are quantified. River bedform variation is implemented by means of roughness scenarios, capturing an extreme range in main channel roughness uncertainty. The defined roughness scenarios should be considered as very extreme. For less extreme roughness scenarios, the bandwidths in water levels on the branches will be smaller. However, these extreme scenarios have shown the potential effects of roughness uncertainty in a river system. It is observed that changes in roughness and subsequent changes in discharge distributions can have varying, possibly opposite, effects on the water levels, due to the feedback mechanisms. It is expected that for less extreme roughness scenarios these water level variations will still be found and can largely be explained by the effects found for the extreme scenarios.

This study has applied an one-dimensional model to obtain the understanding of the propagation of uncertainties in a bifurcating river system as well as to obtain a first quantitative estimation. In future work it may be necessary to apply a two-dimensional model to attain more accurate uncertainty predictions. Furthermore, a two-dimensional model allows the inclusion of the regulation structures at the bifurcation points of the river Rhine and allows the inclusion of the uncertain effects of river engineering works (Berends et al. 2018) on the discharge distribution.

6 Conclusions

In this study, the effect of uncertain river bedforms on the water levels in the bifurcating river Rhine system was estimated. In the presented case study the presence of river bifurcations showed to have a strong influence on the possible water level variation given the roughness uncertainty. The feedback mechanisms between downstream water levels and discharge distributions can cause both an increase and a decrease in bandwidths in possible water levels. For the larger Waal branch, the effect of a high roughness on the water levels is partly compensated by a reduction in the local discharge. Thereby, the maximum bandwidth in possible water levels is decreased. For the smaller Nederrijn and IJssel branches, scenarios exist in which the discharge in the branch increases even though the branch has a high roughness. Therefore, the discharge distribution effect causes a wider range of possible water levels.

The observed differences in roughness effects and discharge distribution effects on the water levels affect the probabilities of water levels along the branches. Therefore, it is important to regard the branches as an interconnected system in which varying discharge distributions may cause an increasing or decreasing uncertainty in water levels. This impacts river management and future planning and assessment of river engineering works.

Acknowledgements This work is part of the Perspectief research programme All-Risk with project number P15-21, which is (partly) financed by NWO Domain Applied and Engineering Sciences, in collaboration with the following private and public partners: the Dutch Ministry of Infrastructure and Water Management (RWS), Deltares, STOWA, HKV consultants, Natuurmonumenten and the regional water authorities Noorderzijlvest, Vechtstromen, it Fryske Gea, HHNK. Dr. F. Huthoff is thanked for his input during the study.

References

Berends KD, Warmink JJ, Hulscher SJMH (2018) Efficient uncertainty quantification for impact analysis of human interventions in rivers. Environ Model Softw 107:50–58

Deltares (2015). SOBEK 3, D-Flow 1D. User manual. Version 3.4.0

Frings RM, Kleinhans MG (2008) Complex variations in sediment transport at three large river bifurcations during discharge waves in the river Rhine. Sedimentoligy 55(5):1145–1171

Gensen MRA, Warmink JJ, Hulscher SJMH (subm.) River dune based roughness uncertainty for the Dutch Rhine branches. Marine and River Dune Dynamics 2019, Bremen, Germany
Gensen MRA (2018) Large-scale uncertainties in river water levels: literature report. University of Twente, Enschede. CE&M research report 2018-001/WEM-001
Merz B, Vorogushyn S, Lall U, Viglione A, Blöschl G (2015) Charting unknown waters—on the role of surprise in flood risk assessment and management. Water Resour Res 51:6399–6416
Sieben J (2008) Taal van de rivierbodem. Rijkswaterstaat
Van Rijn LC (1993) Principles of sediment transport in rivers estuaries and coastal areas. Aqua publications, Blokzijl, Netherlands
Vanoni VA, Hwang LS (1967) Relation between bedforms and friction in streams. J Hydraul Div 93:121–144
Warmink JJ, Booij MJ, Van der Klis H, Hulscher SJMH (2013) Quantification of uncertainty in design water levels due to uncertain bed form roughness in the Dutch river Waal. Hydrol Process 27(11):1646–1663
Wilbers AWE, Ten Brinke WBM (2003) The response of sub-aqueous dunes to floods in sand and gravel bed reaches of the Dutch Rhine. Sedimentology 50(6):1013–1034

Discharge Characteristics of Triangular Weir with Upstream Ramp and Its CFD Modelling Using Ansys CFX Module

Subhojit Kadia, Binit Kumar and Zulfequar Ahmad

Abstract To understand the flow field and discharge characteristics of a triangular weir with an upstream ramp (TW-UR), experimental study as well as computational fluid dynamics (CFD) simulation were performed. The Ansys CFX module and standard k-ε turbulent model were used in the simulation. It was observed that the TW-UR had about 9.8–14.3% higher discharging capacity than a sharp-crested weir of the same height and it was found that about 10–15% higher discharging capacity was estimated in the CFD simulation as compared to the observed data under the same head. The highly active flow field in the upstream side and enhanced velocity along the flow direction due to flow contraction in the vertical plane are helpful in enhancing the hydrodynamic force exerted by the moving fluid and creating the chances of sediment passage as well as increasing the discharging capacity. Existing two equations of the coefficient of discharge for TW-UR were checked for their accuracy using the present and previous experimental data, and it was observed that the equation proposed by Azimi et al. (J Irrig Drain Eng 139(1):75–83, 2013) predicted the coefficient of discharge within ±5, ±10 and ±15% error ranges for 43.0, 72.5 and 92.5% datasets, whereas within those error ranges, equation proposed by Di Stefano et al. (J Irrig Drain Eng 142(10):04016036-1–9, 2016) estimated the coefficient of discharge for 61.7, 92.3 and 100% datasets, respectively. Statistical analysis showed that in most of the cases, the equation of Di Stefano et al. (J Irrig Drain Eng 142(10):04016036-1–9, 2016) showed better precision than Azimi et al. (J Irrig Drain Eng 139(1):75–83, 2013) equation, and overall, the equation proposed

S. Kadia (✉)
Department of WRD & M, Indian Institute of Technology Roorkee, Roorkee, Uttarakhand 247667, India
e-mail: subhojitkadia@gmail.com

B. Kumar · Z. Ahmad
Department of Civil Engineering, Indian Institute of Technology Roorkee, Roorkee, Uttarakhand 247667, India
e-mail: binit.nit2010@gmail.com

Z. Ahmad
e-mail: zulfifee@iitr.ac.in

M. B. Kalinowska et al. (eds.), *Recent Trends in Environmental Hydraulics*, GeoPlanet: Earth and Planetary Sciences,
https://doi.org/10.1007/978-3-030-37105-0_7

by Di Stefano et al. (J Irrig Drain Eng 142(10):04016036-1–9, 2016) is more accurate than the equation proposed by Azimi et al. (J Irrig Drain Eng 139(1):75–83, 2013).

Keywords Discharge characteristics · Triangular weir · Experimentation · CFD · Existing equations

1 Introduction

Water resources engineers and professionals come across problems like land erosion, sediment load calculation, degradation, aggradation, scour, silting of reservoirs and stability of channel, which are related to alluvial rivers and channels (Garde and Ranga Raju 2015). Ganges, Brahmaputra, Damodar, and Kosi are few well-known rivers in India which carry sediment at a mean concentration about 0.12–0.39% by their weight and the annual observed erosion rate is about 0.00041–0.00188 m (Garde and Ranga Raju 2015). A transverse structure like a weir placed across a stream to divert or to store water in the upstream side increases the depth of water and reduces the flow velocity in the upstream side which results in flow and sediment discontinuity (Bai and Duan 2014; Fan and Morris 1992; Tiwari and Sharma 2017). The restriction in sediment passage may results in the deposition in the upstream of a weir and scouring in the downstream side (Garde and Ranga Raju 2015; Kim et al. 2014). The downstream scouring depends on the sediment properties, flow parameters, tailwater depth and the difference in head.

Sharp-crested weirs and weirs of finite crest length are the two groups of overflowing weirs (Azimi and Rajaratnam 2009; Azimi et al. 2013). A triangular weir with an upstream ramp (TW-UR) placed across a stream may accomplish the requirement of sediment passage due to the presence of the ramp. The authors are presently investigating the sediment movement over a TW-UR. However, this paper is about the investigation made on the flow field and discharge characteristics of a TW-UR using experimentation and computational fluid mechanics (CFD) tool.

The coefficient of discharge for a TW-UR can be evaluated using Eq. (1), which is basically used for a free-flowing weir (Di Stefano et al. 2016; Henderson 1966; Shaker and Sarhan 2017).

$$Q = \frac{2}{3} C_d B \sqrt{2gH^3} \tag{1}$$

where Q is the discharge [m^3 s^{-1}], C_d is the coefficient of discharge for a TW-UR, B is the channel width [m], g is the acceleration due to gravity [m^2 s^{-1}], and H is the head over the weir crest [m] measured at a sufficient distance upstream from the crest. Previous experimental studies on the head-discharge relationship of a TW-UR are available in the literature (Abou-Seida and Quraishi 1976; Azimi et al. 2012, 2013; Horton 1907; Shaker and Sarhan 2017). Azimi et al. (2013) analysed the data obtained from earlier investigations (Abou-Seida and Quraishi 1976; Horton 1907) and suggested Eq. (2), and later Di Stefano et al. (2016) proposed Eq. (3) for the estimation of the coefficient of discharge for a TW-UR.

$$C_d = 1.27\left(\frac{H}{L}\right)^{0.11} \Big/ \left(\sqrt{3}\left(\frac{P}{L}\right)^{1/10}\right) \tag{2}$$

$$C_d = 1.058\left(\frac{H}{P}\right)^{0.0839}\left(\frac{0.002}{P}\right)^{-0.0264}\left(1+\frac{L}{P}\right)^{0.1134} \Big/ \sqrt{3} \tag{3}$$

where L is the base length of the weir [m] and P is the height of the weir [m]. Di Stefano et al. (2016) considered 0.002 m crest length along the longitudinal direction. Shaker and Sarhan (2017) compared TW-UR with a sharp-crested weir having the same crest height. For sharp-crested weir, the coefficient of discharge can be calculated using the Rehbock equation (Henderson 1966), Eq. (4).

$$C_{d,sharp} = 0.611 + 0.08\frac{H}{P} \tag{4}$$

where $C_{d,sharp}$ is the coefficient of discharge for a sharp-crested weir having a crest height P [m]. $C_{d,sharp}$ is affected by the geometric parameters of channel and weir, and the flow phenomenon (Kumar et al. 2011). It was observed that a TW-UR was having higher discharging capacity than broad-crested and sharp-crested weirs.

This paper deals with the current experimental study and CFD analysis on the flow field and discharge characteristics of a TW-UR model and checking the accuracy of the available equations for the coefficient of discharge suggested by previous investigators (Azimi et al. 2013; Di Stefano et al. 2016). The ranges of parameters used in the present and previous studies are listed in Table 1. A total of 200 experimental datasets were collected.

Table 1 Summary of the present and previous investigations along with the range of parameters

Investigator	L [m]	P [m]	H [m]	Angle α
1. Abou-Seida and Quraishi (1976) (*group 2 case 1)	0.1415, 0.117, 0.0882	0.0818, 0.117, 0.1527	0.0335–0.1273	30°, 45°, 60°
2. Bazin (*125) (Horton 1907)	0.167	0.5	0.3–0.427	71.6°
3. Shaker and Sarhan (2017)	0.06–0.3	0.06, 0.08, 0.1, 0.12	0.021–0.06	21.8°, 26.6°, 33.7°, 45°
4. USDW (*16) (Horton 1907)	5.63	1.45	0.51–1.27	15.9°
5. Present study	0.189	0.105	0.0425–0.1055	29.1°

Note 1 *indicates the number/set of the experiment

2 Experimental Work

The experiments were conducted in the Hydraulics Laboratory, Dept. of Civil Engg., IIT Roorkee, in a 15 m long glass-walled flume having a cross-section of 0.39 m × 0.52 m. The TW-UR model shown in Figs. 1 and 2 was fabricated by 0.006 m thick sheet. The length (L), height (P) and width (B) of the model were 0.189 m, 0.105 m and 0.39 m, respectively, and the slope angle α was 29.1° to the horizontal axis. In the downstream side, the weir height was 0.14 m. The flume was fed by clear water through an inlet pipe of 0.1 m diameter and the discharge was measured with an ultrasonic flow meter (accuracy ±1%) connected to the inlet pipe. In total, 23 discharge samples were studied, ranging from 0.0073 to 0.02995 $m^3 s^{-1}$. The depth of water and H were measured using a point gauge with a Vernier scale (least count = 0.0001 m), which was placed at 0.2 m upstream from the weir crest, as shown in Fig. 2. The flow condition in the upstream side was steady and subcritical.

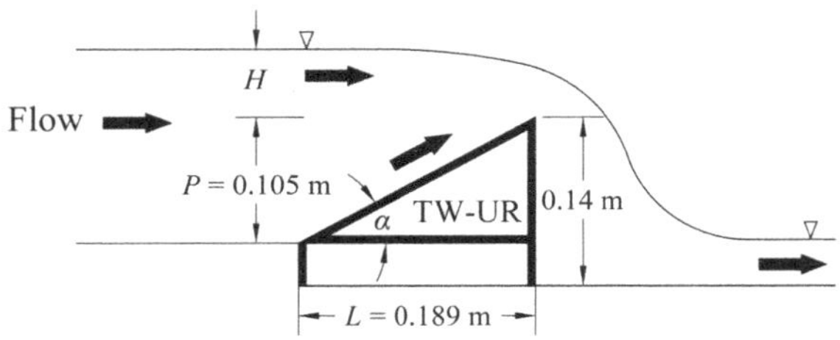

Fig. 1 Triangular weir with an upstream ramp

Fig. 2 Experimental activity

3 CFD Analysis

The Computational Fluid Dynamics (CFD) technique was employed earlier for the analysis of head-discharge relationship for the Side Weir (Azimi and Shabanlou 2015; Hoseini et al. 2013), Labyrinth Weir (Aydin and Emiroglu 2013, 2016), Piano Key Weir (Bremer and Oertel 2017; Crookston et al. 2018; Hu et al. 2018) and it was found to be handy, accurate, time and cost saving tool which can minimise the experimental efforts. The present CFD simulation is on a TW-UR using the Ansys CFX solver within Ansys 19.1 academic research version (ANSYS 2018). The Ansys CFX solver uses the finite volume and vertex-centered methods, whereas Fluent uses the finite volume and cell-centered methods (Acharya 2016; Berggren et al. 2009). The vertex-centered method is useful in reducing the computational space and cost due to a less number of degrees of freedom than the cell-centered method (Acharya 2016; Berggren et al. 2009). In CFD analysis the well-known Reynolds-averaged Navier-Stokes equations (continuity and momentum) are used as the basic equations. The standard k-ε turbulent model (Launder and Spalding 1972, 1974) is a type of eddy-viscosity model that uses the basic equations of turbulent kinetic energy k and dissipation rate ε (Acharya 2016; Bates et al. 2005).

3.1 Geometry, Meshing and Boundary Setup

The current CFD analysis involved the formation of geometry (1:1 with the flume model), meshing, setup's physics, simulation run and post-processing. The 3D geometry was generated in AutoCAD 3D and imported in Ansys workbench. Inlet, free surface, outlet boundaries were fixed thereafter. The domain was 1.5 m long with cross-sections of 0.39 m × 0.3 m in the upstream side and 0.39 m × 0.335 m in the downstream side, respectively. For mesh configuration, the tetrahedron method was adopted and the size of the near-bed faces was kept at 0.00709 m and the maximum size of the element was set at 0.0156 m. A maximum of 10 inflation layers was created near the bed surfaces for better representation of the flow near the bed. Therefore, the whole fluid domain was divided into a number of small finite volumes. In this CFD analysis, 5 discharge values out of 23 experimental values were simulated in the Spatial Sciences Laboratory, Department of WRD & M, IIT Roorkee. For the simulation, the transient flow condition for a total time span of 30 s at a time step of 0.05 s was selected. Gravity, pressure, and buoyancy were defined thereafter. Continuous surface tension was fixed at 0.072 N m^{-1}. The fluid domain was initialised by applying the mean velocity approach. For each of the 5 cases, the mean observed velocity was assigned to the inlet boundary and the turbulence intensity was set at 1% (low). The outlet was defined as a pressure-based outlet. The top opening was configured with relative pressure normal to the plane. The boundary conditions are shown in Fig. 3. The convergence criteria were fixed with the residual type being RMS and the residual target being 0.0001. For better convergence 10 coefficient

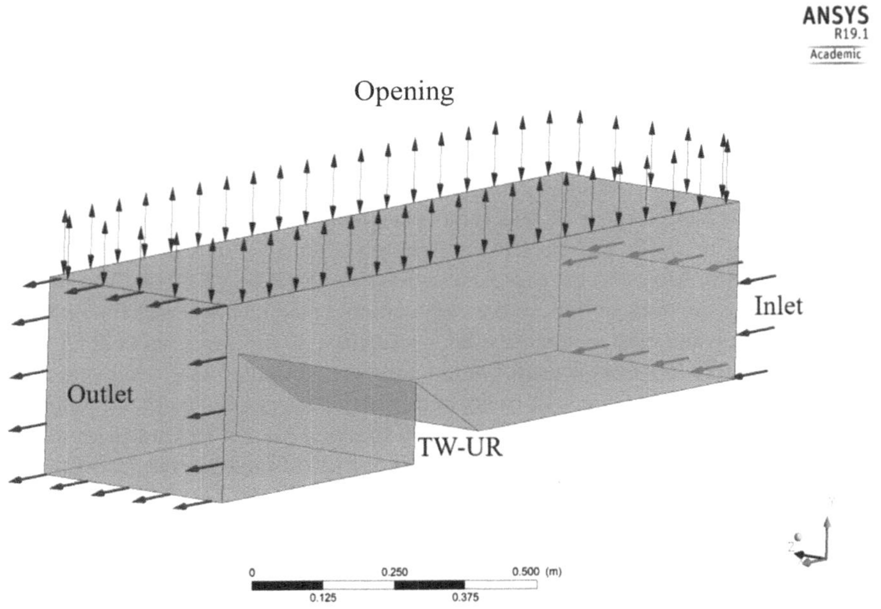

Fig. 3 TW-UR model along with the boundary conditions

loops were considered. The simulation was based on the finite volume technique and Volume of Fluid (VOF) method by adopting multiphase, turbulent, open channel conditions.

4 Results and Discussion

4.1 Flow Field Obtained from CFD Simulation

To understand the difference in flow field near and over a TW-UR as compared to a sharp-crested weir, the flow over a sharp-crested weir with same weir height as the TW-UR model was simulated at 0.02076 $m^3\ s^{-1}$ discharge using the same methodology as described in the earlier section. Figure 4a shows the less active flow field (low-velocity magnitude) below the crest level in the upstream of a sharp-crested weir, whereas the flow field upstream of TW-UR is more active as shown in Fig. 4b. The flow velocity and hydrodynamic force over the ramp increased along the direction of flow due to the vertical contraction in the flow area. Therefore, TW-UR can pass more discharge and increases the possibility of sediment passage as compared to a broad-crested weir.

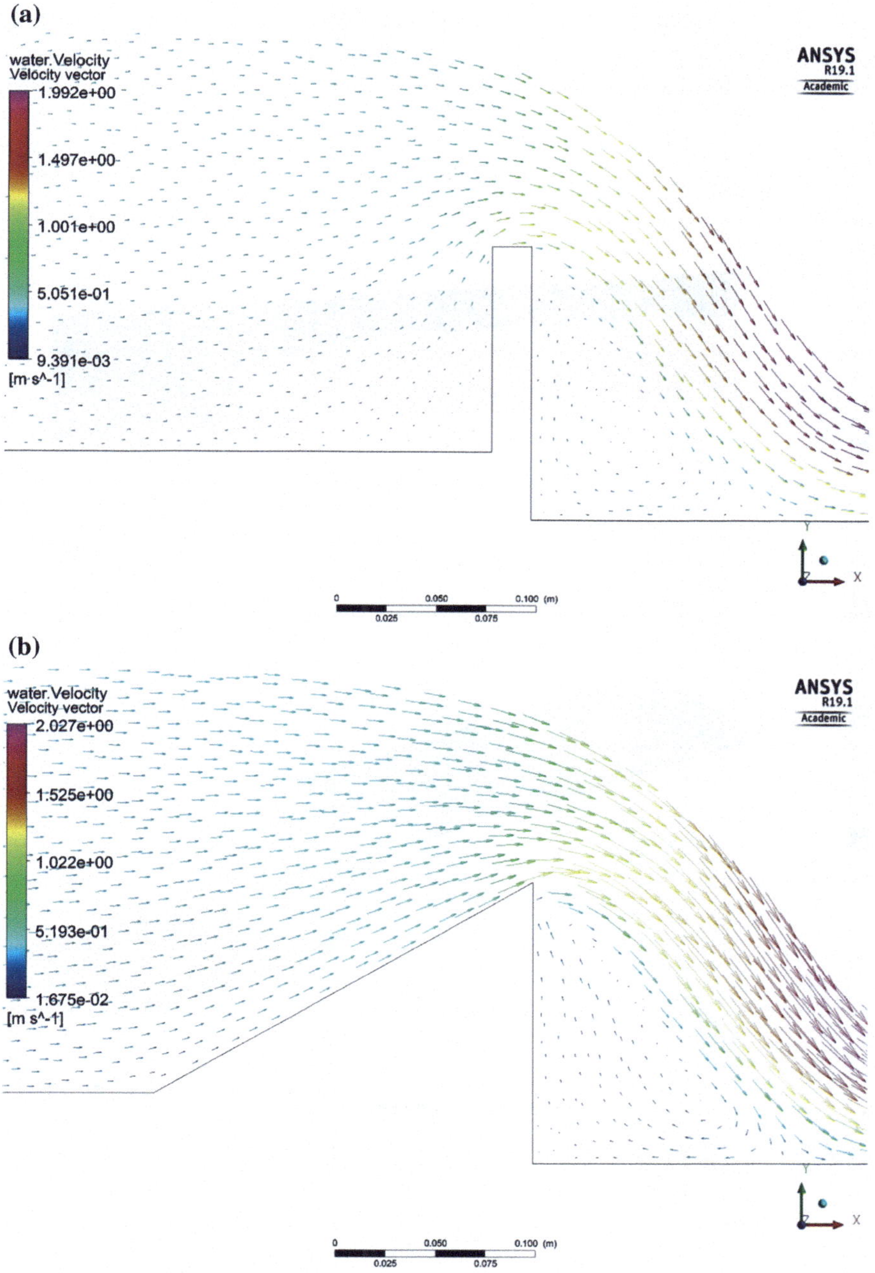

Fig. 4 Simulated flow field at 0.02076 $m^3\ s^{-1}$ discharge for: **a** sharp-crested weir, **b** TW-UR

Figure 5a represents the distribution of air and water mixture along the mid-section of the channel. The simulated free surface was defined by an isosurface having water volume fraction = 0.5, and the same is shown in Fig. 5b.

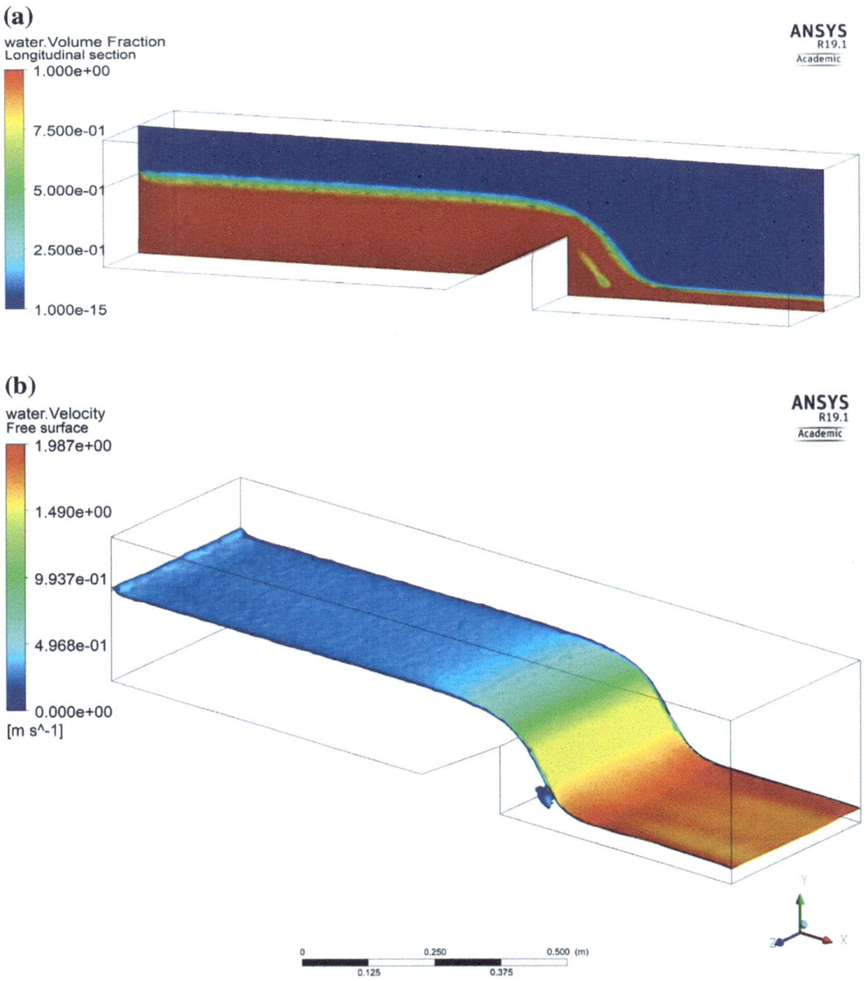

Fig. 5 Simulated flow at 0.02076 $m^3\ s^{-1}$ discharge: **a** air-water mixture, **b** free surface

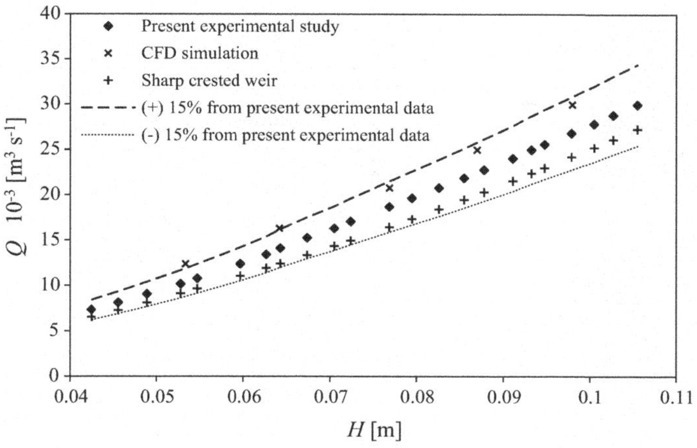

Fig. 6 Head-discharge correlation

4.2 Comparison of the Discharge Obtained from CFD with the Observed Value

The head-discharge characteristics of TW-UR were studied experimentally for 23 discharge values out of which 5 discharge values were simulated using CFD technique. Figure 6 represents the head-discharge variation for both observed as well as simulated flows over a TW-UR model and its comparison with a sharp-crested weir of the same height. The coefficient of discharge for the sharp-crested weir was obtained using Eq. (4). The TW-UR model had a better discharging capacity (9.8–14.3% higher) than a sharp-crested weir under the same head for all head values as shown in Fig. 6. The more active flow field near and over the TW-UR, as shown in Fig. 4b, helped it to pass additional discharge than a broad-crested weir. To pass the same discharge over a TW-UR, the CFD model estimated lesser head requirement as compared to the observed data and it was found that the CFD simulation estimated about 10–15% higher discharge under the same head in comparison to the observed results. It was also noticed that the observed C_d increased initially with an increase in H and attained almost a constant value beyond $H/P \approx 0.65$.

4.3 Checking the Accuracy of Existing Discharge Equations

To determine the accuracy of existing discharge equations suggested by Azimi et al. (2013) and Di Stefano et al. (2016), the collected 200 datasets from previous and present experimental studies were analysed graphically and using the statistical parameters. The range of the parameters used in the present and earlier studies are summarised in Table 1. Out of 200 datasets, the datasets from United States Deep

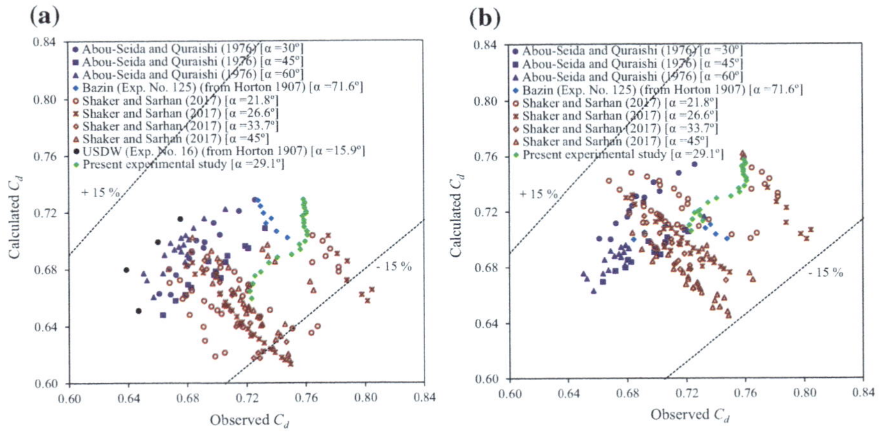

Fig. 7 Calculated C_d versus observed C_d for: **a** Azimi et al. (2013), **b** Di Stefano et al. (2016)

Waterways (USDW) were not considered in the evaluation of Eq. (3) because they were not satisfying the criteria $16.05^{\circ} \leq \alpha \leq 90^{\circ}$ (Di Stefano et al. 2016). After plotting the observed and calculated values of C_d, it was observed that about 92.5% of data were within ±15% error for Eq. (2), whereas all data were well within ±15% for Eq. (3), as shown in Fig. 7a, b. Therefore, the equation of the coefficient of discharge proposed by Di Stefano et al. (2016) appears to be performing better than the equation suggested by Azimi et al. (2013).

For both equations and especially for Eq. (2), a comparatively higher deviation between the calculated and observed C_d was found for the datasets of Shaker and Sarhan (2017) with H <0.03 m and especially for $P = 0.06$ m model. This may be because the surface tension force is not negligible for H <0.03 m (Kabiri-Samani and Javaheri 2012). For the present experimental data, the maximum absolute error in calculated C_d was about 3 and 9% for Eq. (3) and Eq. (2), respectively. While arranging and plotting the datasets in an increasing sequence of absolute error, it was observed that the % of the total number of datasets lay within 5, 10 and 15% absolute error ranges for Eq. (2) was 43.0, 72.5 and 92.5%, respectively. Whereas, for Eq. (3) it was 61.7, 92.3 and 100%, respectively. The maximum absolute error for Eqs. (2) and (3) was 18.2 and 13.8%, respectively. Figure 8 shows the better performance of Eq. (3). Overall, the equation proposed by Di Stefano et al. (2016) is predicting C_d better than the equation suggested by Azimi et al. (2013).

To evaluate the accuracy of the available equations statistically, the observed and calculated C_d were compared using two of the most commonly used statistical parameters; "root mean square error (*RMSE*)" and "mean absolute percentage error (*MAPE*)", respectively (Aydin and Emiroglu 2013, 2016; Crookston et al. 2018). They are expressed in Eqs. (5) and (6).

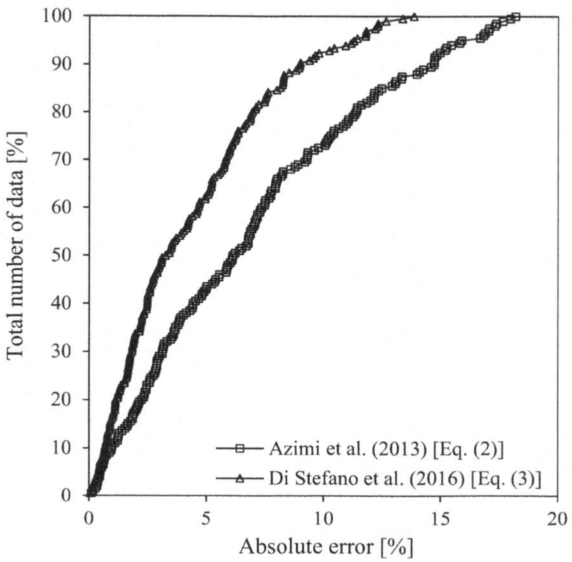

Fig. 8 Variation of the total number of data to the percentage absolute error

$$RMSE = \sqrt{\frac{1}{N}\sum_{i=1}^{N}(C_{do} - C_{dc})^2} \tag{5}$$

$$MAPE = \frac{1}{N}\sum_{i=1}^{N}\left|\frac{C_{do} - C_{dc}}{C_{do}}\right| 100\,\% \tag{6}$$

where N is the number of datasets, C_{do} is the observed coefficient of discharge obtained from the present and previous experimental investigations (Abou-Seida and Quraishi 1976; Horton 1907; Shaker and Sarhan 2017) and C_{dc} is the calculated or predicted coefficient of discharge using Eqs. (2) and (3) (Azimi et al. 2013; Di Stefano et al. 2016). The *RMSE* and *MAPE* values for different cases are shown in Table 2. It was observed that in 7 out of 9 cases the equation proposed by Di Stefano et al. (2016) predicted C_d with greater accuracy than the equation suggested by Azimi et al. (2013). For the present experimental datasets, the calculated C_d had *RMSE* and *MAPE* values of 0.05 and 6.54% for Eq. (2) and 0.012 and 1.41% for Eq. (3), respectively. Overall, the *RMSE* and *MAPE* of calculated C_d were 0.062 and 6.86% for Eq. (2) and 0.04 and 4.3% for Eq. (3), respectively. Therefore, the equation proposed by Di Stefano et al. (2016) is found to be more accurate than the other one.

Table 2 Error in C_d prediction for Eqs. (2) and (3)

Investigator	Angle α	Equation (2) (Azimi et al. 2012)		Equation (3) (Di Stefano et al. 2016)	
		RMSE	*MAPE* [%]	*RMSE*	*MAPE* [%]
1. Abou-Seida and Quraishi (1976) (*group 2 case 1)	30°	0.008	0.91	0.037	5.40
	45°	0.019	2.67	0.009	1.07
	60°	0.020	2.84	0.015	2.05
2. Bazin (*125) (Horton 1907)	71.6°	0.025	2.67	0.028	3.37
3. Shaker and Sarhan (2017)	21.8°	0.066	7.59	0.039	4.57
	26.6°	0.091	10.86	0.050	5.58
	33.7°	0.073	8.98	0.039	4.55
	45°	0.074	8.84	0.059	6.92
4. USDW (*16) (Horton 1907)	15.9°	0.034	4.68	–	–
5. Present experimental study	29.1°	0.050	6.54	0.012	1.41
6. Total data	–	0.062	6.86	0.040	4.30

Note 2 * indicates the number/set of the experiment

5 Conclusions

The head-discharge characteristics of a TW-UR model were studied experimentally for 23 discharges, out of which 5 discharge conditions were simulated in Ansys CFX module using the standard k-ε turbulent model. It was found that the TW-UR model had about 9.8–14.3% higher discharging capacity than a sharp-crested weir of same height and C_d increased initially with an increase in H and attained almost a constant value beyond $H/P \approx 0.65$. About 10–15% higher discharge was estimated in the CFD simulation as compared to the observed data under the same head. The TW-UR may be helpful in passing sediments as compared to the conventional weirs due to the presence of the upstream ramp and a highly active flow field in the upstream side. Increasing velocity along the flow direction over the upstream ramp due to flow contraction in the vertical plane is helpful in enhancing the hydrodynamic force exerted by the moving fluid and increasing the discharge capacity. Authors are presently investigating the sediment transport over the TW-UR. Two existing equations suggested by Azimi et al. (2013) and Di Stefano et al. (2016) were checked for their accuracy after analysing 200 available datasets. Graphical illustration showed that the equation proposed by Azimi et al. (2013) predicted the coefficient of discharge within ±5, ±10 and ±15% error ranges for 43.0, 72.5 and 92.5% datasets, whereas within those error ranges, the equation proposed by Di Stefano et al. (2016) estimated the coefficient of discharge for 61.7, 92.3 and 100% datasets, respectively. The maximum absolute error for Eqs. (2) and (3) was 18.2 and 13.8%, respectively. Statistical analysis has shown that in 7 out of 9 cases, Eq. (3) predicted C_d with greater accuracy than Eq. (2). Considering all datasets, the *RMSE* and *MAPE* of calculated C_d were 0.062 and 6.86% for Eq. (2) and 0.04 and 4.3% for Eq. (3), respectively. Overall, it can be concluded that the equation proposed by Di Stefano et al. (2016) is more accurate than the equation proposed by Azimi et al. (2013) for the used datasets. Further, these 200 datasets can be utilised in the establishment of a more accurate equation.

Acknowledgements The first author would like to thank WBSEDCL for sponsoring his master's study. He would also like to express his gratitude to Prof. S. K. Mishra, WRD & M, IIT Roorkee, for his fruitful advice and encouragement. The second author is extremely thankful to the MHRD, Government of India for sponsorship.

References

Abou-Seida MM, Quraishi AA (1976) A flow equation for submerged rectangular weirs. Proc Inst Civil Eng Part 2(61):685–696

Acharya R (2016) Investigation of differences in Ansys solvers CFX and fluent. Master thesis, KTH Royal Institute of Technology, Stockholm, Sweden

ANSYS (2018) ANSYS academic research mechanical and CFD, Release 19.1, Canonsburg, PA

Aydin MC, Emiroglu ME (2013) Determination of capacity of labyrinth side weir by CFD. Flow Meas Instrum 29:1–8

Aydin MC, Emiroglu ME (2016) Numerical analysis of subcritical flow over two-cycle trapezoidal labyrinth side weir. Flow Meas Instrum 48:20–28

Azimi AH, Rajaratnam N (2009) Discharge characteristics of weirs of finite crest length. J Hydraul Eng 135(12):1081–1085

Azimi AH, Rajaratnam N, Zhu DZ (2012) A note on sharp-crested weirs and weirs of finite crest length. Can J Civ Eng 39(11):1234–1237

Azimi AH, Rajaratnam N, Zhu DZ (2013) Discharge characteristics of weirs of finite crest length with upstream and downstream ramps. J Irrig Drain Eng 139(1):75–83

Azimi H, Shabanlou S (2015) U-shaped channels along the side weir for subcritical and supercritical flow regimes. Flow Meas Instrum 46:170–178

Bai Y, Duan JG (2014) Simulating unsteady flow and sediment transport in vegetated channel network. J Hydrol 515:90–102

Bates PD, Lane SN, Ferguson RI (2005) Computational fluid dynamics: applications in environmental hydraulics. John Wiley & Sons Ltd, Chichester

Berggren M, Ekström SE, Nordström J (2009) A discontinuous Galerkin extension of the vertex-centered edge-based finite volume method. Commun Comput Phys 5(2–4):456–468

Bremer F, Oertel M (2017) Numerical investigation of wall thickness influence on Piano Key Weir discharge coefficients: a preliminary study. In: Proc. of the third international workshop on labyrinth and Piano Key Weirs 2017, Taylor & Francis Group, London, pp 101–108

Crookston BM, Anderson RM, Tullis BP (2018) Free-flow discharge estimation method for Piano Key Weir geometries. J Hydro-Environ Res 19:160–167

Di Stefano C, Ferro V, Bijankhan M (2016) New theoretical solution of the outflow process for a weir with complex shape. J Irrig Drain Eng 142(10):04016036-1–9

Fan J, Morris GL (1992) Reservoir sedimentation. I: delta and density current deposits. J Hydraul Eng 118(3):354–369

Garde RJ, Ranga Raju KG (2015) Mechanics of sediment transportation and alluvial stream problems. Revised 3rd edn. New Age International (P) Ltd., New Delhi

Henderson FM (1966) Open channel flow. Macmillan, New York

Horton RE (1907) Weir experiments, coefficients, and formulas. In: Proc. U.S. geological survey—water supply and irrigation (200), Government Printing office, Washington, DC

Hoseini SH, Jahromi SHM, Vahid MSR (2013) Determination of discharge coefficient of rectangular broad-crested side weir in trapezoidal channel by CFD. Int J Hydraul Eng 2(4):64–70
Hu H, Qian Z, Yang W, Hou D, Du L (2018) Numerical study of characteristics and discharge capacity of Piano Key Weirs. Flow Meas Instrum 62:27–32
Kabiri-Samani A, Javaheri A (2012) Discharge coefficients for free and submerged flow over Piano Key Weirs. J Hydraul Res 50(1):114–120
Kim S, Im J, Lee SO (2014) Assessment of sediment exclusion efficiency for several modified labyrinth weirs. Paddy Water Environ 12(Supp. 1):133–140
Kumar S, Ahmad Z, Mansoor T (2011) A new approach to improve the discharging capacity of sharp-crested triangular plan form weirs. Flow Meas Instrum 22:175–180
Launder BE, Spalding DB (1972) Lectures in mathematical models of turbulence. Academic Press, London, New York
Launder BE, Spalding DB (1974) The numerical computation of turbulent flows. Comput Methods Appl Mech Eng 3:269–289
Shaker AJ, Sarhan AS (2017) Performance of flow over a Weir with sloped upstream face. ZANCO J Pure Appl Sci 29(3):43–54
Tiwari H, Sharma N (2017) Turbulence study in the vicinity of Piano Key Weir: relevance, instrumentation, parameters and methods. Appl Water Sci 7:525–534

Modelling of Velocity Distribution in a Channel Partly Covered by Submerged Vegetation

Monika B. Kalinowska, Kaisa Västilä, Adam Kozioł, Paweł M. Rowiński, Adam Kiczko and Janusz Kubrak

Abstract The presence of vegetation has a significant impact on the flow conditions in streams and rivers by changing the roughness, which has an important effect on flow velocity distribution. The article focuses on modelling and analysing the depth-averaged velocity distribution in a rectangular channel partially covered by submerged grassy vegetation. The possibility of using two selected models (CCHE2D flow model and Shiono and Knight model) has been evaluated using the measurement data from a laboratory experiment. The measurements and modelling have been carried out for different flow conditions. Applied models calibrated for low velocities were found to be extendable to the same vegetative condition during high velocities.

Keywords Submerged vegetation · CCHE2D · Shiono and knight model · Depth-averaged velocity distribution · Roughnes · Laboratory experiment · Streams · Rivers

1 Introduction

The presence of vegetation in open channels significantly affects water flow, transport and mixing processes, water quality as well as morphodynamics and ecological properties of the stream (e.g. Rowiński et al. 2018). Research related to flow and transport processes become very complicated in the channels with vegetation, and although many experimental and numerical studies have already been carried out to determine the effect of vegetation on flow hydrodynamics, there are still many questions and problems that require further analysis (e.g. Kalinowska et al. 2019). The problem is especially important in situations when detailed measurements at different

M. B. Kalinowska (✉) · P. M. Rowiński
Institute of Geophysics, Polish Academy of Sciences, Ksiecia Janusza 64, 01-452 Warsaw, Poland
e-mail: Monika.Kalinowska@igf.edu.pl

K. Västilä
Aalto University School of Engineering, P. O. Box 15200, 00076 Aalto, Finland

A. Kozioł · A. Kiczko · J. Kubrak
Warsaw University of Life Sciences, 159 Nowoursynowska Str, 02-776 Warsaw, Poland

M. B. Kalinowska et al. (eds.), *Recent Trends in Environmental Hydraulics*, GeoPlanet: Earth and Planetary Sciences,
https://doi.org/10.1007/978-3-030-37105-0_8

flow rates are not possible. In such cases, the possibility of using computational models is very beneficial. However, an application of any detailed, most accurate, 3D numerical models is usually limited to academic questions, since in practice there is not enough input data to run 3D models. This study is motivated by the practical needs for predicting the flow velocities in partly vegetated channels during high flows when vegetation is deeply submerged as defined by Nepf (2012). The aim of this paper is to investigate the capabilities of selected simplified models for simulating the lateral distribution of streamwise velocity in a channel consisting of an unvegetated part and a vegetated part imitating maintained grassy floodplain vegetation. We applied two models: the CCHE2D, a two-dimensional (2D) depth-averaged hydrodynamic model, and the quasi 2D Shiono and Knight model.

2 Materials and Methods

2.1 Laboratory Experiment

The experiments were conducted in the 20 m long, 0.6 m wide Environmental Hydraulics Flow Channel at Aalto University as part of the measurement campaigns of Box et al. (2018) and Caroppi et al. (2019). Artificial grass 20 mm in height was covering 38% of the flume width within the 10 m long test channel reach, whereas the rest of the flume bottom consisted of smooth PVC (see Fig. 1). The 0.8 mm wide grass stems were semi-flexible and had a uniform density of ~65 stems per cm^2 with the frontal projected area per unit volume of ≈500 m^{-1} in still water conditions.

In the three test runs differing in discharge (Table 1), the water depth was $h = 0.17$ m, resulting in relative vegetative submergence of 0.12 (vegetation height divided by water depth). Uniform flow conditions (in the cross-sectionally averaged sense) along the 10 m test reach were achieved in all test runs by adjusting the bed slope and the weir at the downstream end of the flume. The water depth was measured with pressure transducers at 80 Hz in six locations along the centerline of the 10 m test region while discharge was measured at 80 Hz with a magnetic flow meter located in the pipe supplying the water. The experimental setting is presented in detail in Caroppi et al. (2019).

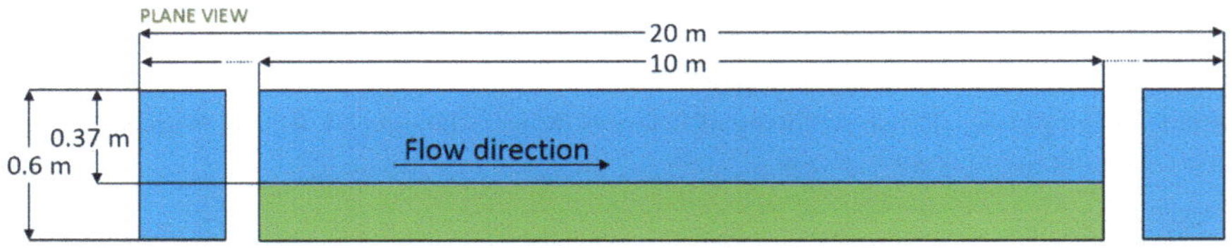

Fig. 1 Schematically presented laboratory channel, vegetated area shown in green color

Table 1 Measured hydraulic properties of the partly vegetated channel

Variant	Vegetation hight (m)	Water depth *h* (m)	Discharge *Q* (m^3 h^{-1})	Slope *I* (–)	Averaged streamwise velocity (m s^{-1})
B1	0.02	0.17	80	0.0005	0.22
B2	0.02	0.17	180	0.0015	0.49
B3	0.02	0.17	300	0.0029	0.81

The flow field was recorded with Acoustic Doppler Velocimeter (ADV) at 200 Hz over 120 s period in the fully developed region in a cross-section located ~7.5 m downstream from the leading edge of the vegetation. The measuring verticals were located in the vegetated part (y = 250 mm and y = 130 mm), at the interface (y = 70 mm) and in the unvegetated part (y = 0 mm and y = −150 mm), with y = 0 at the centerline of the channel and increasing towards the right wall (i.e., the vegetated part). In addition, a transect was measured at a depth of z = 0.6 h. ADV data with signal-to-noise-ratio <17 were filtered out. After measuring all the other points, circular holes with a diameter of 1 cm were cut into the grass mat to measure the points within the grasses. The data was de-spiked using the modified phase-space threshold technique assuming C1 = C2 = 1.48, according to Parsheh et al. (2010) and the median absolute deviation as the characteristic scalar. The filtered data were replaced by the last good value, as recommended by Jesson et al. (2013). The ADV data were processed using the Velocity Signal Analyser software (v1.5.64) developed by Jesson et al. (2015).

2.2 CCH2D Model

CCHE2D model is a two-dimensional depth-averaged hydrodynamic model of unsteady turbulent free-surface flows (Jia and Wang 2001; Ye and McCorquodale 1997). This model has been developed by NCCHE—National Center for Computational Hydroscience and Engineering of the University of Mississippi, USA. To obtain a well-posed initial-boundary value problem, and consequently to run the model, a set of initial and boundary conditions, as well as simulation parameters, are required. The laboratory channel geometry (Sect. 2.1) has been used to prepare the computational mesh using the CCHE-MESH generator (Zhang 2005). The generated, rectangular mesh has been schematically presented in Fig. 2. The grid sizes were equal to $\Delta x = 0.02$ m and $\Delta y = 0.01$ m in x and y-direction, respectively. To obtain an accurate solution the grid was thickened to $\Delta y = 0.001$ m close to the walls (see Fig. 2). Finally, the number of grid nodes was equal to 40,581 (I = 81 × J = 501). Information about the channel geometry, the initial bed elevation, the initial water level and the bed roughness has been specified based on the laboratory experiment described above. The bed elevation has been prepared using the

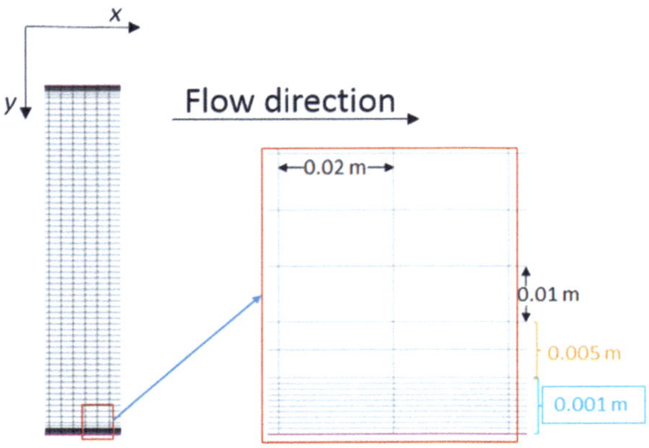

Fig. 2 Fragment of a computational mesh, with marked grid sizes

information about the slope for each assumed discharge separately. To characterize the bed roughness, the Manning coefficient was used. Final simulations were performed for the time step equal to $\Delta t = 0.1$ s, and simulation time equalled to 200 s (sufficient to receive the steady-state solution). The mixing length model was used as the first-order turbulence closure hypothesis. The simulations have been done for different values of the wall slipness coefficient (w_p), as well as using the logarithmic law of the wall (von Kármán 1930) to provide more accurate wall effect simulations, although, as mentioned above, this approach requires fine grid near the walls. For each simulation variant, the total discharge value (Q) of each laboratory experiment variant was used as an inlet boundary condition. As an outlet boundary condition, we used the measured water surface level of 0.17 m for all variants.

2.3 Shiono and Knight Model

The Shiono and Knight Method (SKM) is derived from the Navier-Stokes equations and in fact is a combination of the depth-averaged longitudinal momentum equation together with the continuity equation, expressed for steady flow as follows (Shiono and Knight 1990, 1991; Rameshwaran and Shiono 2007; Knight et al. 2010; Shiono and Rameshwaran 2015):

$$\underbrace{\frac{\partial\left[H(\rho UV)_d\right]}{\partial y}}_{I} = \underbrace{\rho g H S_0}_{II} - \underbrace{\rho\frac{f}{8}U_d^2}_{III} + \underbrace{\frac{\partial}{\partial y}\left[\sigma\lambda H^2\sqrt{\frac{f}{8}}U_d\frac{\partial U_d}{\partial y}\right]}_{IV}, \quad (1)$$

where: H is the local water depth; U and V are velocity components in the x and y directions, respectively; x is the streamwise coordinate parallel to the channel bed; y is the lateral coordinate; the subscript d refers to a depth-averaged value; ρ is the fluid density; g is the acceleration due to gravity; S_0 is the channel bed slope;

f is the Darcy-Weisbach friction factor; and λ is the dimensionless eddy viscosity. The term (I), which includes the streamwise and lateral components of velocity, is called the secondary flow term. The terms (II), (III) and (IV) describe the weight component, the component of the friction force determined based on the bed shear stress and the resistance component determined based on the transverse Reynolds stresses, respectively. Studies have shown that the addition of the secondary flow term was found to make a significant improvement in the boundary shear stress prediction (Knight et al. 1994).

In many solutions of Eq. (1) the secondary flow term $\partial\left[H(\rho UV)_d\right]/\partial y$ is expressed as a single parameter Γ, representing the lateral gradient of the advective term. This assumption causes that Eq. (1) becomes a second-order linear differential equation which can be solved analytically. The approach used in SKM consisted in subdividing the channel cross-section into various subareas (panels) with either constant depth domains or sloping side slope domains, then solving the boundary conditions between adjacent panels and between the edge panels and their boundary, assuming no-slip condition (Tang and Knight 2009). Three parameters, the Darcy-Weisbach friction factor (f), the dimensionless eddy viscosity (λ) and the secondary flow term (Γ), are required for each panel. In many works, values of these parameters are a priori given, determined in cross-sections of natural and artificial channels (e.g. Tang and Knight 2009; Shiono et al. 2012; Kordi et al. 2015).

For a constant depth panel, if local friction factor (f), dimensionless eddy viscosity (λ) and secondary flow term (Γ) are known, the depth-averaged velocity (U_d) can be calculated as follow:

$$U_d = \left[A_1 e^{\gamma y} + A_2 e^{-\gamma y} + k\right]^{1/2}; \tag{2}$$

where A_1 and A_2 can be obtained through the application of the appropriate boundary conditions (e.g. Knight et al. 2007). The values of constants γ and k are determined from the following equations:

$$\gamma = \sqrt{\frac{2}{\lambda}}\left[\frac{f}{8}\right]^{1/4}\frac{1}{H}, \tag{3}$$

$$k = \frac{8gHS_0}{f}\left(1 - \frac{\Gamma}{\rho g S_0}\right). \tag{4}$$

The values of the dimensionless friction factor (f) in the channel or its part can be estimated in different ways. The most common way to calculate the friction factor (f) is to use the Darcy-Weisbach equation $f = 8gR_hS_0U_i^{-2}$, where R_h is the hydraulic radius, U_i is the section mean velocity at known average water flow velocity in a single cross-section or particular parts of the cross-section of a compound channel. With the known Manning roughness coefficient (n) in the cross-section of the channel or its part, the equivalent sand roughness height (k_s) is calculated from the equation $n = k_s^{1/6}(8.25\ g^{1/2})^{-1}$ (Rameshwaran and Shiono 2007; Ackers 1991) and then the local friction factor (f) is calculated from the Colebrook-White equation (e.g. Kozioł et al. 2016). In the literature, calibrated forms of the Colebrook-White equation for the friction factor (f) for smooth channels and vegetated floodplains can be found (e.g., Rameshwaran and Shiono 2007). Determination of the values of eddy viscosity (λ) and secondary flow term (Γ) requires the calibration of the model. The results of the performed hydraulic tests were used to calculate the lateral distribution of the depth-averaged velocities and the water flow. In the previous studies, the authors used SKM to predict velocity distributions in other cases of both laboratory channels and actual Ihme River in Germany and the results already turned out to be promising but sensitive to the assumed method of the evaluation of roughness height (Kubrak and Rowiński 2007).

3 Results and Discussion

3.1 Parameters of the Calibrated Models

Numerous simulations were performed to obtain correct model parameters to replicate the measured lateral velocity distributions of the variant B1 (Table 1) in the best possible way. The parameters calibrated for the B1 variant were then used to simulate the B2 and B3 variants. In case of CCHE2D model, the measured values of the velocity distribution in the x-direction at 0.6 h (0.095 m) were taken into account to reproduce the results close to the walls most accurately. Assuming the constant value of the Manning coefficient, for the main channel and grassy part of the channel, the best results were obtained for $n = 0.015$ and 0.030, respectively. The initial values for the Manning coefficients were estimated based on the laboratory measurements. In case of SKM application, the cross-section area was divided into two panels based on the roughness of the bottom (smooth PVC, artificial grass). The same values of Manning coefficients as in case of CCHE2D model were used. The SKM model parameters were determined using Monte Carlo sampling. Values of the eddy viscosity coefficient were uniformly sampled in the range of 0.005–4.5 and the secondary flow values from -5 to 5. For combinations of these two parameter values, the velocity distribution and the flow rate was calculated. The best parameter set was selected using the criterion of the lowest residuals (the difference between the measured and the calculated values) of the velocity (dU) and flow rate (dQ). Finally, the obtained value of the eddy viscosity coefficient was $\lambda_v = 0.129$, and the values of the secondary flow terms were $\Gamma_1 = 0.532$ and $\Gamma_2 = -1.054$ for the main channel and the grassy part of the channel, respectively.

3.2 Performance of the Models and Considerations for Practical Applications

The results of the calibrated models in the selected cross-section where the flow was fully developed are presented in Figs. 3, 4 and 5 for the B1, B2 and B3 variants. Distributions of depth-averaged longitudinal velocities obtained from the two models are juxtaposed with experimental data. The red dots represent the measured depth-averaged values, while the yellow diamonds the values measured at depth 0.6 h (0.095 m). Green triangles and blue quadrants represent the modelled results, with CCHE2D and SKM, respectively. Please note that the values at 0.6 depth noticeably differ from the depth-averaged values within the vegetation and in the shear layer.

Appropriate identification of parameters of the Shiono and Knight model allowed to calculate the distribution of the depth-averaged velocity in the rectangular cross-section of the channel with non-uniform roughness of the bottom, consistent with the observed values (Figs. 3, 4 and 5). Distributions of the depth-averaged velocity that reasonably agree with observations can be obtained for the same values of coefficients for the vortex viscosity and secondary flow. Obviously, the best agreement between the simulated and measured results was obtained for the B1 case, for which the model was calibrated. For the B2 and B3 variants the agreement is a bit less accurate. The lateral distribution of the velocity was more accurately reproduced with the CCHE2D, including the near wall region. In the case of the CCHE2D physical flow model, there is no so large discrepancy in the agreement between the simulated and measured results for different variants.

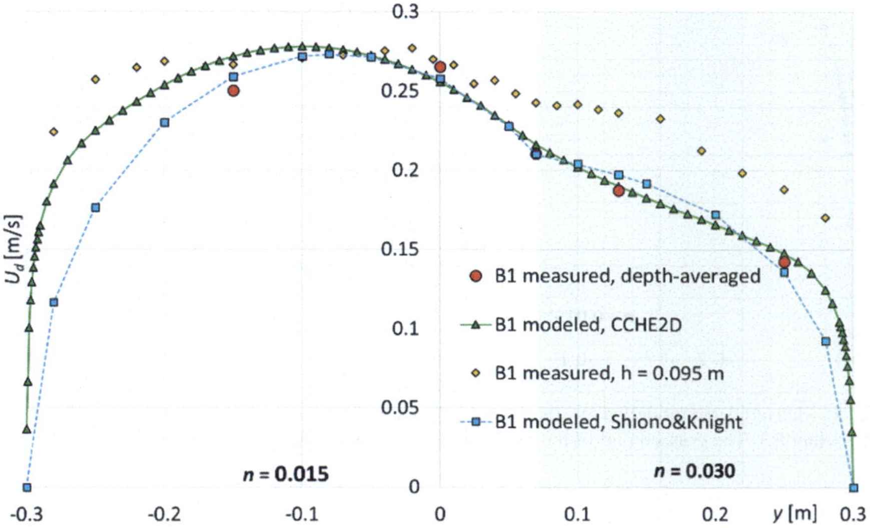

Fig. 3 Measured and modelled results of the depth-averaged velocity distribution in the x-direction for variant B1. The grassed part of the cross-section is shown in light green colour

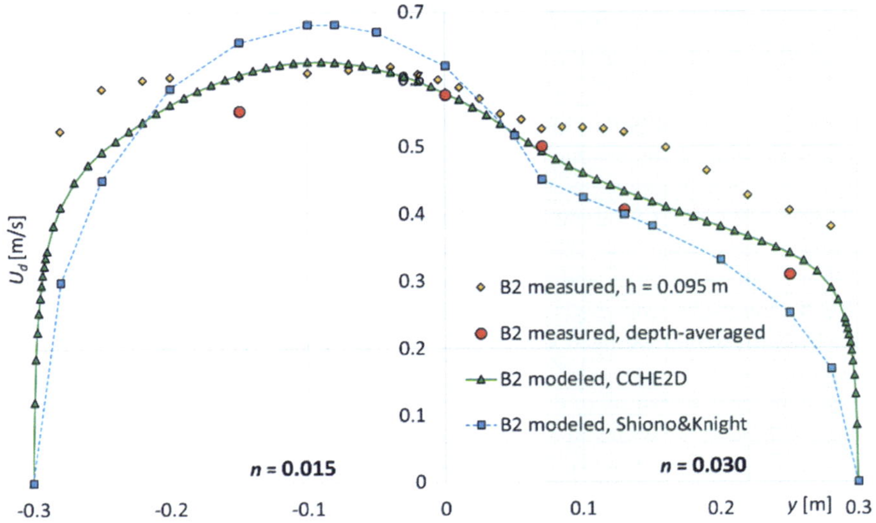

Fig. 4 Measured and modelled results of the depth-averaged velocity distribution in the x-direction for variant B2. The grassed part of the cross-section is shown in light green colour

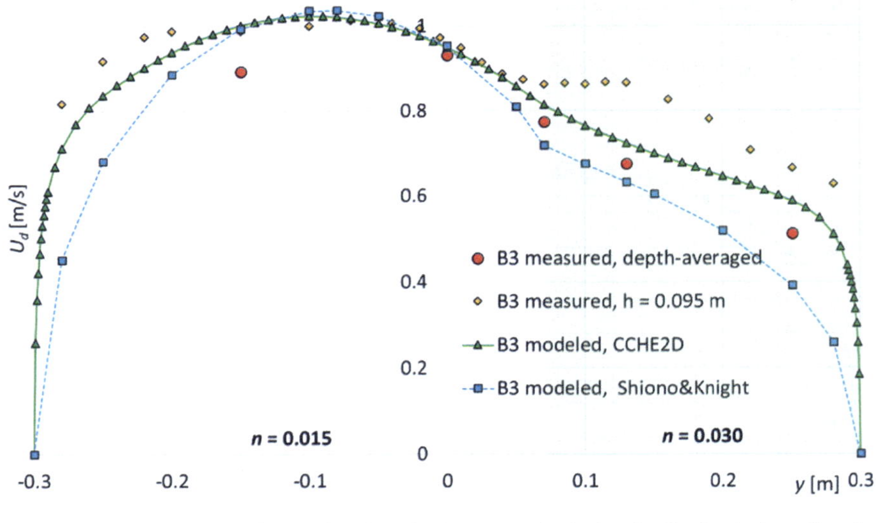

Fig. 5 Measured and modelled results of the depth-averaged velocity distribution in the x-direction for variant B3. The grassed part of the cross-section is shown in light green colour

If these models are planned to be applied in practice, one should note that the simulations and the calibration procedure were much more time consuming with the CCHE2D model than with the SKM model. Using the CCHE2D model, we should also be aware of the influence of the grids steps selection: in such narrow channel they must be very dense, especially close to the wall. Also, the proper wall slip parameter must be chosen in the CCHE2D model. The selected results, presented in the paper were computed using the logarithmic law of the wall. For practical applications aimed at modeling real floodplains, it has to be considered that the vegetation often consists of a grassy understory and taller woody shrubs. Future research should focus on ensuring the performance of 2D models for more complex floodplain vegetation, e.g. through implementing the advanced vegetation parameterization proposed by Västilä and Järvelä (2018). For the SKM model, investigations determining the dependency of the vortex viscosity and secondary flow parameters on vegetation properties are needed.

4 Conclusions

Our computations showed that both the CCHE2D and the SKM models calibrated at low velocities were extendable to the same vegetative condition during high velocities. Following this approach, both models may be used to reasonably predict the depth-averaged velocity profile in a partly vegetated channel with deeply submerged vegetated cover. The SKM model is more suited to simple practical applications due to its more straightforward and fast calibration process whereas CCHE2D is more reliable in predicting the lateral distribution of the streamwise velocity across the entire flow domain including low velocity areas near the channel walls. Application of these models to flows with more complex, vertically varying vegetation structures requires further investigations.

Acknowledgements This study was supported within statutory activities No. 3841/E-41/S/2019 of the Ministry of Science and Higher Education of Poland, by Maa- ja vesitekniikan tuki ry (No. 33271) and by Maj and Tor Nessling Foundation (No. 201800045). We are grateful for the doctoral student Gerardo Caroppi for conducting the measurements.

References

Ackers P (1991) Hydraulic design of straight compound channels. vol 1—summary and design method, vol 2—appendices. SR Report 281. HR Wallingford, UK

Box W, Västilä K, Järvelä J (2018) Transport and deposition of fine sediment in a channel partly covered by flexible vegetation. In: Paquier A, Rivière N (eds) River Flow 2018. https://doi.org/10.1051/e3sconf/20184002016

Caroppi G, Västilä K, Järvelä J, Rowinski P, Giugni M (2019) Turbulence at water-vegetation interface in open channel flow: experiments with natural-like plants. Adv Water Resour 127:180–191. https://doi.org/10.1016/j.advwatres.2019.03.013
Jesson MA, Sterling M, Bridgeman J (2013) Despiking velocity time-series-optimisation through the combination of spike detection and replacementmethods. Flow Meas Instrum 30:45–51. https://doi.org/10.1016/j.flowmeasinst.2013.01.007
Jesson MA, Bridgeman J, Sterling M (2015) Novel software developments for the automated post-processing of high volumes of velocity time-series. Adv Eng Softw 89:36–42. https://doi.org/10.1016/j.advengsoft.2015.06.007
Jia Y, Wang SSY (2001) CCHE2D: two-dimensional hydrodynamic and sediment transport model for unsteady open channel flows over loose bed. Technical report no. NCCHE-TR-2001-1, National Center for Computational Hydroscience and Engineering, The University of Mississippi, USA
Kalinowska M, Västilä K, Rowiński PM (2019) Solute transport in complex natural flows. Acta Geophys 67:939–942. https://doi.org/10.1007/s11600-019-00308-z
Knight DW, Omran M, Tang X (2007) Modelling depth-averaged velocity and boundary shear in trapezoidal channels with secondary flows. J Hydraul Eng 133(1):39–47
Knight DW, Yuen KW, Al-Hamid AA (1994) Boundary shear stress distributions in open channel flow. In: Beven KJ, Chatwin PC, Millibank JH (eds) Mixing and transport in the environment. Wiley, pp 51–87
Knight DW, Tang X, Sterling M, Shiono K, McGahey C (2010) Solving open channel flow problems with a simple lateral distribution model. In: River flow 2010, Proceedings of the international conference on fluvial hydraulics, Braunschweig, Germany, pp 8–10
Kordi H, Amini R, Zahiri A, Kordi E (2015) Improved Shiono and Knight method for overflow modeling. J Hydrol Eng 20(12):04015041
Kozioł A, Kubrak J, Kubrak E, Krukowski M, Kiczko A (2016) Distributions of velocity in compound channels with high vegetation on floodplains. Acta Scientiarum Polonorum Formatio Circumiectus 15(4):227–241. https://doi.org/10.15576/ASP.FC/2016.15.4.227
Kubrak J, Rowiński PM (2007) Effects of variation of banks roughness in open channels on flow conveyance. Publ Inst Geophys Pol Acad Sci E-7(401):137–148
Nepf HM (2012) Flow and transport in regions with aquatic vegetation. Annu Rev Fluid Mech 44:123–142. https://doi.org/10.1146/annurev-fluid-120710-101048
Parsheh M, Sotiropoulos F, Porté-Agel F (2010) Estimation of power spectra of acoustic-doppler velocimetry data contaminated with intermittent spikes. J Hydraul Eng 136:368–378. https://doi.org/10.1061/(ASCE)HY.1943-7900.0000202
Rameshwaran P, Shiono K (2007) Quasi two-dimensional model for straight overbank flows through emergent. J Hydraul Res 45(3):302–315
Rowiński PM, Västilä K, Aberle J, Järvelä J, Kalinowska MB (2018) How vegetation can aid in coping with river management challenges: a brief review. Ecohydrol Hydrobiol 18(4):345–354. https://doi.org/10.1016/j.ecohyd.2018.07.003
Shiono K, Knight DW (1990) Mathematical models of flow in two or multi stage straight channels. In: International conference on river flood hydraulics, pp 229–238
Shiono K, Knight DW (1991) Turbulent open-channel flows with variable depth across the channel. J Fluid Mech 222:617–646
Shiono K, Rameshwaran P (2015) Mathematical modelling of bed shear stress and depth averaged velocity for emergent vegetation on floodplain in compound channel. In: E-Proceedings of the 36th IAHR World Congress, 28
Shiono K, Takeda M, Yang K, Sugihara Y, Ishigaki T (2012) Modeling of vegetated rivers for in bank and overbank flows. In: River flow 2012—Proceedings of the international conference on fluvial hydraulics, vol 1, pp 263–269
Tang X, Knight DW (2009) Lateral distributions of streamwise velocity in compound channels with partially vegetated floodplains. Sci China Ser E: Technol Sci 52(11):3357–3362. https://doi.org/10.1007/s11431-009-0342-7

Västilä K, Järvelä J (2018) Characterizing natural riparian vegetation for modeling of flow and suspended sediment transport. J Soils Sediments 18(10):3114–3130. https://doi.org/10.1007/s11368-017-1776-3

von Kármán T (1930) Mechanische Ähnlichkeit und Turbulenz, Nachrichten von der Gesellschaft der Wissenschaften zu Göttingen. Fachgruppe 1 (Mathematik) 5:58–76 (also as: "Mechanical Similitude and Turbulence", Tech. Mem. NACA, no. 611, 1931)

Ye J, McCorquodale JA (1997) Depth-averaged hydrodynamic model in curvilinear collocated grid. J Hydraul Eng ASCE 123(5):380–388

Zhang Y (2005) CCHE2D mesh generator users' manual—version 2.50. NCCHE technical report, NCCHE-TR-2005-01

Habitat Structure Changes of the Wisłoka River as a Result of Channel Restoration

Leszek Książek, Agnieszka Woś, Maciej Wyrębek and Andrzej Strużyński

Abstract The economic use of rivers in many cases has led to the degradation of their beds, the interruption of ecological continuity, and its defragmentation. In Poland, the effect of restoration activities in the mountain river bed has rarely been analysed. The article presents an analysis of hydraulic conditions and habitat structure, as well as a hydromorphological assessment of the river in selected sections of the Wisłoka River after completion of activities improving fish habitats. On a section of the Wisłoka River, about 30 km long (from the weir in Mokrzec to Pustków), ten sections were selected, which were subjected to restoration activities. These activities included spatial differentiation of water flow conditions in the channel to improve the habitat conditions that were obtained by (i) laying boulder clusters, (ii) depositing a bedload in the river bed, and (iii) forcing a change of the horizontal river system through controlled lateral erosion of the banks. As part of the restoration activities, the diversity of habitat types on the sections examined has significantly increased, which, in the future, should lead to an increase in the diversity of benthic fauna and macro-invertebrates.

Keywords River restoration · Boulder clusters · Morphological units · Hydromorphological quality

1 Introduction

Natural changes in the century scale of morphology in riverbeds and basins include climate changes and tectonic movements. Among anthropogenic reasons, we distinguish the anthropopressure on the environment of river valleys by agricultural development (Wyżga 2007) and the growing industry and growing cities. To increase the safety of flooding and the stabilization of the bottom and banks, training works

L. Książek (✉) · A. Woś · M. Wyrębek · A. Strużyński
Department of Water Engineering and Geotechnics, Faculty of Environmental Engineering and Land Surveying, University of Agriculture in Krakow, Al. Mickiewicza 24/28, 30-059 Krakow, Poland
e-mail: rmksiaze@cyf-kr.edu.pl

M. B. Kalinowska et al. (eds.), *Recent Trends in Environmental Hydraulics*, GeoPlanet: Earth and Planetary Sciences,
https://doi.org/10.1007/978-3-030-37105-0_9

were undertaken which resulted in the straightening of the channels, their narrowing, and a change of the longitudinal profile. The river bed stabilization achieved as a result of the regulation of the horizontal stabilization of the river was accompanied by decreasing river bed reaching up to 3.8 m in the 20th century. The deep incision of rivers leads to the degradation of riverbeds (Korpak 2007; Michalik and Książek 2009) and is also a significant threat to the organic world of watercourses and their riverside areas as they disturb the hydrodynamic continuity and biological rivers and their fragmentation. River valleys are linear structures and are a link between ecosystems extending crosswise to them. One of the effects of losing the naturalness of rivers is the disappearance of places conducive to the existence and reproduction of ichthyofauna. An example of a change in the ichthyofauna composition is Wisłoka River where at the turn of the 19th and 20th century 35 fish species occurred, in the 1970s 25 species of fish were recorded, and at the turn of the 20th and 21st century 22 fish species (Bartnik et al. 2011) were found.

The action opposite to the above-described process is restoration, aimed at restoring the river to a state close to the natural state, that is, the state existing before the regulation or occurring in nature against human interference. The studies investigating the effect of river restoration on hydromorphology found a significant positive effect on richness and abundance of macroinvertebrates and fish (Lorenz et al. 2012; Schmutz et al. 2014). In the river channel, restoration activities consist, among other things, of restoring the continuity of sediment transport along the channel's length, rebuilding the horizontal system of the channel by diversifying its route, recreating oxbow lakes, and correcting the longitudinal profile. To increase the diversity of habitat conditions, it is no less important to diversify local longitudinal slopes of the bottom (Zając et al. 2018) and the shape and dimensions of cross-sections as well as the presence of local obstacles (e.g., fallen trees, oversized boulders).

On the studied section of the Wisłoka River, restoration activities were carried out (Bartnik et al. 2015), as a result of which it is assumed that the habitats of the lithophyll fish species will be restored and anadromous fish such as salmon, trout, zarte, and sturgeon will potentially be spawning. Habitat requirements include a gravel bottom, clean oxygenated water, and fast flow. The diversity of habitat conditions in the riverbed was obtained by laying oversized boulders, supplementing the bottom substrate washed out in the process of bed erosion, and forcing the change of the horizontal river system through controlled lateral erosion of the banks. Because the monitoring is a crucial point in a restoration process (Roni et al. 2018) after completion of the channel restoration, the hydraulic habitat conditions of ichthyofauna were assessed in the areas of habitats created on the basis of the hydromorphological assessment of river quality. This assessment supports the biological evaluation of ecological status in accordance with the requirements of the WFD (2000), including hydromorphological elements and morphological conditions (Hawryło and Leja 2012).

The aim of this article is to present the results of restoration activities aimed at improving the morphology of the channel section and hydraulic conditions of water flow, thus improving the habitat structure for the Wisłoka fish from the weir in Mokrzec to Pustków one year after the reconstruction of the bed.

2 Materials and Methods

2.1 Study Area

The Wisłoka River is a right-bank tributary of the Vistula River with a length of approximately 164 km and a catchment area of over 4.1 thousand km^2. The section covered by the study, from the weir in Mokrzec to Pustków, has a length of about 30 km (km 43 through 73). In this section, ten locations for restoration were selected (Fig. 1) and one reference section, as summarized in Table 1.

In this area, the river flows over agricultural areas. Wisłoka's sandy-gravel bottom decreased in the second half of the 20th century to the 1990s at a rate of 5–9 cm/year, while in the last 20 years, the intensity of this process was reduced (Galarowski and Klimek 1991; Wyżga 2007). The entire section was rated as heavily modified Uniform Surface Water. This was determined by the transversal and longitudinal hydraulic structures limiting the diversity of habitats for fish and invertebrates as

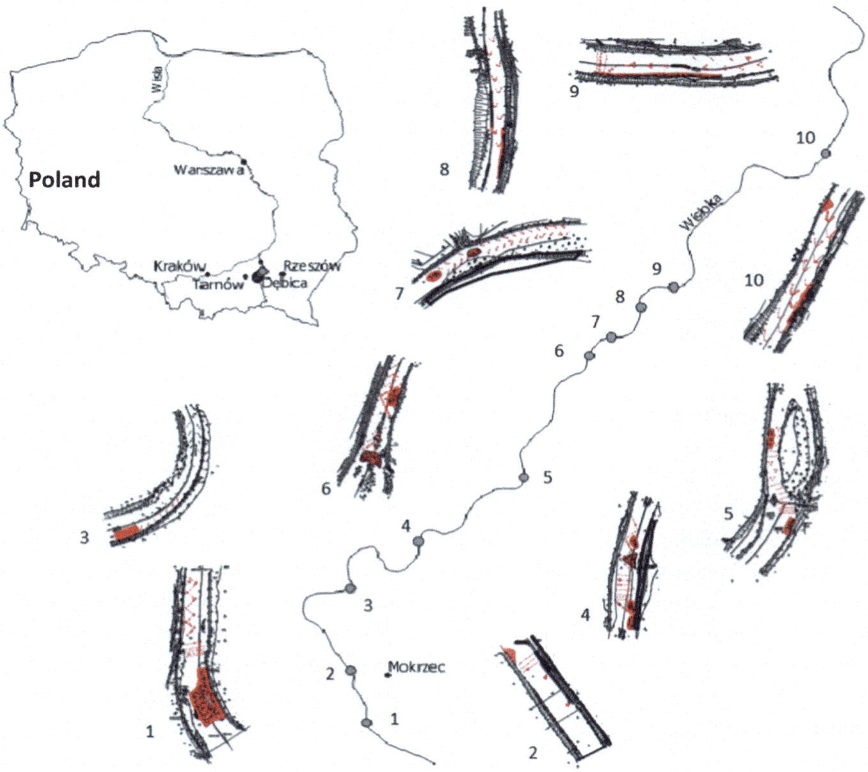

Fig. 1 Location of the sections with designed restoration program (in red) at Wisłoka River

Table 1 Description of the restoration program on the Wisłoka River sections

No.	Chainage	Description	Boulder cluster arrangement	Gravel volume (m^3)	Deflector
01	73110–72923	Weir, riffie-pool sequence, varied flow conditions	Openwork + grouped	1100	–
02	71886–71382	Wide channel, uniform flow conditions	Openwork + grouped	1000	+
03	68381–68084	Riffie-pool sequence	Grouped	800	–
04	65015–64772	Riffie-pool sequence, varied flow conditions	Openwork + grouped	900	–
05	61612–61380	Riffie-pool sequence, varied flow conditions	Openwork + grouped	900	–
06	57723–57526	Weir, varied flow conditions	Grouped	1200	–
07	57207–56874	Riffie-pool sequence	Grouped	1000	–
08	55850–55512	Wide channel, uniform flow conditions	Openwork cluster	900	–
09	54750–54386	Riffie-pool sequence, varied flow conditions	Openwork + grouped	900	–
10	48397–48082	Wide channel, uniform flow conditions	Openwork cluster	1250	+

well as embankments of the riverbed, water management related to energy and the excess flow of mean low flow. However, about 43% of the length of the section belongs to the Natura 2000 area.

At the bottom of the channel, fine and small gravels dominate. The average value of the mean diameter is $d_m = 0.022$ m (0.010 m $< d_m < 0.035$ m) and the maximum diameters (dimension b) are in the range $d_{max} = 0.079$ through 0.157 m. With such a construction of the bottom, conditions for the development of invertebrate organisms are fulfilled; however, conditions for the development of ichthyofauna are not good enough for species historically existing in Wisłoka. The requirement for improving the ecological quality is the restoration of the habitat structures (cobbled sections and boulders), the presence of which depends on increase in the number of large species and fish.

2.2 Morphology Change of the Sections of the Wisłoka

Diversification of water flow conditions was achieved by placing boulders, depositing bed material in the river channel, and forcing the change of the horizontal river system through controlled lateral erosion of the banks described in Książek et al. (2018).

1. The impact of individual boulders on the bottom and the streamlines (Książek et al. 2017; Strutyński et al. 2013) is increased by grouping them together, and several variants are used:
   - openwork arrangement over the entire width of the trough,
   - grouped arrangement of boulders which differentiates local flow conditions,
   - openwork cluster (deflector for decade flows) which directs the mainstream line and generates conditions for creating the bends.
2. Deposition of bottom material to increase the diversity of the granulation of the material forming the bottom cover; a coarse bottom substrate was supplemented, which will be transported during the passage of the flood. The effects of these operations should be moving gravel bars, which, after passing the flood, change their geometry.
3. Forcing the change of the horizontal river system to initiate the lateral erosion process by directing the thalweg to the opposite bank by the use of deflectors.

The most advantageous solution was adapted to the local morphology of the channel using the possibilities of modeling two-dimensional water flow conditions. In models of this type, the hydraulic values are averaged at depth. It uses the finite element method to solve equations (Jia and Wang 1998). The model has been subjected to the calibration process by comparing the calculated and measured water surface levels at cross-sections distributed in intervals about channel width along longitudinal profile. The calibration was performed for low discharges when roughness coefficients strongly determine water surface and flow conditions. The values of roughness coefficient assumption were based on the grain size composition, e.g. Stri-clear formula. The calculated and measured water surface levels do not exceed 5%, which allows to evaluate the models as very good (Książek et al. 2010). Restoration structures were implemented in the model as real objects in the bottom by adjusting the computational grid.

2.3 Assessment of Hydromorphology

The hydromorphology quality of the river was made by identifying hydromorphological units in the channel and carrying out hydromorphological assessment of the sections. Identification of hydromorphological units before restoration was made on the basis of the types identified in the Mesohabitat Evaluation Model (Hauer et al. 2008). This model, using the water flow velocity, depth, and shear stresses on the bottom, allows for the distinguishment of six hydromorphological units corresponding to different water flow conditions: fast current, rapid flow, normal flow, calm flow, stagnation, and shallow water (Książek and Bartnik 2009). Flow parameters are determined locally on the basis of numerical modeling. After restoration, the identification of hydromorphological units was made by unit defining (mapping) with support of water depth and velocity measurements.

The hydromorphological assessment of the Wisłoka River was carried out in accordance with the EN 14614 standard with the comments of Wyżga et al. (2010). The degree of naturalness was assessed for ten categories of features, some of them concerning the channel (1–6) and some of them the floodplains (7–10).

3 Results and Discussion

As a result of the restoration process, the differentiation of water flow conditions, gravel bars movement, and the change of the horizontal layout of the river were obtained. Morphological unit mapping was carried out on 6 and 7 November 2014 at the flow which changed with the increase in catchment area in the range of $Q = 7.8$ through 13.6 $m^3\ s^{-1}$.

3.1 Spatial Velocity Distribution

Changes in the spatial flow velocity distribution in selected cross-sections are shown in Fig. 2a–e. Figure 2a shows the velocity distribution immediately after the row of openwork boulders. In the right part of the cross-section 01-03, zones with an increased (about 1–1.5 ms^{-1}) and with a reduced velocity (about 0.4 ms^{-1}) can be distinguished, which alternate as a series of small streams.

In a pool at the left bank the boulders were not placed because they have no influence to the flow due to complete submergence. Figure 2b shows the velocity distributions behind the deflector reaching 24% of the width of the channel—section 02-09. The narrowing of the active cross-section causes acceleration of the water velocity (up to 0.8 ms^{-1}) in relation to the cross-section without a narrowing. At the same time, the backwater area behind the deflector is visible. Distributions of velocities behind grouped boulders (clusters of 3 grains) at the section 09-03 is shown in Fig. 2c. The effect of putting boulders in a triangular group is the creation of two large, wide zones with reduced speed (about 0.4 ms^{-1}). Cross-section 10-01 in Fig. 2d shows the velocity distributions passing through the arms of the openwork cluster. The maximum velocity of about 1 ms^{-1} is inside the cluster, while its arms reduce the flow rate. In addition, the openwork cluster activates the central part of the cross-section. Behind the full deflector, a backwater zone formed, which disappeared at a distance of 4 deflector width.

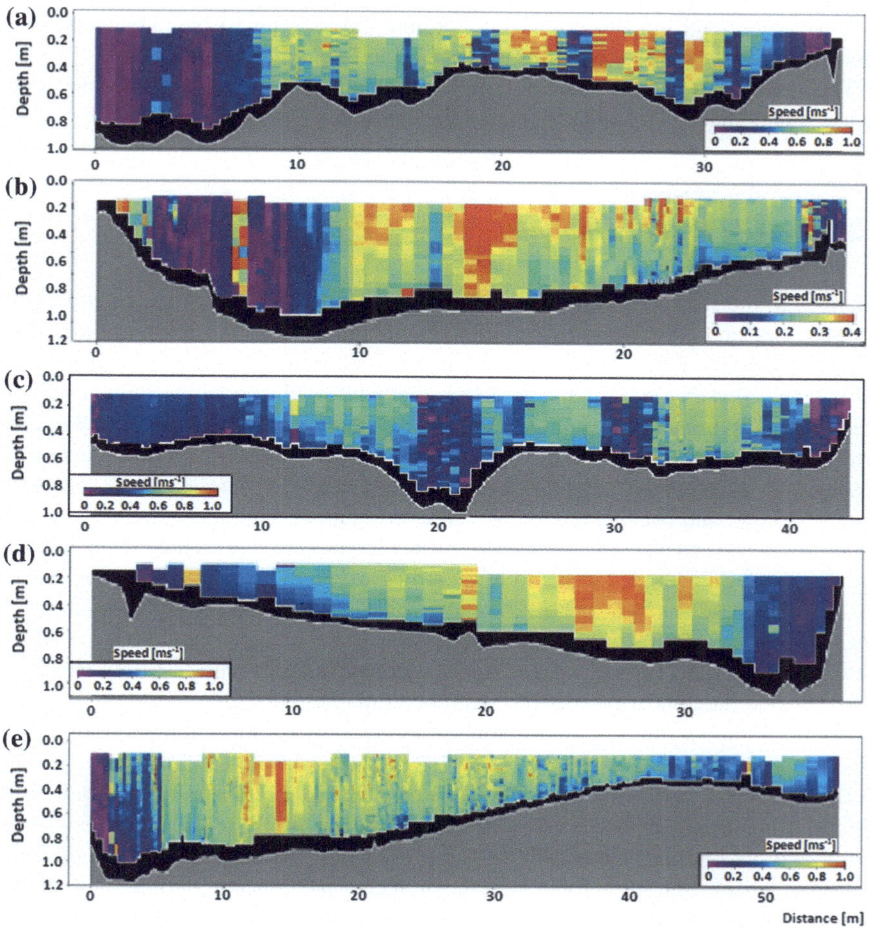

Fig. 2 Spatial velocity flow distribution, the Wisłoka River, date of measurement 6–7 Nov 2014, cross-sections: **a** 01–03 km 73 + 090 right after the boulder cluster—openwork arrangement for discharge $Q = 7.8\ \text{m}^3\ \text{s}^{-1}$, **b** 02–09 km 71 + 309 downstream the deflector for discharge $Q = 8.9\ \text{m}^3\ \text{s}^{-1}$, **c** 09–03 km 54 + 482 downstream boulder cluster groups for discharge $Q = 11.9\ \text{m}^3\ \text{s}^{-1}$, **d** 10–01 km 48 + 250 through openwork deflector for discharge $Q = 13.6\ \text{m}^3\ \text{s}^{-1}$, and **e** 10–11 km 48 + 072 after the deflector outside the impact zone for discharge $Q = 13.6\ \text{m}^3\ \text{s}^{-1}$. The spatial velocity flow distribution, water depth, channel geometry as well as discharge have been measured using Acoustic Doppler Current Profiler

The selected arrangement of boulders has produced the expected changes in morphology. Three years after completing the work in the channel, the local scour was about 0.15–0.20 m (Książek et al. 2017).

The deflectors were applied on two straight sections, 02 and 10. On the day of measurements, the deflector reduced the cross-section width by 24% per section 02 and 16% on section 10. This narrowing resulted in an increase in mean velocity in the cross-section by 130% and 350%, respectively. On the opposite banks, there were signs of erosion (loss of vegetation).

The assessment of the impact of boulders on water flow conditions was preceded by a detailed analysis by Książek et al. (2017).

3.2 Distribution of Habitats

Changes in the riverbed in the form of placement of boulder clusters, construction of deflectors, and gravel supplementation resulted in the diversification of the distribution of habitats. Often, the homogeneous riverbed turned into a system of various habitat types that change with the flow. In many sections, the part of individual habitat types in the total surface area of the riverbed has changed, as summarized in Table 2. Prior to restoration, the distribution of habitats on all sections was, on average, as follows: riffie—0.2% of the water area at decade, fast run—7.4%, run—55.6%, shallow water—25.0%, and backwater—11.9%. The structure of habitats completely lacked structures described as pool. After restoration, all types are represented in the habitat structure and, on average, are as follows: riffie—0.8% of the water area, fast run—18.1%, run—33.2%, pool—23.5%, shallow water—13.9%, and backwater—10.5%. There is a noticeable decrease in the contribution of low energy habitats (i.e. shallow water and backwater) from 36.9 to 24.4%, increase in areas with fast flow from 9.6 to 18.9% and the appearing of areas with medium energy (pool).

Figure 3 shows the spatial distribution of habitats on the tenth section of the Wisłoka River with a flow $Q = 13.6\ m^3\ s^{-1}$ as an example of changes on all ten sections covered by restoration activities. There was a large diversity of habitat distribution: the dominant normal flow—run pattern was partly replaced with shallow water, and pool behind the boulders. The presence of the deflector caused the formation of a backwater directly behind the deflector and concentrated the flow in the right part of the channel, where the fast run dominates.

Characterizing changes in habitats in 2011 and 2014 on the remaining sections (from 01 to 09), it should be noted that in section 01, habitats from the riffle and fast run have transformed into a run and pool, and shallow water in the vicinity of a gravel pile. In section 02, the largest part in the distribution of habitats has a pool and run, while behind the boulder clusters and behind the deflector a backwater has been identified. The new structures in section 03 have formed backwaters at the left bank, a riffle has appeared in the middle of the riverbed, and a pool on the right bank. In section 04, on the right bank, a backwater was created, while the presence of boulder structures has transformed the riffle into a fast run. At the estuary of the Ostra River (section 05), after the changes in the Wisłoka River, a fast run was identified; while in 2011, it was a run in the right arm of the channel that was previously dominated by shallow water and backwater, and the left bank showed a pool. Below the weir in Dębica (section 06), the area with a fast run and riffle increased, as well as a backwater and shallow water on the right bank near the gravel pile. In section 07, the distribution of habitats diversified, the mainstream transformed from a run into a

Table 2 Mesohabitats structure area before restoration, the Wisłoka River, Nov 2011, Nov 2014

No.	Mesohabitat area content in the wetted area (%)											
	Riffle		Fast run		Run		Pool		Shallow water		Backwater	
	2011	2014	2011	2014	2011	2014	2011	2014	2011	2014	2011	2014
01	2.0	0.0	21.4	0.0	56.8	49.8	0.0	25.7	7.0	15.6	12.8	8.9
02	0.0	0.0	0.1	0.0	0.1	42.4	0.0	36.0	81.8	3.1	18.0	18.6
03	0.0	0.0	0.1	40.7	67.1	28.3	0.0	26.9	28.3	0.0	4.6	4.0
04	0.0	0.0	44.7	29.6	29.7	10.5	0.0	24.2	12.4	17.6	13.1	18.1
05	0.2	0.0	2.3	29.6	60.4	10.5	0.0	24.2	24.1	17.6	12.9	18.1
06	0.3	6.4	23.8	39.8	67.0	23.3	0.0	0.0	5.3	22.0	3.6	8.4
07	0.0	0.0	0.0	39.6	60.0	27.3	0.0	27.3	31.0	5.8	9.0	0.0
08	0.3	0.0	2.5	39.5	59.9	18.8	0.0	35.8	7.7	5.8	29.6	0.0
09	0.0	0.0	0.1	0.0	53.1	60.2	0.0	18.4	40.8	21.4	6.0	0.0
10	0.0	0.0	0.2	18.4	82.8	35.3	0.0	20.5	5.9	19.7	11.2	6.0

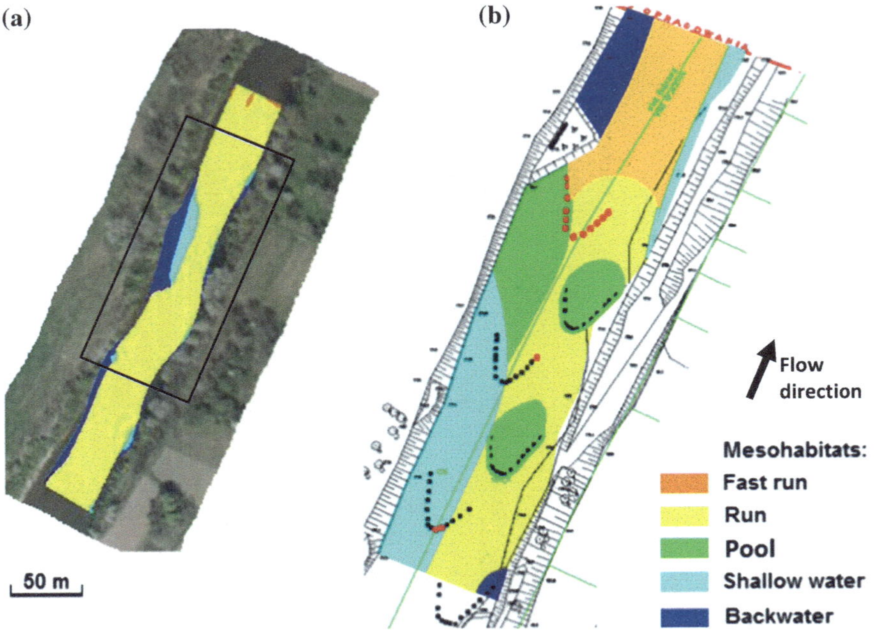

Fig. 3 Mesohabitats structure in the section number 10 of the Wisłoka River. **a** Before restoration, the 2D model for discharge $Q = 7.95\ m^3\ s^{-1}$, **b** after restoration, discharge $Q = 13.6\ m^3\ s^{-1}$ (*source* Projekt 2013, modified); black and red symbols—restoration design

fast run, additionally behind structures, and a pool was identified on the right bank. In section 08, the run still dominates the riverbed; there was also a pool on the left side of the channel and behind the boulders. The backwater in section 09 has been transformed into a run and pool in the left part of the channel, and behind boulders, there was shallow water at the right bank.

3.3 Hydromorphological Assessment

After completing the work, the structure of hydromorphological units has been changed by diversifying the spatial velocity distribution, caused by the breakdown of the high-velocity zone into smaller areas between which the pool and backwater units were created. This can also be seen in a better assessment of parameters describing the condition of the main channel. Figure 4 shows the comparison of hydromorphological assessments evaluate sections of the Wisłoka River in 2011 (Hawryło and Leja 2012) and in 2014 i.e. before and after restoration.

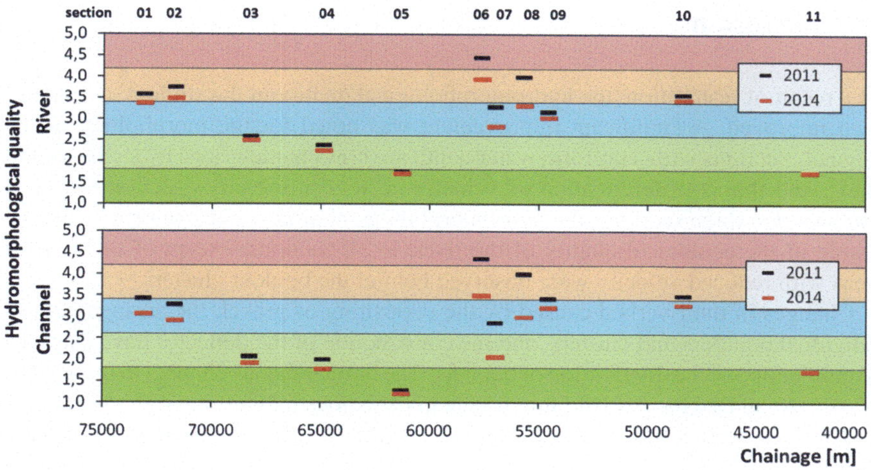

Fig. 4 River hydromorphological quality before (2011) and after (2014) restoration in comparison with hydromorphological quality class: green—excellent (I), light green—good (II), blue–moderate (III), orange—weak (IV), red—bad (5), the Wisłoka River; 11—reference section

In three section (01, 06, and 08), the hydromorphological quality class has changed by one degree. These sections represented bad or poor hydromorphological conditions mainly due to existing hydraulic structures that disturbed the supply of gravel and the hydrological regime. Low grades were also obtained for the diversification of cross-sections that had a trapezoidal shape. As a result of the use of gravel piles and appropriate boulder clusters, the grain composition of the bottom has been improved, creating favourable conditions for variation of water depth and velocity. The sections with poor quality (i.e. 02 and 10) are straight sections with very little differentiation in the bed morphology. The purpose of improvement was to use a deflector, which directs the current to the opposite bank, causing lateral erosion—the first signs of erosion in the form of vegetation loss were recorded. Location 05 is on the edge of classes 1 and 2, with a similar value to the reference location 11.

The assessment of naturalness includes ten categories of features, six of which concern the main channel and the remaining flood plains. Since the restoration covered only the main channel, the changes in the hydromorphological quality are the highest in locations 06, 07, and 08—the quality decreased by approximately 0.9 points, which translates to an increase in the grade by one class. In seven locations their channel grade increased by one class. The average assessment of the main channel in 2011 was 3.02, and in 2014 was 2.60 (with flood terraces 3.25 and 3.00, respectively). Taking into account only the main channel to class 1, location 04 and 05 can be qualified.

Restoration did not include changes of bank protection. Within the channel, restoration activities improved the morphological quality of the channel but did not cause a significant change of the river valley.

4 Conclusions

As a result of restoration, the hydromorphological quality in the studied section has been improved. A significant improvement was noted for the morphology of the channel. Sections with a uniform water depth are now characterized by shallows and pools, and thus a differentiation of velocity vectors in the cross-section has been obtained. On three sections, the hydromorphological quality class improved by one degree of the ecological quality of the river. In all sections, except of 08, distinct areas with reduced velocity were observed behind the boulder clusters.

Changes in the riverbed caused by the depositing of gravel, the arrangement of individual boulders and clusters, and the construction of the deflector have caused a diversification of the distribution of habitats. The homogeneous channel in section 02, 03, 07, 09, and 10 has evolved into a system of various habitat types.

The presented results indicate that the requirement for improving the channel hydromorphology quality is the restoration of habitats. The presence of different hydromorphic units helps to increase the number of species and amount of fish individuals. Restitution of biodiversity depends on physical habitats variability as well as support the restoration of the ecological continuity of the Wisłoka River for fish and other organisms.

After the work has been completed, the hydromorphological structure of the channel has improved due to the differentiation of water flow conditions, mainly velocity distributions caused by the breakdown of the high-velocity zone into smaller areas between which the slow flow areas will be created. There is a noticeable reduction in the part of low energy habitats (i.e. shallow water and backwater), increased areas with fast flow, and the appearance of areas with medium energy (pool). An interdisciplinary approach will allow a full assessment of the effects of restoration in the future.

Acknowledgements The study has been financially supported by the PIOS-05.02.00-00-182-09-00 project. The authors would like to thank Wojciech Bartnik and Małgorzata Leja for their help with the field measurements and computation analyses.

References

Bartnik W, Epler P, Jelonek M, Klaczak A, Książek L, Mikołajczyk T, Nowak M, Popek W, Sławińska A, Sobieszczyk P, Szczerbik P, Wyrębek M (2011) Fisheries management with relation to the restoration of connectivity of the little and the Upper Vistula River basins. In: Piotr E, L Książek (eds) Infrastructure and ecology of rural areas, p 13

Bartnik W, Książek L, Jelonek M, Sobieszczyk P, Florek J, Hawryło A, Leja M, Strużyński A, Strutyński M, Wałęga A, Wyrębek M, Wiśniewolski W, Parasiewicz P, Prus O, Adamczyk M (2015) Conditions of restoring fish habitat structure in the Wisłoka River section between km 73 + 200 and 42 + 600. Gospodarka Wodna 5:147–152

EN 14614 (2004) Water quality—guidance standard for assessing the hydromorphological features of rivers. CEN

Galarowski T, Klimek K (1991) Funkcjonowanie koryt rzecznych w warunkach zagospodarowania. Dynowska I, Maciejewski L (eds) Dorzecze górnej Wisły, cz. I, PWN
Hauer C, Tritthart M, Habersack H (2008) Computer-aided mesohabitat evaluation, Part I: background, model concept, calibration and validation based on hydrodynamic numerical modeling. In: Altinakar M, Kokpinar M, Darama Y, Yegen B, Harmancioglu N (eds) International conference on fluvial hydraulics—river flow, pp 1967–1974
Hawryło A, Leja M (2012) Hydromorphological assessment on the example of a section of the Wisłoka River from Mokrzec to Pustków. Episteme 14:85–91
Jia Y, Wang SS (1998) Capability assessment of CCHE2D in channel flow simulations. Proc Adv Hydro-sci Eng 3
Korpak J (2007) The influence of river training on mountain channel changes (Polish Carpathian Mountains). Geomorphology 92:166–181
Książek L, Bartnik W (2009) Using hydraulics conditions for mesohabitat assessment in a river channel. Nauka Przyr Technol 3(89):1–8
Książek L, Wałęga A, Bartnik W, Krzanowski S (2010) Calibration and verification of computational model of the Wisłok River by means of flood wave transformation. Infrastruct Ecol Rural Areas 8(1):15–28
Książek L, Woś A, Roche G (2017) Boulder cluster influence on hydraulic microhabitats distribution under varied instream flow regime. Acta Sci Polonorum Form Circumiectus 16(4):139–153
Książek L, Wyrębek M, Strutyński M, Woś A (2018) Numerical modeling of water flow conditions with spatial distribution of boulders in main channel. In: Kalinowska M, Mrokowska M, Rowiński P (eds) Free surface flows and transport processes. GeoPlanet: Earth and Planetary Sciences. Springer, Cham, pp 271–280. https://doi.org/10.1007/978-3-319-70914-7_17
Lorenz A, Korte T, Sundermann A, Januschke K, Haase P (2012) Macrophytes respond to reach-scale river restorations. J Appl Ecol 49:202–212
Michalik A, Książek L (2009) Dynamics of water flow on degraded sectors of polish mountain stream channels. Polish J Environ Stud 18:665–672
Projekt (2013) Żwirowanie, kubatury i usytuowania deflektorów na rzece Wisłoce z uwzględnieniem optymalizacji warunków hydraulicznych przepływu wody. PUP Algebar
Roni P, Åberg U, Weber C (2018) A review of approaches for monitoring the effectiveness of regional river habitat restoration programs. North Am J Fish Manag 38:1170–1186
Schmutz S, Kremser H, Melcher A, Jungwirth M, Muhar S, Waidbacher H, Zauner G (2014) Ecological effects of rehabilitation measures at the Austrian Danube: a meta-analysis of fish assemblages. Hydrobiologia 729:49–60
Strutyński M, Strużyński A, Kulesza K (2013) The Influence of large roughness elements on natural morphological changes in a mountain river bed. In: P. Rowiński (ed) Experimental and computational solutions of hydraulic problems. GeoPlanet: earth and planetary sciences, pp 237–247
WFD (2000) Water framework directive, directive of the European Parliament and the Council 2000/60/EC establishing a framework for community action in the field of water policy. Off J Eur Communities 22.12.2000 L 327/1
Wyżga B (2007) A review on channel incision in the Polish Carpathian rivers during the 20th century. Dev Earth Surf Process 11:525–553. https://doi.org/10.1016/S0928-2025(07)11142-1
Wyżga B, Zawiejska J, Radecki-Pawlik A, Amirowicz A (2010) A method for the assessment of hydromorphological river quality and its application to the Czarny Dunajec, Polish Carpathians, Cultural landscapes of river valleys. Publishing House of the University of Agriculture in Krakow, pp 145–164
Zając K, Florek J, Zając T, Adamski P, Bielański W, Ćmiel AM, Klich M, Lipińska AM (2018) On the reintroduction of the endangered thick-shelled river mussel *Unio crassus*: the importance of the river's longitudinal profile. Sci Total Environ 624:273–282. https://doi.org/10.1016/j.scitotenv.2017.11.346

Theoretical Analysis of the Reduction of Pressure Wave Velocity by Internal Circular Tubes

Michał Kubrak and Apoloniusz Kodura

Abstract This study presents a theoretical analysis of the influence of the insertion of circular tubes into a pipeline on the rapid water hammer phenomenon. Three types of circular tubes are distinguished: a thin-walled tube, a thick-walled tube and a solid cylindrical tube. For each case, by applying the work-energy principle, a formula for the pressure wave velocity is derived. The differences between the derived formulas are analyzed. It is shown that the insertion of a tube with a low bulk elastic modulus may have a damping effect on the water hammer phenomenon, i.e., it reduces the pressure wave velocity and maximum pressure increase.

Keywords Water hammer · Pressure wave velocity · Pressure increase · Elastic tube

1 Introduction

The water hammer phenomenon is a rapid pressure change generated by a sudden variation of the water velocity in closed pipelines. From a physical point of view, this phenomenon deals with a pressure wave that propagates in compressible liquid flowing in an elastic pipeline (Mitosek and Szymkiewicz 2016). This pressure wave can cause failures of pipeline systems. Its velocity can be determined by the Korteweg formula (Wylie and Streeter 1993):

$$c = \frac{\sqrt{\frac{K}{\rho}}}{\sqrt{1 + \frac{KD}{Ee}}} \tag{1}$$

M. Kubrak (✉) · A. Kodura
Faculty of Building Services, Hydro and Environmental Engineering, Warsaw University of Technology, Nowowiejska 20, 00-653 Warsaw, Poland
e-mail: michal.kubrak@pw.edu.pl

M. B. Kalinowska et al. (eds.), *Recent Trends in Environmental Hydraulics*, GeoPlanet: Earth and Planetary Sciences,
https://doi.org/10.1007/978-3-030-37105-0_10

where ρ—water density [kg m^{-3}], c—pressure wave velocity [m s^{-1}], K—bulk modulus of elasticity of water [Pa], D—inner diameter of the pipeline [m], e—pipeline wall thickness [m] and E—Young's modulus of the pipeline wall [Pa].

The maximum pressure increase caused by a rapid valve closure can be calculated from the Joukovsky equation:

$$\Delta p = \rho c v_0 \tag{2}$$

where Δp—maximum pressure increase [Pa] and v_0—initial water flow velocity [m s^{-1}].

Since the maximum pressure increase is directly proportional to the pressure wave velocity, the efforts of the designers are aimed at reducing it. One of the methods to reduce the pressure wave velocity is to lower the effective bulk modulus of the system by inserting a flexible tube into the pipeline (Tijsseling et al. 1999). The aim of this study is to analyze the potential benefits of inserting different types of circular tubes into the pipeline in order to dampen the pressure wave during the occurrence of the water hammer phenomenon.

2 Derivation of the Formula for the Pressure Wave Velocity

2.1 General Assumptions

Three types of circular tubes are distinguished in this study: a thin-walled tube, a thick-walled tube and a solid cylindrical tube (Fig. 1). In practice, the division into thin-wall and thick-wall pressure vessels depends on the thickness of the vessel wall in relation to its radius. When the thickness of the vessel is less than about one tenth of its mean radius, it is deemed a thin-walled vessel (Grote and Antonsson 2009). In this work, the formulated division is used to distinguish different ways of expressing

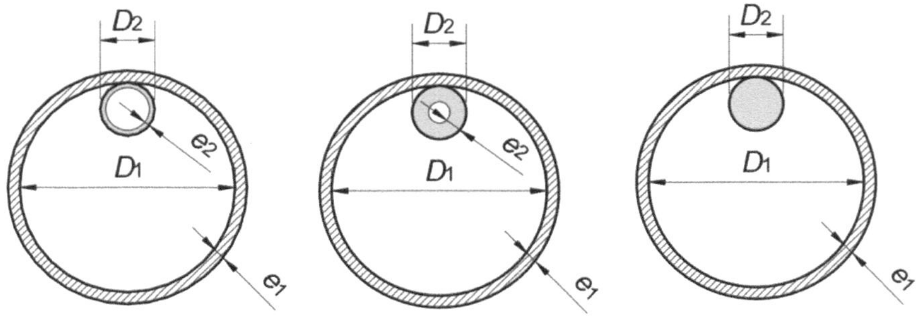

Fig. 1 Scheme of the thin-walled, thick-walled and solid cylindrical tubes inserted into the pipeline

the change of the tube's circumference caused by the pressure increase. Therefore, no specific limits of e/D ratios for individual types of tubes are given.

The formulas for the pressure wave velocity were derived using the work-energy principle. Perfect elastic behavior of the liquid, pipeline and tube was assumed. It was also assumed that the pipeline is thin-walled and, therefore, only hoop stress was taken into account (Bergant et al. 2008). Regardless of the type of the tube, the cross-sectional area of the liquid stream is equal to:

$$A = A_1 - A_2 \tag{3}$$

where A—cross-sectional area of the liquid stream [m^2], A_1—cross-sectional area of the pipeline with the inner diameter of D_1 [m^2] and A_2—cross-sectional area of the tube with the outer diameter of D_2 [m^2].

The kinetic energy before the pressure wave occurs must be converted into elastic energy in compressing the liquid, stretching the pipeline and compressing the tube (Streeter 1958). The principle of energy conservation may be written as follows:

$$E_k = E_l + E_p + E_t \tag{4}$$

where E_k—change in the kinetic energy of the liquid stream [J], E_l—increase of the elastic energy of the liquid stream [J], E_p—increase of the elastic energy of the pipeline [J] and E_t—increase of the elastic energy of the tube [J].

The change in the kinetic energy of the liquid stream is given by:

$$E_k = \frac{1}{2} A L v_0^2 \rho \tag{5}$$

where L—length of the pipeline [m].

The work of compression of the liquid is:

$$E_l = \frac{1}{2} A L \frac{\Delta p^2}{K} \tag{6}$$

The work done in stretching the pipeline walls is the product of average force exerted in the pipeline wall and the additional strain or extension of the pipe-wall circumference:

$$E_p = F_a \Delta l_1 \tag{7}$$

where F_a—average force exerted in the pipeline wall [N] and Δl_1—change of pipeline's circumference [m].

The average force in the pipeline wall due to the water hammer is:

$$F_a = \frac{\Delta p D_1}{4} L \tag{8}$$

where D_1—inner diameter of the pipeline [m].

The forces resulting from the increase of stress in the pipeline walls are balanced by the increase of water pressure on the longitudinal section of the pipeline:

$$\Delta\sigma_1 2e_1 L = \Delta p D_1 L \tag{9}$$

where $\Delta\sigma_1$—lateral stress of the pipeline wall caused by the pressure increase [Pa].

The extension of the pipeline wall circumference can be calculated using Hooke's law, and combining this with Eq. (9) gives:

$$\Delta l_1 = l_1 \frac{\Delta\sigma_1}{E_1} = \pi D_1 \frac{\Delta\sigma_1}{E_1} = \pi \frac{\Delta p D_1^2}{2E_1 e_1} \tag{10}$$

where l_1—pipeline's circumference [m], E_1—Young's modulus of the pipeline-wall material [Pa] and e_1—thickness of the pipeline wall [m].

Hence, using Eqs. (7), (8), and (10), the increase of the elastic energy of the pipeline is:

$$E_p = \frac{1}{2} A_1 L \frac{D_1 \Delta p^2}{E_1 e_1} \tag{11}$$

E_k, E_l and E_p do not depend on the type of inserted tube. The formula for the pressure wave velocity will depend on how the increase of the elastic energy of the tube E_t is expressed. For each of the considered tubes, the formula for E_t takes a different form.

2.2 Thin-Walled Tube

Assuming that the tube is thin-walled, its increase in the elastic energy can be expressed in the same way as the increase in the elastic energy of the pipeline—Eq. (11). Denoting the increase of the elastic energy of the thin-walled tube as E_{t_n}, one obtains:

$$E_{t\_n} = \frac{1}{2} A_2 L \frac{D_2 \Delta p^2}{E_2 e_2} \tag{12}$$

where D_2—outer diameter of the tube [m], E_2—modulus of the elasticity of the tube-wall material [Pa] and e_2—thickness of the tube wall [m].

Hence, the expression for conversion of the kinetic energy to the work of compression of the liquid and the tube and expansion of the pipeline is equal to:

$$\frac{1}{2} A L v_0^2 \rho = \frac{1}{2} A L \frac{\Delta p^2}{K} + \frac{1}{2} A_1 L \frac{D_1 \Delta p^2}{E_1 e_1} + \frac{1}{2} A_2 L \frac{D_2 \Delta p^2}{E_2 e_2} \tag{13}$$

After simplifying, substituting Δp with the Joukovsky equation, Eq. (2), and denoting the pressure wave velocity in a pipeline with inserted thin-walled tube as c_n, one obtains:

$$c_n = \sqrt{\frac{\frac{K}{\rho}}{1 + \frac{A_1}{A}\frac{KD_1}{E_1 e_1} + \frac{A_2}{A}\frac{KD_2}{E_2 e_2}}} \tag{14}$$

From Eq. (14), it follows that the pressure wave velocity in a pipeline with inserted thin-walled tube depends on the outer diameter, wall thickness and Young's modulus of the tube as well as the corresponding properties of the pipeline.

2.3 Thick-Walled Tube

The fact that the inserted tube is thick-walled was taken into account when expressing the change of the circumference caused by the increase in water pressure. In order to determine the elastic energy of the thick-walled tube, the average change of its circumference was taken into account, which was expressed as the arithmetic mean on the external and internal circumferences of the tube. Assuming that the tube operates in the area of elastic deformations, the average change of its circumference can be expressed as follows:

$$\Delta l_2 = \pi (D_2 - e_2)\frac{\Delta \sigma_2}{2E_2} \tag{15}$$

where $\Delta \sigma_2$—lateral stress of the tube wall caused by the pressure increase [Pa].

By expressing $\Delta \sigma_2$ by Eq. (9), one obtains:

$$\Delta l_2 = \pi D_2 \frac{\Delta p}{2E_2}\left(\frac{D_2}{e_2} - 1\right) \tag{16}$$

Further derivation of the formula for the pressure wave velocity can be completed in the same way as for the thin-walled tube. By expressing the work done in compressing the thick-walled tube with Eqs. (7), (8), and (16), and solving Eq. (4) for c_k, one obtains:

$$c_k = \sqrt{\frac{\frac{K}{\rho}}{1 + \frac{A_1}{A}\frac{KD_1}{E_1 e_1} + \frac{A_2}{A}\frac{K}{E_2}\left(\frac{D_2}{e_2} - 1\right)}} \tag{17}$$

where c_k—pressure wave velocity in pipeline with thick-walled tube [m s^{-1}].

2.4 Solid Cylindrical Tube

In order to derive the formula for the pressure wave velocity in a pipeline with the inserted solid cylindrical tube, in Eq. (17) it was assumed that $e_2 = D_2/2$. Thus:

$$c_c = \sqrt{\frac{\frac{K}{\rho}}{1 + \frac{A_1}{A}\frac{KD_1}{E_1 e_1} + \frac{A_2}{A}\frac{K}{E_2}}} \tag{18}$$

where c_c—pressure wave velocity in pipeline with solid cylindrical tube [m s^{-1}].

Equation (18) is the same as the formula for the pressure wave velocity in a pipeline with an inserted fiber optic cable, which was derived using the mass balance method in (Kubrak et al. 2018).

3 Computational Analysis of Derived Equations for Pressure Wave Velocity

In order to illustrate the differences between the values of the velocity of the pressure wave calculated from the formula for pipeline with the inserted thin-walled and thick-walled tubes, using Eqs. (14) and (17), the c_n/c_k ratios were calculated. It was assumed that the pipeline has an outer diameter of $D_1 = 0.1$ m, a wall thickness of $e_1 = 0.003$ m, its Young's modulus is $E_1 = 200 \times 10^9$ Pa, the wall thickness of the tube is $e_2 = 0.001$ m, the density of water is $\rho = 1000$ kg m^{-3} and its bulk modulus of elasticity is $K = 2 \times 10^9$ Pa. Calculations were made for different values of the tube's outer diameter D_2 and its modulus of elasticity E_2. The results of the calculations are shown in Fig. 2 as a function of D_2/D_1.

Figure 2 shows that the values of the pressure wave velocity c_k calculated from Eq. (17) are higher than c_n calculated from Eq. (14). These differences depend on the D_2/D_1 ratios. In the case where $E_2/E_1 > 0.01$, the c_k/c_n ratio does not exceed 1.01. The highest value of ~5.6% was obtained for $E_2/E_1 = 0.0001$ and for diameter $D_2 = 0.07\ D_1$.

To show the differences between the values of pressure wave velocity calculated from the formula for the pipeline with the inserted thick-walled tube and for the pipeline without the inserted tube, using Eqs. (1) and (17), the c_k/c ratios were calculated. The same values of geometric and material parameters of the pipeline were assumed. In order to explain the influence of the wall thickness of the tube on the pressure wave velocity, calculations were made for different e_2/D_2 values. It was also assumed that the Young's modulus of the tube is $E_2/E_1 = 0.0001$ (which corresponds to the modulus of elasticity of the material with high deformability, such as silicone). The results of the calculations are shown in Fig. 3 as a function of D_2/D_1. It should be noted that in practice, due to the increase of pressure losses, the outer diameter of the tube would not exceed 0.20 of the inner diameter of the pipeline ($D_2/D_1 \leq 0.20$).

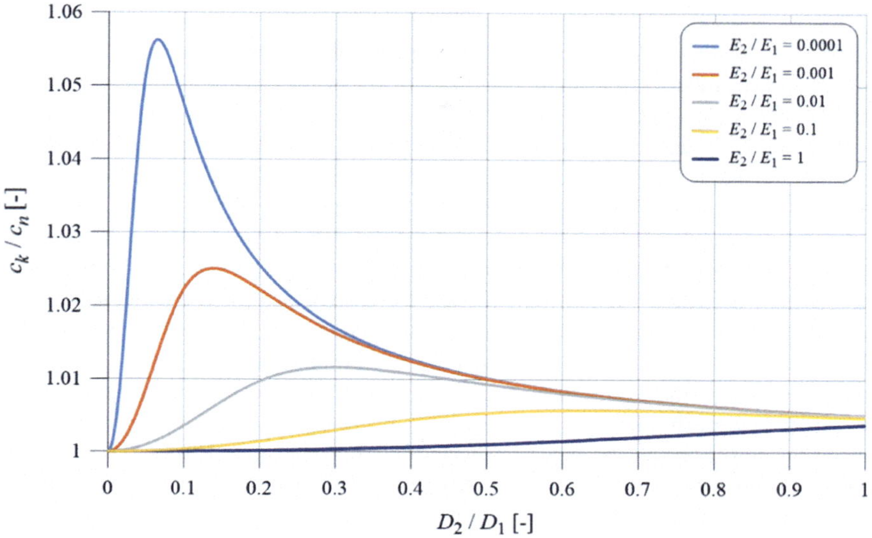

Fig. 2 Comparison of pressure wave velocity values calculated from the ratio of Eq. (14)–(17)

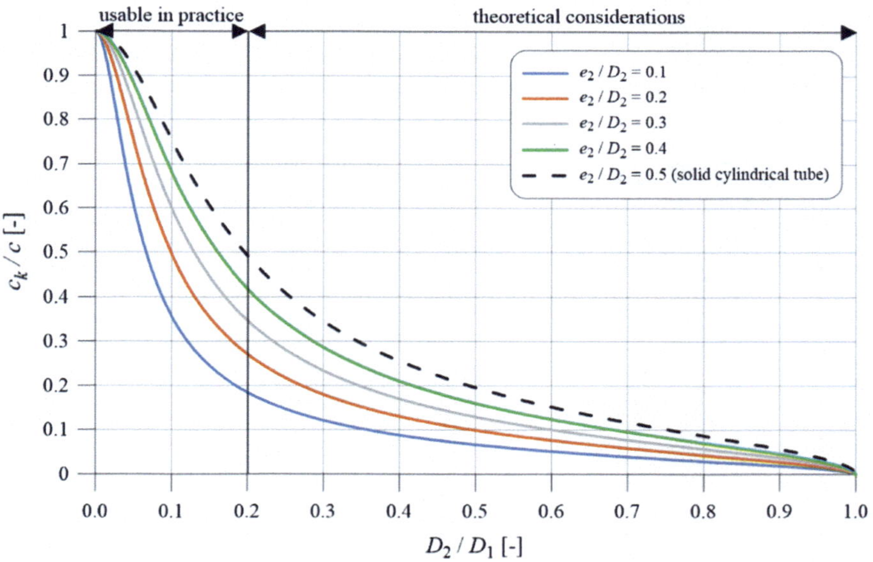

Fig. 3 Comparison of pressure wave velocity values calculated from the ratio of Eqs. (1) and (17)

Figure 3 shows that the insertion of the thick-walled (e.g. silicone) tube can significantly reduce the pressure wave velocity of the system. The larger the outer diameter of the tube and the smaller its wall thickness, the greater its ability to reduce the pressure wave velocity.

It should be noted, however, that the insertion of the tube into a pipeline reduces its diameter, and increases the average water flow velocity. According to Eq. (2), this causes a higher pressure increase during the water hammer phenomenon.

In order to examine how the insertion of the elastic tube into a pipeline affects the maximum pressure increase in the event of a rapid water hammer, $\Delta p_k/\Delta p$ ratios were calculated, where Δp_k—maximum pressure increase in the pipeline with the inserted thick-walled tube and Δp—maximum pressure increase in the pipeline without the tube. The maximum pressure increase Δp for a pipeline without the inserted tube was calculated using Eqs. (1) and (2). The maximum pressure increase Δp_k for the pipeline with the inserted thick-walled tube was also calculated from Eq. (2), but taking into account the pressure wave velocity determined from Eq. (17) and different (higher) average water flow velocities. The calculations were made for the same assumptions as in the previous considerations. The results of calculations for different e_2/D_2 ratios are shown as a function of D_2/D_1.

Figure 4 indicates that inserting the thick-walled tube with a low Young's modulus ($E_2/E_1 = 0.0001$; $E_2 = 0.02 \times 10^9$ Pa) can significantly attenuate pressure changes that occur during the water hammer phenomenon. The damping properties of the tube are higher the lower its Young's modulus and wall thickness are. For example, assuming that outer diameter of the thick-walled tube would be 0.10 of the inner

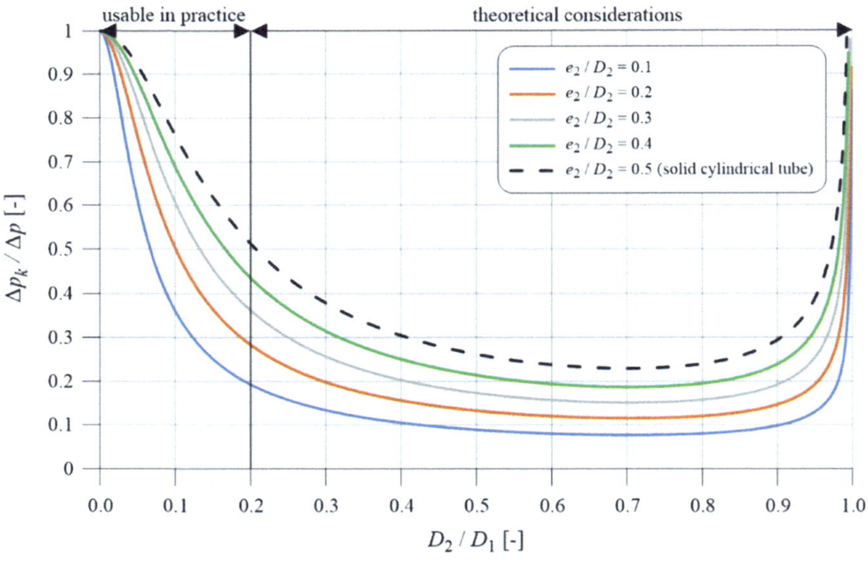

Fig. 4 Ratio of pressure increase in a pipeline containing a thick-walled tube to a pressure increase with no thick-walled tube

diameter of the pipeline ($D_2 = 0.10D_1$) and that the thickness of the wall would be $e_2 = 0.2D_2$, it would theoretically reduce pressure increase by 50% in relation to the pipeline without the tube (orange line in Fig. 4).

4 Conclusions and Summary

In the theoretical analysis of the influence of inserting a circular tube in a pipeline on the propagation of the pressure wave, three types of tubes were distinguished: thin-walled tube, thick-walled tube and solid cylindrical tube. For each of these cases, using the work-energy principle, a formula for calculating the pressure wave velocity was derived. It was shown that the values calculated for thin-walled tubes are smaller than those calculated for thick-walled tubes. However, for the assumed range of calculations, these differences are negligibly small. The formula for the pressure wave velocity in a pipeline with a thick-walled tube has a wider application because by substituting $e_2 = D_2/2$, it transforms into a formula for the velocity of pressure wave with a solid cylindrical tube. It was shown that insertion of a tube with low bulk elastic modulus, may have a damping effect on the water hammer phenomenon, i.e., it reduces the pressure wave velocity and maximum pressure increase. The damping properties of the tubes are higher when the Young's modulus and wall thickness are lower.

References

Bergant A, Tijsseling A, Vitkovsky J et al (2008) Parameters affecting water-hammer attenuation, shape and timing—Part 1: mathematical tools. J Hydraul Res 46(3):377

Grote K, Antonsson E (2009) Springer handbook of mechanical engineering, 1st edn. Springer, Berlin, p 947

Kubrak M, Kodura A, Imiełowski S (2018) Analysis of pressure wave velocity in a steel pipeline with inserted fiber optic cable. Free Surface Flows and Transport Processes/GeoPlanet: Earth and Planetary Sciences, pp 281–292

Mitosek M, Szymkiewicz R (2016) Reservoir influence on pressure wave propagation in steel pipes. J Hydraulic Eng 132(8):1–5

Streeter VL (1958) Fluid mechanics. McGraw-Hill Companies, Columbus, OH

Tijsseling AS, Kruisbrink ACH, Pereira da Silva A (1999) The reduction of pressure wavespeeds by internal rectangular tubes. In: Proceedings of 3rd ASME/JSME joint fluids engineering conference, San Francisco

Wylie EB, Streeter VL (1993) Fluid transients in systems. Prentice-Hall, Upper Saddle River, NJ

An Experimental Investigation of Reaeration and Energy Dissipation in Hydraulic Jump

Serhat Kucukali and Sevket Cokgor

Abstract A hydraulic jump is characterized by a highly turbulent flow with macro-scale vortices and some kinetic energy dissipation. Reaeration is defined as the transfer of oxygen from air towards water. Aeration performance of a classical hydraulic jump has been investigated by means of energy dissipation in a 0.5 m wide flume with upstream Froude and Reynolds numbers ranging from 2.3 to 6.4 and 1.4 to 5.4 $\times 10^4$, respectively. Dissolved Oxygen (DO) measurements were conducted simultaneously upstream and downstream of the hydraulic jump with a manual oxygen meters. A strong correlation was found between the reaeration and energy dissipation rate indicating the macro turbulence dominant role in the process. In addition, the turbulence quantities were collected by an Acoustic Doppler Velocimeter (ADV) at low Froude numbers. Turbulent kinetic energy and Reynolds shear stress took their maximum values at the toe of the jump and exhibited a longitudinal decay. We related macro-scale turbulent length scale with upstream flow depth. The average energy dissipation rate has been correlated with the 1.5 power of the maximum turbulent kinetic energy in a dimensionally homogeneous form.

Keywords Hydraulic jump · Aeration · Oxygen transfer · Energy dissipation · Turbulence

1 Introduction

Hydraulic jump constitutes at the transition from supercritical regime to subcritical regime and characterized by highly turbulent flow, macro-scale vortices, kinetic energy dissipation and bubbly two-phase flow due to air entrainment. In the literature, there are lots of studies concerned with the hydraulic jump flow patterns. However, the self-aeration aspect of hydraulic jumps took less attention (Kucukali and Cokgor

---

S. Kucukali (✉)
Civil Engineering Department, Hacettepe University, 06800 Beytepe, Ankara, Turkey
e-mail: kucukali78@gmail.com

S. Cokgor
Civil Engineering Department, Istanbul Technical University, Istanbul, Turkey

M. B. Kalinowska et al. (eds.), *Recent Trends in Environmental Hydraulics*, GeoPlanet: Earth and Planetary Sciences,
https://doi.org/10.1007/978-3-030-37105-0_11

2007). The term self-aeration means transfer of oxygen from air into water and it has important environmental and ecological implications for polluted streams. Aeration efficiency is calculated with the equation suggested by Gameson (1957)

$$E = \frac{C_d - C_u}{C_s - C_u} = 1 - \frac{1}{\exp(K.t)} \tag{1}$$

In Eq. (1), E denotes aeration efficiency, which has a range between 0 for no aeration, to 1 for total downstream saturation. C_u and C_d are dissolved oxygen concentrations up- and downstream of a hydraulic structure, respectively, C_s is the saturation concentration of dissolved oxygen for a given ambient condition, K is the reaeration rate (1/s), and t is the bubble residence time. At different temperatures, the physical and chemical properties of water change. This in turn will influence the air-water gas transfer rate. Self-aeration efficiency therefore converted to a standard temperature. In this study Gulliver et al. (1990) formula was used to adjust aeration efficiency to 20 °C and it is defined as E_{20}. On the basis of a small-eddy model, Lamont and Scott (1970) correlated the reaeration rate with the energy dissipation rate and their result suggested that the gas transfer efficiency is controlled locally by the energy dissipation. One can calculate the average energy dissipation rate per unit mass ε_a along a reach as follows,

$$\varepsilon_a = \frac{\Delta H.Q.\gamma}{V.\rho} \tag{2}$$

where ΔH (m) is head loss, Q ($m^3\ s^{-1}$) is the discharge, V is the volume of the reach (m^3), γ is the specific weight and ρ is the density of water. Melching and Flores (1999) collected data sets from 166 streams in USA in order to develop empirical equations to predict reaeration rates of streams. The data of Melching and Flores (1999) were re-analyzed in Fig. 1 and the reaeration rates were plotted as a function of the average energy dissipation rate. The relationship was best fitted by a linear function as

$$K_{20} = 1102 \times \varepsilon_a + 4.4 \tag{3}$$

Equation (3) has a sound physical basis, because air entrainment takes place when the turbulence kinetic energy overcomes the surface tension forces (Ervine and Falvey 1987). For non-uniform flow conditions, Moog and Jirka (1998) related the gas transfer efficiency of macro roughness elements with the energy dissipation rate on the basis of small-eddy model. Kucukali and Cokgor (2008) investigated the aeration performance of boulder structures at sudden expansions and plunging jets, where they found an interrelation between the aeration efficiency and the energy dissipation. Indeed, hydrodynamic processes which ensure the self-aeration mechanism such as: (1) hydraulic jump, (2) plunging jet or water fall, and (3) stepped channels have other common property: they are also used as energy dissipaters at hydraulic structures. From this point of view, one can expect to find a positive correlation between the

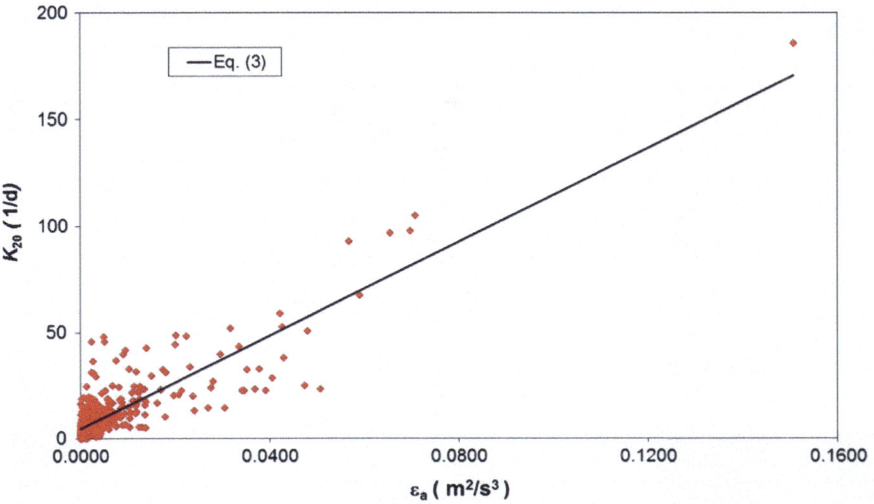

Fig. 1 Reaeration K_{20} versus average energy dissipation rate per unit mass ε_a for uniform flow conditions. *Data source* Melching and Flores (1999)

aeration efficiency and the energy dissipation. It could be suggested that the macro-scale eddies, which are responsible for the mixing and energy dissipation, could also be responsible for the suction of air into water (Hoyt and Sellin 1989). In this reasoning, Vischer and Hager (1995) defined the energy dissipation as a mixing process, which is originated from the vortex flow and turbulence. For instance, in fish passage design standards, energy dissipation rate is used as an indicator for turbulence (Kucukali and Hassinger 2018). The objective of this experimental study is to investigate reaeration rate of hydraulic jumps in terms of energy dissipation and to reveal the important role of turbulence in the process.

2 Experimental Set-Up and Methods

The hydraulic jump is a rapidly varied flow occurring at the transition from the supercritical to the subcritical flow regime. Figure 2 shows the main variables, U_1 and U_2, as the up- and downstream average flow velocities; d_1 and d_2 are the corresponding flow depths, x is the streamwise distance from the toe of the hydraulic jump and, L_j is the jump length. The upstream Froude number is the main parameter of a hydraulic jump, $Fr_1 = U_1/(gd_1)^{0.5}$, where g is the acceleration of gravity. The experiments were carried out in a horizontal rectangular flume, 0.50 m wide, 0.45 m deep and 13.15 m long, with glass sidewalls and a concrete bottom (Fig. 3). The discharge was measured with a V-notch weir to ±2%. The flow depths were measured by

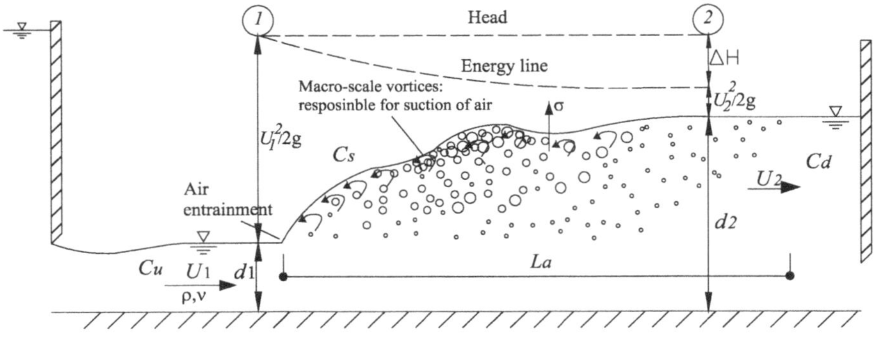

Fig. 2 Definition sketch for reaeration in a hydraulic jump

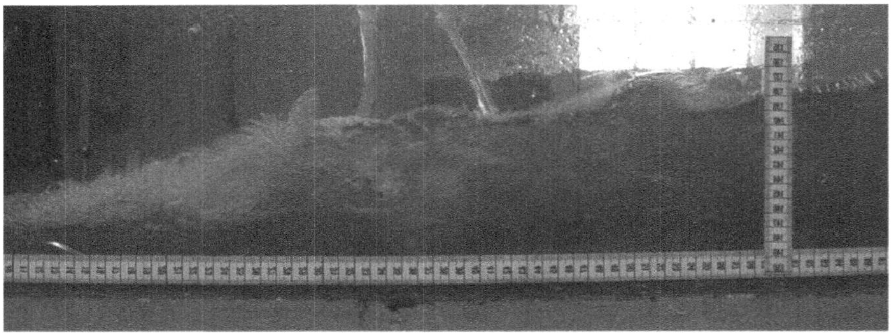

Fig. 3 Macro-scale vortices in a hydraulic jump for: $Fr_1 = 4.85$; $Re = 5.3 \times 10^4$. For oxygen transfer, the energy containing large scale eddies should overcome surface tension forces by creating surface disturbances

pointer gages with an accuracy of ±1 mm. The jump lengths were determined visually and, beyond the end of jump, the flow is practically not perturbed by the jump. The hydraulic jumps were generated with the aid of sluice and tail gates (Fig. 2). The longitudinal distance from the sluice gate to the jump toe varied in the range of 0.1–0.94 m, resulting in partially developed flow conditions. The jumps were grouped into seven different unit width discharges q in the range of 0.020–0.053 m^2/s. The head loss ΔH of the hydraulic jump was calculated from

$$\Delta H = \left(d_1 + \frac{U_1^2}{2g}\right) - \left(d_2 + \frac{U_2^2}{2g}\right) \tag{4}$$

and the average energy dissipation rate per unit mass for the hydraulic jump equals to

$$\varepsilon_a = \frac{\Delta H.Q.g}{(d_1 + d_2).0.5.L_j} \tag{5}$$

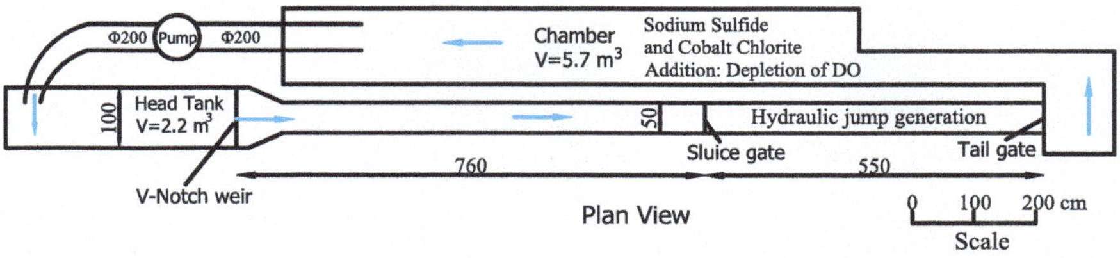

Fig. 4 Plan view of the experimental set-up

Dissolved Oxygen (DO) measurements were conducted simultaneously upstream and downstream of the hydraulic jump with manual oxygen meters (Model WTW Oxi 330i). The air calibration technique was used in the experiments with C_s values measured in the air. The uncertainties during the measurements were ±0.5% for the oxygen concentration and ±0.1% K for water temperature. To drop the DO concentration of the approaching flow, Sodium Sulfide and Cobalt Chlorite were added into the chamber (Fig. 4). The two minutes time-averaged values of C_u and C_d were measured 0.50 m up- and downstream of the hydraulic jump at the central flow depths, respectively. The measurement distance was sufficiently far from downstream for turbulent mixing to ensure the uniformity of flow conditions. The deficit ratios were calculated for each experiment from Eq. (1) and formula of Gulliver et al. (1990) was used to adjust the deficit ratio to a standard temperature.

The turbulence quantities were collected by a Nortek 10 MHz type Acoustic Doppler Velocimeter (ADV) at 25 Hz during a sampling time of 2 min, with velocities from about 1 mm/s to 2.5 m/s with an accuracy of ±1%. From the time-series of the velocity measurements, the root-mean-square of the turbulent fluctuation velocities in the longitudinal u′, lateral v′, and vertical w′ directions, respectively, were calculated at the each measurement point. Then, the turbulent kinetic energy per unit mass k is calculated from

$$k = \frac{1}{2}(\overline{u'^2} + \overline{v'^2} + \overline{w'^2}) \tag{6}$$

The distribution of the turbulent kinetic energy in the flow field is important, because the energy dissipation is resulted from the turbulence generation. Also, the presence of turbulence in the flow domain plays an important role in conversion of flow kinetic energy to turbulent kinetic energy. For the turbulent kinetic energy analysis, also the data of Liu et al. (2004) were used. The flow conditions of the turbulence measurements are given in Table 1, with Re = q/υ the Reynolds number and k_{max} the maximum turbulent kinetic energy per unit mass measured at any streamwise location in the hydraulic jump. Moreover, Reynolds shear stress is calculated by using Eq. (7)

$$\tau = -\rho\overline{u'v'} \tag{7}$$

Table 1 The flow conditions of the hydraulic jump turbulence measurements

Reference	Fr_1	Re	U_1 (m s^{-1})	d_1 (mm)	d_2 (mm)	ε_a (m^2 s^{-3})	k_{max} (m^2 s^{-2})
Liu et al. (2004)	2	118,570	1.67	71	172	0.235	0.060
	2.5	147,680	2.08	71	216	0.493	0.096
	3.3	86,100	2.10	41	165	0.864	0.102
Present study	1.9	34,500	1.05	33	70	0.173	0.038
	2.3	34,500	1.19	29	80	0.207	0.048

The Reynolds shear stress is a transport effect resulting from turbulent motion induced by velocity fluctuations with its subsequent increase of momentum exchange and of mixing (Piquet 2010).

Both the DO and the turbulence quantities collection methods and related calculations are used by several researches in the literature. Making measurements by using manual oxygen meters and ADV is relatively cheaper and easier. Hence, we selected those reliable and proven test methods. Moreover, the ADV technique has a limitation in bubbly flow; therefore, the turbulence measurements were applied only to the low Froude numbers $1.9 < Fr_1 < 2.3$ (Table 1).

3 Experimental Results and Discussion

In this study, it is expected to find a positive correlation between the aeration efficiency and energy dissipation. For this purpose, E_{20} values were plotted against head loss ΔH including the oxygen transfer data at the V-notch weir of the experimental set-up. Figure 5 depicts the E_{20} dependence on ΔH but data were scattered around a linear line suggesting an influence of another parameter on the process. This parameter was thought to be a unit discharge.

Moreover, the analysis of the measured data indicated a direct proportion between the reaeration and the average energy dissipation rate of a hydraulic jump. As ε_a increases from 0.14 to 0.29 m^2 s^{-3}, the reaeration rate increased by a factor of two. Figure 6 shows the hydraulic jump reaeration rate K_{20} versus the average energy dissipation rate per unit mass ε_a with the best fit ($R^2 = 0.97$)

$$K_{20} = 56873.\varepsilon_a^{1.08} \quad for \;\; 2.4 \le Fr_1 \le 6.5 \tag{8}$$

Equation (8) suggests that the gas transfer efficiency of a hydraulic jump is controlled by the energy dissipation rate. The comparison of Eqs. (8) and (3) suggests that the re-aeration rate of a hydraulic jump has about 50 times higher value as compared to uniform flow condition at the same energy dissipation rate.

The turbulent kinetic energy has its maximum value at the jump toe and decays along the hydraulic jump (Fig. 7), because of the energy cascade process in which

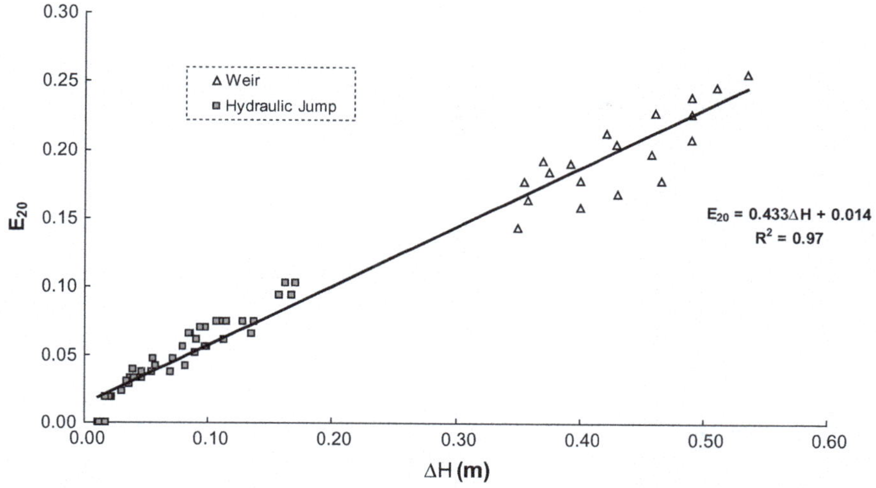

Fig. 5 Variation of aeration efficiency as a function of head loss at a hydraulic jump and weir

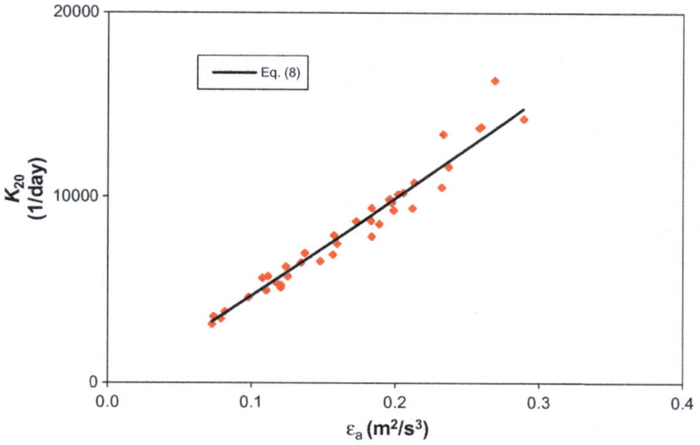

Fig. 6 Reaeration K_{20} versus average energy dissipation rate per unit mass ε_a for hydraulic jump

turbulent kinetic energy is transformed into micro-scale vortices. This resulting correlation was explained with the equation:

$$\frac{k^{1/2}}{U_1} = -0.012x/d_2 + 0.2\ (R^2 = 0.90) \tag{9}$$

where x is the longitudinal distance from the jump toe.

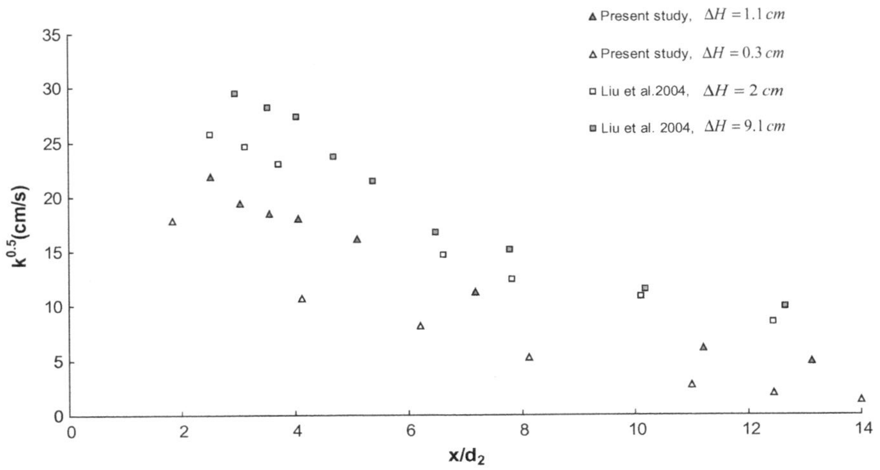

Fig. 7 Distribution of dimensional maximum turbulence kinetic energy in a hydraulic jump

Moreover, Fig. 8 shows the relationship between ε_a and the $k_{\max}$, given that

$$\varepsilon_a = 19.52.k_{\max}^{1.5}\ (R^2 = 0.90) \tag{10}$$

Equation (10) suggests that the energy dissipation rate was under the control of the turbulent kinetic energy. It was attempted to adjust Eq. (10) to the *Prandtl-Kolmogorov* formula. Therefore, the average of the upstream flow depths (Table 1) $d_{avr} = 49$ mm was taken as the integral turbulent length scale, in agreement

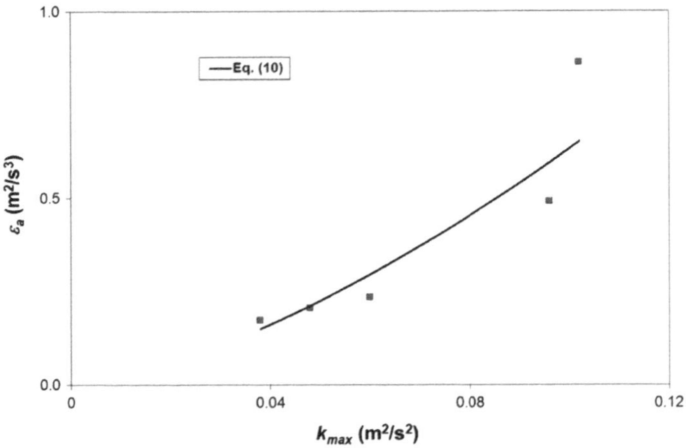

Fig. 8 Variation of the average energy dissipation rate per unit mass ε_a as a function of the maximum turbulent kinetic energy per unit mass k_{max}

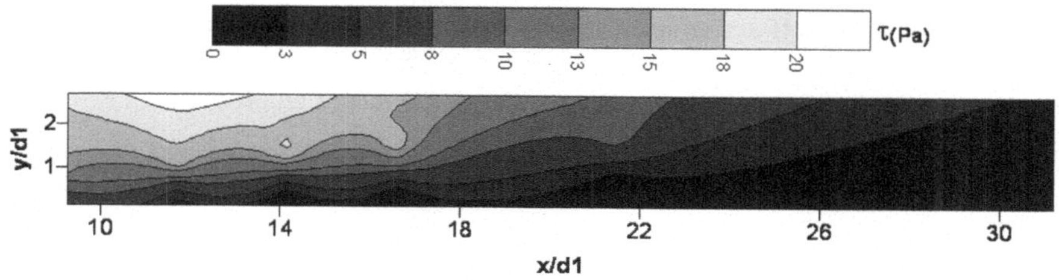

Fig. 9 Reynolds shear stress contour for $Fr_1 = 3.35$; Re = 86,100

with Kucukali and Chanson (2008) who related the macro-scale turbulent length scale L with upstream flow depth. Substituting this value in Eq. (10) leads to the dimensionally homogenous equation

$$\varepsilon_a = 0.97.\frac{k_{\max}^{1.5}}{L} \quad (11)$$

Equation (11) has an exponent consistent with the formula *of Prandtl-Kolmogorov*, although this formula deals with local energy dissipation and turbulent kinetic energy. Moreover, in Fig. 9, Reynolds shear stress (τ) variation through the hydraulic jump is shown. Figure 9 depicts the intensive vertical mixing and momentum transfer near the free-surface of the hydraulic jump.

4 Conclusions

The reaeration rate of hydraulic jumps was experimentally investigated in terms of the average energy dissipation rate per unit mass. 40 test series were conducted to relate the gas transfer efficiency to the energy dissipation rate, supporting the small-eddy model of Lamont and Scott (1970). The reaeration and the average energy dissipation rate of the hydraulic jump may be described with Eq. (8). The proposed equation predicts and captures the system dynamics. Equation (8) has the following features: (i) it is a single parameter equation, (ii) it may be tested for other rapidly varied flow conditions like plunging jets and sudden expansions. In the experiments, viscous scale effects might exist; hence, the validity of Eq. (8) should be tested at higher Reynolds numbers in a field study. It is discussed that a functional relationship between reaeration and energy dissipation rate exists both for uniform and nonuniform flow conditions. At site conditions, measuring energy dissipation is much easier than measuring turbulence quantities. The functional dependence between the reaeration and energy dissipation rate can provide engineers to estimate gas transfer coefficients in the field. The authors believe that the present results bring a fresh perspective towards better understanding of hydraulic jump reaeration process. The overall contribution of this study is thought to be that we showed the main role of macro turbulence in self-aeration and energy dissipation mechanisms. Consequently, experimental findings suggest that hydraulic jump can be used as a self-aerator device in waste water treatment plants to enhance the DO levels of an effluent.

References

Ervine DA, Falvey HT (1987) Behaviour of turbulent water jets in the atmosphere and in plunge pools. Proc Inst Civ Eng 2:295–314

Gameson ALH (1957) Weirs and aeration of rivers. J Inst Water Eng 11(5):477–490

Gulliver JS, Thene JR, Rindels AJ (1990) Indexing gas transfer in self-aerated flows. J Environ Eng 116(3):503–523

Hoyt JW, Sellin RHJ (1989) Hydraulic jump as mixing layer. J Hydraul Eng 115:1607–1613

Kucukali S, Chanson H (2008) Turbulence measurements in the bubbly flow region of hydraulic jumps. Exp Thermal Fluid Sci 33(1):41–53

Kucukali S, Cokgor S (2007) Fuzzy logic model to predict hydraulic jump aeration efficiency. Proc Inst Civ Eng Water Manag 160(4):225–231

Kucukali S, Cokgor S (2008) Boulder-flow interaction associated with self-aeration process. J Hydraul Res 46(3):415–419

Kucukali S, Hassinger R (2018) Flow and turbulent structure in baffle-brush fish pass. PI Civ Eng-Water Manag 171(1):6–17

Lamont JC, Scott DS (1970) An eddy cell model of mass transfer into the surface of a turbulent fluid. AIChE J 16:513–519

Liu M, Rajaratnam N, Zhu DZ (2004) Turbulence structure of hydraulic jumps of low Froude numbers. J Hydraul Eng 130(6):511–520

Melching CS, Flores HE (1999) Reaeration equations derived from U.S. Geological Survey database. J Environ Eng 125(5):407–414

Moog DB, Jirka GH (1998) Stream re-aeration in nonuniform flow: macro-roughness enhancement. J Hydraul Eng 125(1):11–16

Piquet J (2010) Turbulent flows. Models and physics. Springer, Berlin

Vischer DL, Hager WH (1995) Energy dissipators, hydraulic structures design manual. IAHR, Balkema

Flow and Turbulence Structure in a Vertical Slot–Brush Fish Pass

Serhat Kucukali and Reinhard Hassinger

Abstract The flow characteristics of a new vertical slot-brush type fish pass were investigated experimentally in a 2 m wide rectangular flume which has a bed slope of 4% for discharges ranging between 0.16 and 18 $m^3\ s^{-1}$. The physical model results showed that by placing flexible bristle elements in a vertical-slot pool, the flow energy is effectively dissipated by the vibrations and bending of the bristles leading an about 17% reduction in the spatially averaged turbulence kinetic energy and the measured maximum velocity is reduced by 20%. Moreover, turbulent jet and recirculation regions disappeared and deceleration zones reduced after the installation of brush blocks in the pool.

Keywords Vertical slot fish pass · Flexible hydraulic elements · Turbulence · Velocity field

1 Introduction

This paper focuses on vertical-slot fishway which is the most common fish pass type around the world. Vertical-slot fishway consists of a series of pools which are separated by long and short baffles at regular intervals (DWA 2014). However, vertical slot fish passes are designed for large fish which have economic value, such as salmon and trout. But in rivers there are small-bodied and weak-swimming capacity fishes which are also ecologically important (Hassinger 2007; Kucukali and Hassinger 2015, 2018). For instance, in Turkey there are 382 freshwater fish species and about 1/3 of these species are endemic. Some of those endemic fish species have a body length of 5 cm. Accordingly, it is not possible for those endemic fish species to climb classical pool fish passes. The European Union Water Framework Directive requires to ensure the continuity of those species in rivers in order to ensure good

S. Kucukali (✉)
Civil Engineering Department, Hacettepe University, 06800 Beytepe, Ankara, Turkey
e-mail: kucukali78@gmail.com

R. Hassinger
Civil and Environmental Engineering Department, Kassel University, Kassel, Germany

M. B. Kalinowska et al. (eds.), *Recent Trends in Environmental Hydraulics*, GeoPlanet: Earth and Planetary Sciences,
https://doi.org/10.1007/978-3-030-37105-0_12

ecological status. In this context, it is necessary to make a structural modification to existing vertical slot fish passes to allow the passage of such small-bodied and weak swimming capacity fish (Mallen-Cooper 2015). In this context, efforts have been made to facilitate the passage of fish by characterizing turbulence through various configurations. In the previous studies, impermeable hydraulic elements were placed in the pools to dissipate the flow energy and prevent turbulence (Tarrade et al. 2008; Santos et al. 2016). However, this configuration has caused two different turbulence fields around the obstacles and has not been particularly beneficial for the passage of small fish species. Another technical solution is using cylindrical flexible hydraulic elements as bristles to absorb flow energy which is proposed in this paper. In this design, with the help of the bristle, stagnation intervals are established to transform the hydraulic power of the flowing water and to form resting areas. In the proposed vertical slot-brush fish pass, the flow velocities and turbulent kinetic energies are very low behind the brushes; accordingly, there is no need to provide resting pools for fish which reduces the construction costs. For instance, Fig. 1 shows the application of the vertical slot-brush fishway concept in Çataloluk, Turkey. The function of the slot brush is to reduce the maximum velocity in the pool, jet brush is to dissipate the jet energy and to reduce the turbulent kinetic energy in the pool, and wall brush is to prevent the occurrence of short-circuit flow (Fig. 1). It should be noted that, in

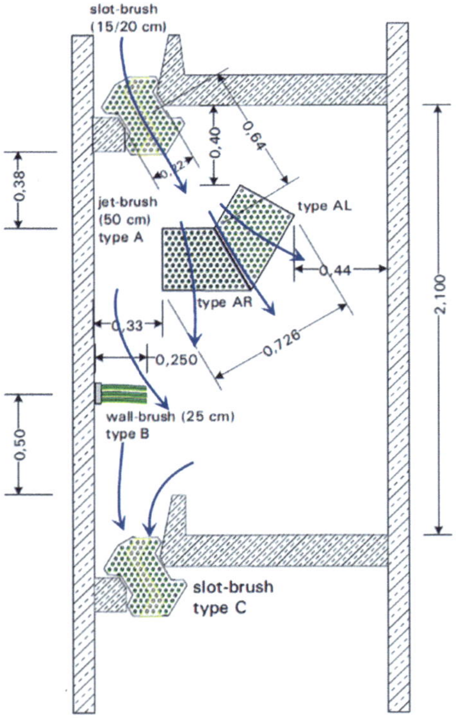

Fig. 1 Application of the vertical slot-brush fishway concept in Çataloluk, Turkey. Dimensions are in m

almost all brush-furnished passages, organic material will be entered into the bristle elements. In real operation, the gaps inside the brush blocks will become partly filled with organic debris and the velocity will be uniform over the water column behind the brush blocks (Kucukali and Hassinger 2015). This material within the brush-blocks is of minor importance for the function as long as the slots remain open. It is recommended that the bristle elements should be combed with a rake at least two times a year (Hassinger 2015). Considering the larger driftwood, however, many observed cases show that the tendency of clogging is lower in bristle system than in the fish way with solid built-in components. The experimental evaluations of Rahn (2011) have also shown that degree of coverage of the bristles with floating objects do not significantly affect the flow and energy dissipation characteristics.

2 Physical Model of the Vertical Slot-Brush Fish Pass and Measurements

Figure 2 shows the general view of the physical model of baffle-brush fish pass which was constructed at a scale of virtually 1:1 in the hydraulics laboratory and testing facilities of Kassel University Department of Civil and Environmental Engineering. The vertical slot-brush fish pass was installed in a rectangular flume with width, height, and length being 2, 0.80, and 8.40 m, respectively, and the flume has a bed slope of 4%. A pump was used to supply the header tank from the laboratory lower reservoir and the discharge was measured by means of a magnetic inductive flowmeter with an accuracy of 0.05% installed in the supply pipe. The discharge was controlled by a valve in the supply pipe with an electric actuator. The water surface elevation was measured by an ultrasonic displacement meter (Sonic Joker) that was mounted above the flow and scanned downwards the free-surface flow. The uncertainty in the measured flow depth was less than 5 mm. An acoustic Doppler velocimeter (ADV) (NortekTek 10 MHz) was used to measure the 3D instantaneous velocity

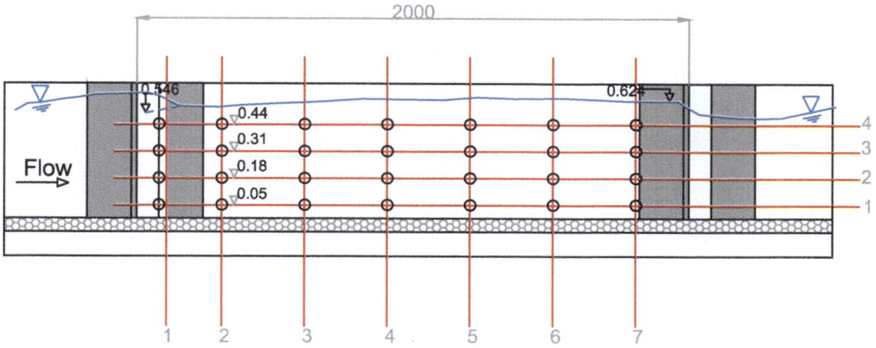

Fig. 2 Water surface profile along the vertical slot fish way and measuring points at *x-z* plane

fields for flow rates of 0.16 and 0.18 $m^3 s^{-1}$. The employed ADV had three sensors and measured the flow velocity in a control volume of in front of its sensors. The turbulence quantities were collected at a frequency of 10 Hz over a sampling time of 60 s. Primary tests showed that 60 s of sampling was sufficient and yielded stationary results for all measuring points. The ADV used measures flow velocities from about 1 mm s^{-1} to 2.5 m s^{-1} with an accuracy of ±1%. The velocity measurements were taken at each grid point with a grid spacing between two points in the lateral (y) and vertical (z) directions, respectively. The position accuracy is approximately 2 mm. The measuring grid was not distributed uniformly over the base area, since it was to be expected that larger velocity gradients occur in the area of the jet and in the vicinity of the wall. As shown in Fig. 3 a grid with 7 cross sections and 7 longitudinal sections as well as 4 vertical planes were selected. In the variant with bristles, some of the measurement loops cannot be approached since the measurement points lay within the bristle elements. In this case, a point was approached which was at the same level in the direction of flow directly below the bristle element. The height of brush play very important role when taking into account the hydraulic conditions of the water flow through the fish pass. The velocity profile can be divided into two parts: lower velocity profile below brushes level and higher velocity profile above the brushes (Bartnik et al. 2012). However, in the present physical model the velocity

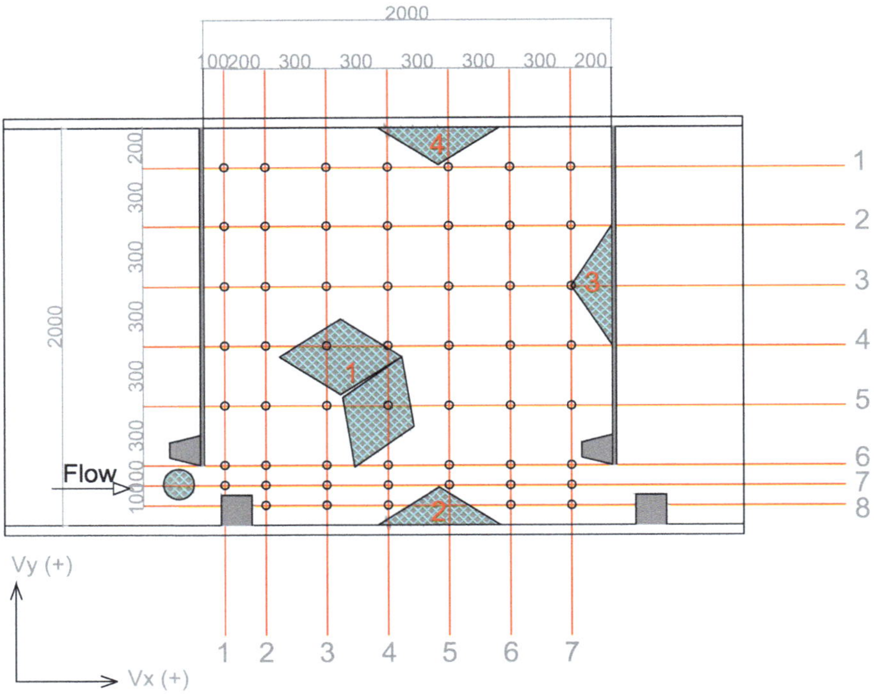

Fig. 3 Top view of the velocity measurement grid. Dimensions are in mm

data were collected only for bellow the brushes level of h_b= 0.47 m (Fig. 2) and we focused on 2D flow structures in the horizontal plane.

Local velocity measurement grid at x-y plane is depicted in Fig. 3. In the figure, circles show the velocity measurement points taken by the acoustic Doppler. From those measurements, the local power velocity V_{pm} is calculated using Eq. (1)

$$V_{pm} = \sqrt[3]{\frac{\sum |V_i^3|}{n}} \text{ in which } V = \sqrt{u^2 + v^2 + w^2} \tag{1}$$

where n is the number of velocity samples. Furthermore, the convective acceleration in the flow direction is computed from

$$a_x = V_x \frac{\Delta V_x}{\Delta x} \tag{2}$$

where $\Delta V_x/\Delta x$ is the longitudinal velocity gradient. Convective acceleration is important because it accounts for the effect of the fluid particle moving to a new location in the flow, where the velocity field is different. From the time-series of the velocity measurements, the root-mean-square of the turbulent fluctuation velocities in the longitudinal u', lateral v', and vertical w' directions, respectively, were calculated at the each measurement point. Then, the horizontal and vertical Reynolds shear stresses are computed from

$$\tau_{xy} = -\rho \overline{u'v'} \tag{3a}$$

$$\tau_{xz} = -\rho \overline{u'w'} \tag{3b}$$

where ρ is density of water. The Reynolds stresses are a transport effect resulting from turbulent motion induced by velocity fluctuations with its subsequent increase of momentum exchange and of mixing (Piquet 2010).

Moreover, the turbulent kinetic energy per unit mass k is calculated using Eq. (4)

$$k = \frac{1}{2}(\overline{u'^2} + \overline{v'^2} + \overline{w'^2}) \tag{4}$$

The distribution of the turbulent kinetic energy in the flow field is important, because the energy dissipation is resulted from the turbulence generation. Also, the presence of turbulence in the flow domain plays an important role in conversion of flow kinetic energy to turbulent kinetic energy. Hence, it is suggested that a better understanding of turbulence arising from the geometry of the fish pass will improve the understanding of fish behavior characteristics.

3 Experimental Results and Discussion

The flow conditions of the model tests are summarized in Table 1. The power density (energy dissipation per unit volume) ΔP is calculated from:

$$\Delta P = \rho g Q \Delta h/(B L h_m) \tag{5}$$

where g is the acceleration due to gravity. The physical model is based on the Froude similitude and values of the Reynolds number ($Re = 8 \times 10^4$) demonstrate that the results are not likely to be affected by Reynolds number effects, i.e., viscous scale effects.

In Fig. 4, the velocity fields with and without brush blocks are compared. The flow characteristics start with an intensely turbulent flow in the slot region and change continuously to the calm and less turbulent flow pattern behind the main brush block creating a refuge for fish As can be clearly seen form Figs. 4 and 5, when the bush blocks are installed in the pool, jet and recirculation regions disappear and deceleration zones reduced.

The spatial distribution of the V_{pm} through the fish pass is shown in Fig. 6. The high speed level without bristles and the moderate speeds in the variant with bristles can be clearly seen in Fig. 6. The resultant velocity is reduced to approx. 2/3 by the bristles. Since the energy content is related to the square of the velocity, it can be inferred from this that the energy content in the flow in the variant with bristles is only nearly half as high as in the conventional design. This means that slightly more

Table 1 The experiments flow conditions and test results

Parameter	Notation	Unit	Without bristles	With bristles
Pool width	B	m	2	2
Pool length	L	m	2	2
Bed slope	S_o	–	0.04	0.04
Discharge	Q	$m^3\ s^{-1}$	0.182	0.160
Elevation difference between pools	Δh	m	0.158	0.157
Water depth at basin entrance	h_o	m	0.546	0.546
Water depth at the end of the tank	h_u	m	0.624	0.624
Average flow depth in pool	h_m	m	0.585	0.585
Bristle height	h_b	m	–	0.47
Volume of pool	V	m^3	2.315	2.315
Power density	p	$W\ m^{-3}$	121.8	107.1
Slot width	s	m	0.165	0.165
Theoretical maximum velocity	v_{max}	$m\ s^{-1}$	1.76	1.755
Measured maximum velocity[a]	v_{ss}	$m\ s^{-1}$	1.934	1.649

[a]Measured at the slot 5 cm above the channel bottom

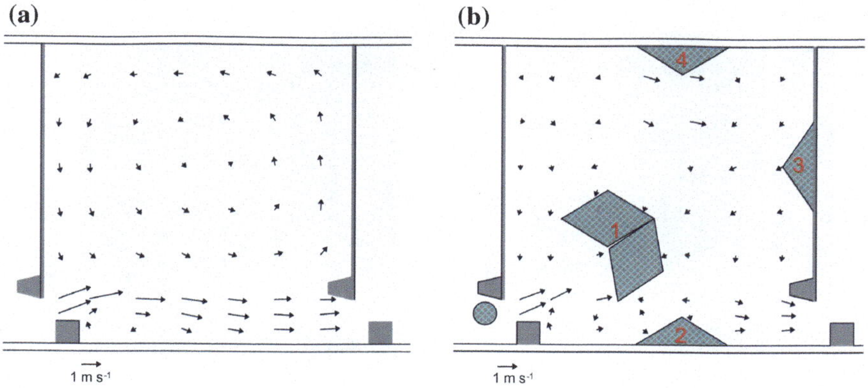

Fig. 4 Velocity field in horizontal plane at $z/d = 0.32$: **a** without brushes, and **b** with brushes

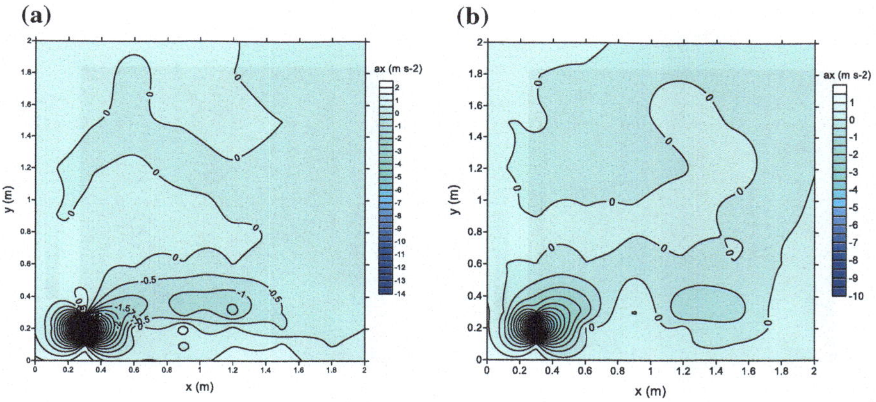

Fig. 5 Distribution of convective acceleration in x direction at $z/d = 0.32$: **a** without brushes, and **b** with brushes

than half of the flow energy is converted from the incoming flow within the bristle elements. From this result, it is also justifiable that the calculated energy conversion (conversion of kinetic energy by turbulence in one volume) can be halved in the case of fish passages with bristle elements. Since the bristle elements occupy space, it is not necessary to derive a demand for a drastic reduction of the fish passages, especially as the basin size is to be selected with regard to large fish.

The experimental results showed that by placing bristle elements in the pool, the flow energy is effectively dissipated by the vibrations and bending of the bristles leading to about 17% reduction in the turbulence kinetic energy (Fig. 7). Also, this reduced turbulent kinetic energy zone creates different migration corridors and favorable hydraulic conditions for different fish species.

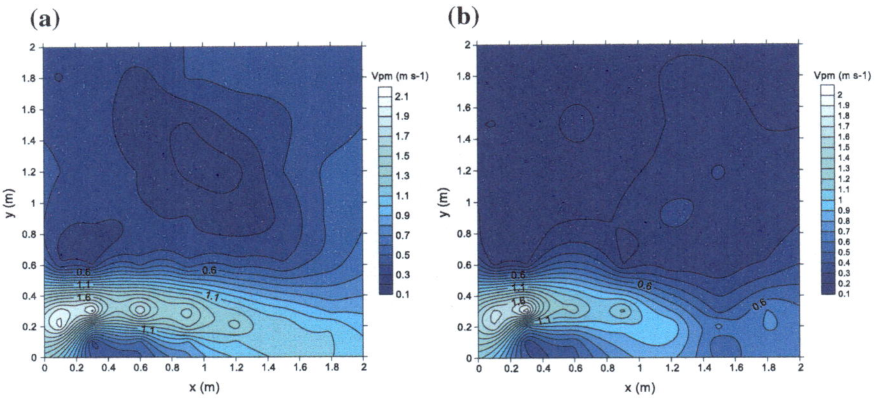

Fig. 6 Power velocity distribution in horizontal plane at $z/d = 0.32$: **a** without brushes, and **b** with brushes

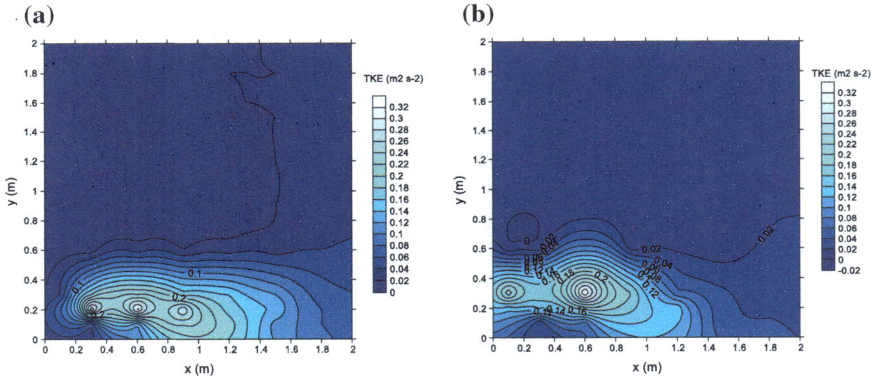

Fig. 7 Turbulence kinetic energy distribution in horizontal plane at $z/d = 0.32$: **a** without brushes, and **b** with brushes

Figure 8 shows relative part of volume in the pool considering the given power velocity. Figure 8 reveals that the power velocity level is reduced considerably by the bristles to such an extent that there is much more space available for small-bodied fish. For Also, the graph shows that the weak-fledged fishes are particularly benefiting from bristles because the space available to them is more than doubled. This means that, with decreasing performance, the fish are clearly increased in the amount of space available, and that the gain for the weaker fishes is increasing. The comparison of the curves (with and without brushes) shows that fish with small and medium-sized swimming pools in the fish pass with bristles are considerably more volume than without bristles (Fig. 8).

The flow-bristle interaction leads to 46% reduction in spatially averaged vertical Reynolds shear stress. However, the horizontal Reynolds shear stress (τ_{xy}) increased

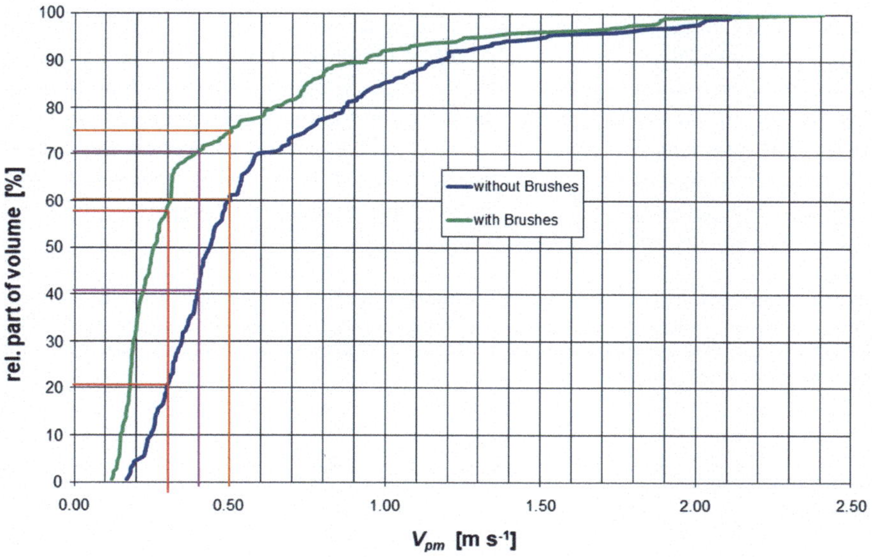

Fig. 8 Spatial distribution of power velocity which is equal and less than in the pool volume

by 6.6% which can be due to the horizontal mixing in the vicinity of brush blocks. This finding may have important implications, because Enders et al. (2017) hypothesized that the horizontal turbulent momentum exchange which matches the body undulation of the swimming fish can be more beneficial for fish passage than the vertical momentum exchange. Moreover, Wilkes et al. (2017) hypothesized that fish would occupy positions with energetically hydrodynamic conditions that are likely to minimize swimming costs.

4 Conclusions

Vertical slot-brush fish pass fulfill the requirements of an efficient fish passage by providing tranquil flows, sufficient flow depths, and different migration corridors with no obstructions for different fish species. A significant amount of energy dissipation takes place in brush blocks with the vibration and bending of bristles which gives rise to energy transfer from the main flow to the bristles. The flow characteristics start with an intensely turbulent flow in the slot region and change continuously to the calm and less turbulent flow pattern behind the main brush block creating a refuge for fish. The main advantage is that all the parts of this fish pass type can be scaled-up in order to adjust it to almost any demand based on the model test results without extraordinary cost. The bristles as energy absorbers represent a cost-effective option, which can be retrofitted at any time existing structures, to improve the hydraulic conditions in fish passages. In the case of a sufficient number of bristle elements, it can be assumed that approximately half of the energy conversion takes place in the bristle elements. Only half the specific capacity is left to the residual volume. This gives a certain saving potential in the basin volume and clear spaces for large fish are still sufficient. Moreover, it is to be expected that macroinvertebrates, which represent a food supply, settle in the brush blocks. It is very likely that various fish species will be permanently present in the fish pool due to the supply of food and the reduced velocity and turbulence levels.

Acknowledgements This work is partially supported by the Scientific and Technical Research Council of Turkey under Scientific and Technological Research Projects Funding Program (1001 TUBITAK) grant with agreement number 117M151. The authors thank Meliha Gamze Ekren for preparing the contour diagrams.

References

Bartnik W, Książek L, Leja M, Wyrębek M, Strutyński M (2012) Bedload equilibrium measurements along canoe fishways. Infrastruct Ecol Rural Areas 11(4):5–16 (In Polish)

DWA (2014) Merkblatt DWA-M 509: Fischaufstiegsanlagen und fischpassierbare Bauwerke – Gestaltung, Bemessung. Qualitätssicherung, Hennef (in German)

Enders EC, Castro-Santos T, Lacey RWJ (2017) The effects of horizontally and vertically oriented baffles on flow structure and ascent performance of upstream-migrating fish. J Ecohydraul 2(1):38–52

Hassinger R (2007) Steigerung der energieumwandlung in schlitzpässen mit borstenelementen. Report on Laboratory Tests on a 1: 1 Scale, Kassel University (in German)

Hassinger R (2015) Brush-furnished fishway and Canoe-fishway: description of the state-of-the-art technology. Kassel University, Germany

Kucukali S, Hassinger R (2015) Hydraulic model test results of baffle-brush fish pass. PI Civ Eng-Water Manag 168(4):189–194

Kucukali S, Hassinger R (2018) Flow and turbulent structure in baffle-brush fish pass. PI Civ Eng-Water Manag 171(1):6–17

Mallen-Cooper M (2015) Fish passage in Australia: universal lessons. Proc., Fish Passage 2015 Conference, Groningen, Netherlands

Piquet J (2010) Turbulent flows. Models and physics. Springer, Berlin, Germany

Rahn S (2011) Hydraulic investigation of the flow conditions in bristle fish passes with triple bars. Master thesis in Faculty of Civil and Environmental Engineering, University of Kassel (in German)

Santos JM, Branco P, Amaral S, Silva A, Ferreira, T, Pinheiro A, Viseu T, Katopodis C (2016) The FISHMOVE project—development of mitigation measures for small instream obstacles to fish migration in Portuguese streams. In: Fish passage 2016 conference, Massachusetts, USA

Tarrade L, Texier A, David L, Larinier M (2008) Topologies and measurements of turbulent flow in vertical slot fishways. Hydrobiologia 609:177–188

Wilkes MA, Enders EC, Silva AT, Maddock I (2017) Position choice and swimming costs of juvenile Atlantic salmon *Salmo salar* in turbulent flow. J Ecohydraul 2(1):16–27

Flow Between the Sub-basins of Charzykowskie Lake—Modeling and Measurements

Artur Magnuszewski and Barbara Nowicka

Abstract Charzykowskie Lake belongs to ribbon type lakes with a flow-through of Brda River. Lake is separated into four sub-basins acting as individual water masses. The long time discussion in the literature is about the possibility of flow between Southern, Central and Northern-2 sub-basins. This problem has been approached with the use of hydrodynamic CCHE2D model and field measurements used to verify modeling results. Model represent steady flow conditions and calm weather which were observed during the field measurements of 2015 08 28 with the use of River-Pro ADCP. It was shown that there is a flow between the Central and Northern-2 sub-basins confirmed by ADCP measurement of the moving boat method. A circular pattern of river water flow is observed at the mouth of Brda inflow to the Northern-1 sub-basin of the lake. This pattern was confirmed with section-by-section ADCP measurement performed at 8 profiles. Study had shown that it is possible to use hydrodynamic modeling to calculate the flow in a wide and deep not thermally stratified lake sub-basins. ADCP measurements provide information on velocity magnitude and orientation of velocity vectors.

Keywords Charzykowskie Lake · ADCP measurements · Hydrodynamic model CCHE2D

1 Introduction and Study Site Description

One of the problems in the studies of deep ribbon lakes fed and drained by the rivers is the calculation of the water inflow and outflow as well as internal circulation. The flow between separated basins of the lake may be difficult to estimate using

A. Magnuszewski (✉)
Department of Hydrology, Faculty of Geography and Regional Studies, University of Warsaw, Krakowskie Przedmieście 30, Warsaw, Poland
e-mail: asmagnus@uw.edu.pl

B. Nowicka
Department of Hydrology, Institute of Meteorology and Water, Management—National Research Institute, Podleśna 61, 01-673 Warsaw, Poland

M. B. Kalinowska et al. (eds.), *Recent Trends in Environmental Hydraulics*, GeoPlanet: Earth and Planetary Sciences,
https://doi.org/10.1007/978-3-030-37105-0_13

only inflow and outflow rivers discharge measurements. With the improvement of hydrodynamic models it is possible to approach that problem using CFD methods. The complete description of the lake dynamics can be achieved by the application of Navier-Stokes-Fourier-Fick equation. These partial differential equations describe fields of velocity vector, pressure, temperature, and mass concentration of different tracers. The simplifications are introduced by restricting calculations to specific process. The density is often assumed to be constant for the whole water body, making the thermal diffusion negligible as compared to convection (Hutter 1986). Gradient in the free surface elevation drives the system by the gravity force. Important in lake modeling is the action of external force of wind. There is a constant progress of using hydrodynamic models in limnology and hydrobiology; an overview of different approaches can be found in Avalos-Cueva et al. (2016), Bhagowati and Ahamad (2018). The external fluxes of water control both the hydrological balance of the lake and biological processes providing nutrients and transporting products of biological production (Munara et al. 2018).

In this study, for showing the flow between the sub-basins of Lake Charzykowskie (Fig. 1) we apply 2D hydrodynamic numerical model CCHE2D developed at the National Center for Computational Hydroscience and Engineering (NCCHE) at the University of Mississippi, USA. The model solves the depth-averaged Navier-Stokes equations utilizing the control volume approach and the efficient element method and has been tested successfully to simulate flow in natural channels in hydraulic research (Jia et al. 2002). Indispensible step in hydrodynamic modeling is the verification of the model results we use for the field measurements performed by RiverPro ADCP instrument. ADCP instruments that are designed for discharge measurements at rivers and other open channels by the method of moving boat or section by section method; both techniques have been tried at the Charzykowskie Lake to be used as a hydrodynamic modeling verification data. Similar approach is reported in the study of Avalos-Cueva et al. (2016) where ADCP measurements are used to verify the thermal structure and circulation of Lake Chapala in Mexico.

Charzykowskie Lake is a flow-through glacial ribbon lake located in the Tuchola Forest Biosphere Reserve in North-West part of Poland. Furthermore, it is part of the Zaborski Landscape Park (ZPK), in the buffer zone of the Tuchola Forest National Park (PNBT). The main river supplying the lake is Brda River; there are also tree smaller rivers with their mouths in the lake (Fig. 1). The size of their catchments and characteristic discharges is given in Table 1. The Charzykowskie Lake morphometry is shown in Fig. 1; there are visible underwater sills separating the whole lake into the following sub-basins: Northern-1, Northern-2, Central and Southern. Bathymetry data for creating DTM of the lake bottom have been taken from EULAKES research project accomplished by the Institute of Meteorology and Water Management-National Research Institute (IMGW-PIB). Using DTM resolution of 20 × 20 m and average water level of 120.3 m a.s.l., bathymetric characteristics of the Charzykowskie Lake bowl were calculated in four sub-basins (Table 2).

Charzykowskie Lake was a subject of many limnological and hydrobiological studies. In many papers, a separation of the lake bowl into distinctive sub-basins has been confirmed, as proved by their morphometry, physical and chemical properties

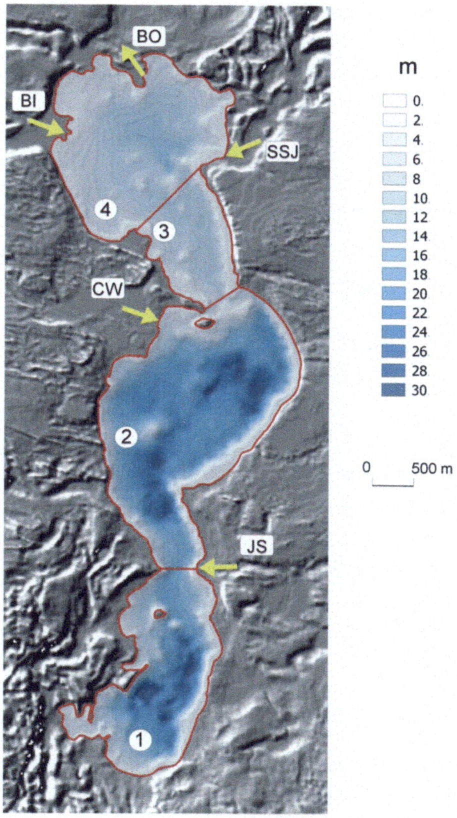

Fig. 1 Bathymetric map of Charzykowskie Lake and delineation of sub-basins: 1—Southern, 2—Central, 3—Northern-2, 4—Northern-1, BI—Brda inflow, BO—Brda outflow, JS—Jarcewska Struga, CW—Czerwona Woda, SSJ—Struga Siedmiu Jezior

of the water, ice thickness, phytoplankton distribution. List of the papers and related parameters confirming the subdivision of the lake bowl into separate sub-basins is shown in Table 3. The Northern-1 and Northern-2 sub-basins are under strong influence of Brda River inflow: they are shallow and without the epilimnion.

In Fig. 2, a pattern of Lake Charzykowskie bottom sediments with organic matter after Stangenberg (1950) shows the differences between the sub-basins. This differentiation of sub-basins bottom sediments was confirmed recently by discrete measurements performed by Nowicka et al. (2013).

The Brda River at the inflow to the lake builds a delta of sediments with a dense aquatic vegetation on the deposition cone. This process of building the delta and sedimentation of organic sediments in Northern-1 sub-basin is important for understanding ecological process and water quality controls in this part of Charzykowskie Lake. One of crucial problems is if there exists water flows between the sub-basins of Charzykowskie Lake.

Table 1 Parameters of the rivers in Charzykowskie Lake catchment

River (symbol)	Catchment area [km^2]	Discharge [m^3 s^{-1}]			
		2015 08 21 field measurement	Mean low flow	Average	Mean high flow
Inflow to Charzykowskie Lake					
Brda inflow (BI)	668	2.74	2.07	6.16	12.3
Czerwona Woda (CW)	78.8	0.05	0.00	0.20	0.41
Jarcewska Struga (JS)	58.0	0.12	0.11	0.42	0.64
Struga Siedmiu Jezior (SSJ)	41.0	0.26	0.05	0.21	0.33
Outflow from Charzykowskie Lake					
Brda outflow (BO)	920	3.59	2.23	7.78	15.6
Inflow—outflow difference					
Difference BO-(BI + CW + JS + SSJ)	36.8	0.85	0.16	1.62	3.30

Table 2 Charzykowskie Lake sub-basins morphometric characteristics calculated from DTM

Sub-basin	Area [km^2]	Volume [km^3]	Average depth [m]
Southern	2.98	0.038	13
Central	5.49	0.083	15
Northern-2	1.36	0.010	7
Northern-1	3.69	0.026	7
Total	13.5	0.157	12

2 Data and Description Methods

The outflow from the lake is lower Brda River collecting all water from the lake and supplying rivers. The outflow is controlled by the Sworne Gacie gauge, while Brda inflow is controlled by Ciecholewy gauge both operated by the IMGW-PIB. IMGW-PIB gauge is located also near Charzykowy and controls water level of Charzykowskie Lake. Flow from Struga Siedmiu Jezior is controlled by the gauge operated by PNBT, flow from Czerwona Woda and Jarcewska Struga was estimated from field measurements performed in the years 1999–2018 by the authors and

Table 3 Features justifying division of the Charzykowskie Lake into sub-basins according to different authors

Features differentiating Charzykowskie Lake into sub-basins	Source
Morphology, bottom sediments	– Stangenberg (1950) – Nowicka et al. (2013, 2017)
Fluvial inputs of nutrients and water	– Goszczyński and Jutrowska (1997) – Barańczuk and Borowiak (2010) – Nowicka (2012)
Ice cover thickness	– Choiński et al. (2006)
Chemical composition of bottom sediments (spatial distribution)	– Stangenberg and Żemoytel (1952) – Nowicka et al. (2013)
Physical and chemical characteristics of water	– Goszczyński and Jutrowska (1997)
Phytoplankton (the concentration of chlorophyll, biomass, spatial distribution)	– Wiśniewska and Luścińska (2012) – Bresciani and Giardino (2012) – Pęczuła (2012)

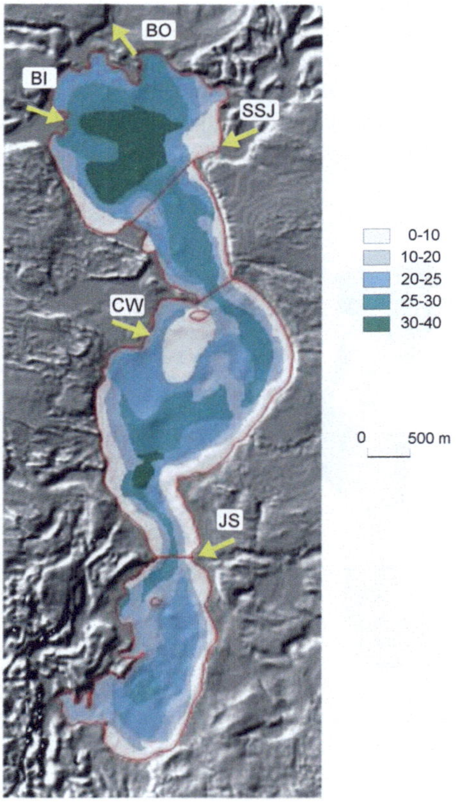

Fig. 2 Charzykowskie Lake—distribution of bottom sediments with different organic matter concentration in % (Stangenberg and Żemoytel 1952)

staff of ZPK. The Brda River inflow is a major flux of the water supplying Lake Charzykowskie.

Hydrometric measurements performed on August 28, 2015 (Table 1) shows that the difference between the outflow and inflow of Brda River is 0.85 $m^3 s^{-1}$ while a sum of discharges of three tributaries shows the inflow of 0.43 $m^3 s^{-1}$. This difference cannot be explained by the rivers discharge measurements uncertainty which is in the range of 7–10% (Bajkiewicz et al. 1993), but by the additional discharge of ground water at the littoral zone of the lake (Nowicka et al. 2017). Taking into account water balance we may expect that water supplying Southern and Central sub-basins should create a flow in direction toward the Brda River outflow.

In this study, to show the water flow between the sub-basins of Charzykowskie Lake, a hydrodynamic modeling with the use of CCHE2D model has been applied. The data to prepare the computing mesh was DTM of 20 × 20 m resolution containing 108035 (x, y, z) coordinates. Computing mesh had $i = 100$ by $j = 100$ lines, and average spacing of lines creating computing mesh 20 × 40 m. The upper boundary condition were steady discharges measured in the field, the lower boundary condition being a water level at the Brda outflow $z = 120.3$ m a.s.l. During field measurements, weather was calm, with no wind, so this force was not specified in the model.

Results of velocity field at Charzykowskie Lake calculated by CCHE2D model is shown in Fig. 3. It reveals the flow over the sill separating sub-basins Central and Northern-2. Calculated velocities are very small, with maximum value $v_h = 0.02$ m s^{-1} at sill dividing sub-basins.

In this study, to get the reference data for hydrodynamic model verification, an ADCP Doppler-profiler type RiverPro made by Teledyne was used (Fig. 4).

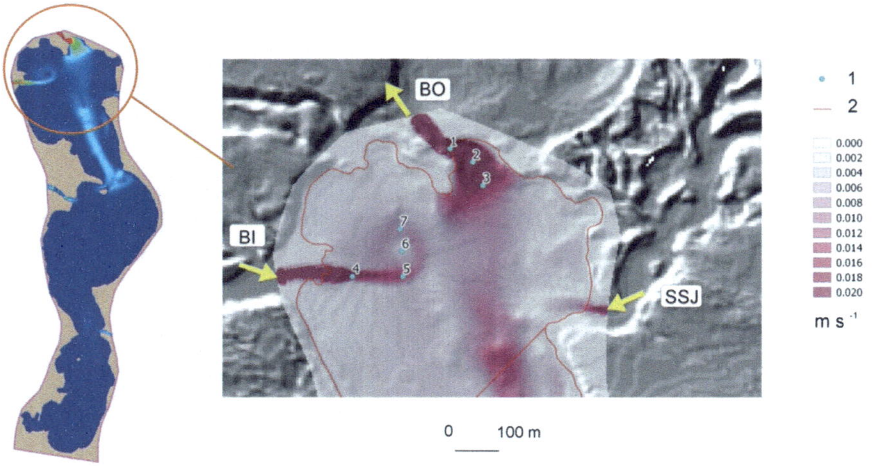

Fig. 3 Velocities calculated by CCHE2D model and ADCP point measurements in Northern-1 sub-basin of Charzykowskie Lake on 2015 08 28: 1—location of boat anchorage, 2—coast line and boundary of the sub-basin

Fig. 4 ADCP RiverPro by Teledyne mounted on a surveying boat ready for measurements at Charzykowskie Lake on 2015 08 28

Flow between sub-basins Central and Northern-2 was measured 4 times, but one measurement was not recorded properly so it was neglected. The ADCP measurement results are shown in Table 4; they confirm very slow velocities obtained from CCHE2D model which creates the problem during recording them from a moving boat.

The study of Avalos-Cueva et al. (2016) with the use of ADCP for velocity measurements at Lake Chapala report that the boat speed of 1.39 m s^{-1} is practical for measurements. In our study from 3 repetitions of measurement we had taken profile No. 9003 with the smallest boat speed of 0.906 m s^{-1} which gave proper values of water vector velocity and orientation. Such a low boat velocity was possible to

Table 4 Results of ADCP measurements over the sill separating sub-basins Central and Northern-2 of Charzykowskie Lake performed on 2015 08 28

Passage No.	Width	Total area	Top Q	Total Q	Flow velocity	Boat speed
	[m]	[m^2]	[m^3 s^{-1}]	[m^3 s^{-1}]	[m s^{-1}]	[m s^{-1}]
9002	431.26	1612.34	0.026	3.216	0.005	0.974
9003	436.45	1638.29	0.858	1.475	0.013	0.906
9004	433.89	1588.13	0.152	1.634	0.004	1.025

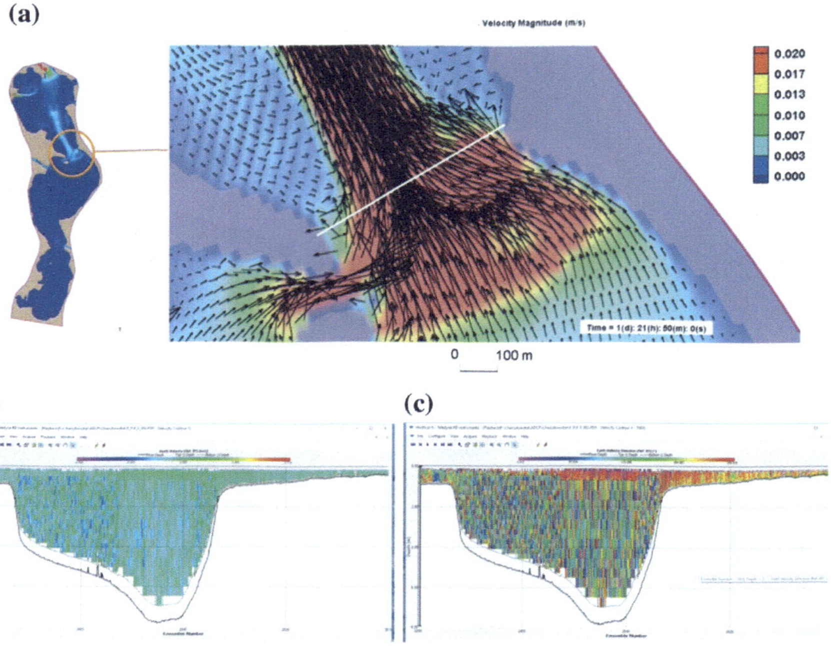

Fig. 5 Selected ADCP profile No. 9003 at sill separating sub-basins Central and Northern-2 compared to: **a** velocity field calculated by CCHE2D model, **b** velocity magnitude by ADCP measurement in geographic N direction, and **c** geographic flow directions by ADCP measurement

maintain due to a favorable weather conditions during measurement, with no wind. Comparison of the ADCP profile No. 9003 and CCHE2D model results is shown at (Fig. 5).

The ADCP measurements show a main flux with the direction to the north (Brda outflow) located in the deepest center of the cross-section and above shallow area close to eastern bank. CCHE2D model shows reverse flow near the western bank and this pattern is confirmed by ADCP measurements by lower magnitude of velocity and directions of flow toward the south. Reverse flow can be explained by the expansion of the cross-section and return of the flow near lake banks in circulation induced by the main stream. Measurements of vertical temperature profiles in the Charzykowskie Lake had shown that there is no thermal stratification of the water in Northern-1 and Northern-2 sub-basins, but thermal stratification occurs in Central and Southern sub-basins with the thermocline at 6 m. Water velocity distribution at ADCP profile do not indicate the vertical differences which probably could occur at the profile with thermal stratification.

The distance between inflow leaving the delta cone and outflow of Brda River in Northern-1 sub-basin is only 1300 m. It would be natural to expect the direct flow of river water between these two points, but results of velocity distribution in that area calculated by CCHE2D model revealed the circulation pattern of Brda River water,

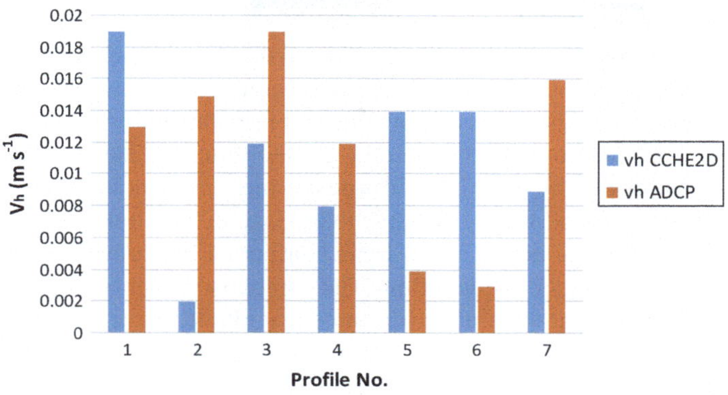

Fig. 6 Velocity in verticals—calculated by CCHE2D model and measured by ADCP in a field

heading to north and west direction. For the verification of the CCHE2D model results showing circulation pattern we designed a pattern of 8 points for velocity measurements in a section by section mode. To localize the measurement points, their geographical coordinates were given and boat was positioned on a two anchors, bow and aft. In given points, average velocity in vertical was measured. The ADCP instrument makes it also possible to measure the velocity vector direction, but in the low flow velocities we could not obtain a reliable value of current direction. The comparison of velocities calculated by CCHE2D model and measured by ADCP is shown in Fig. 6. The average difference of the performed ADCP measurements and CCHE2D results is only 0.002 m s^{-1}, which confirms the proper estimation of model calculated velocities.

3 Discussion of the Results

The Charzykowskie Lake is subject of many studies in which the problem of water exchange between sub-basins is still discussed. Comparison of river inflow and outflow indicates that there is a water flux between the sub-basins but other biological and hydro-chemical characteristics show isolation of the sub-basins (a list of papers confirming that is shown in Table 3). Performed modeling and ADCP verification of flow pattern of Charzykowskie Lake has confirmed a water flux between the sub-basins. The fact that Northern-1 and Northern-2 sub-basins are not thermally stratified confirms the existence of flows and makes it possible to apply the CCHE2D model where water properties are uniform in a whole computing domain. In a moving boat method, the ADCP instrument properly shows the magnitude of velocities and the pattern of reverse flow. This property of ADCP instrument makes it possible to use the measurement results for hydrodynamic model verification and helps to understand complex pattern of flow in a lake with separated sub-basins.

Discharge between sub-basin Central and Northern-2 measured by ADCP is 1.48 $m^3 s^{-1}$ and is higher than that calculated from the sum of discharges CW + JS = 0.17 $m^3 s^{-1}$. It has been discussed before that the current meter measurements made at the mouth profiles of the rivers do not represent fully the inflow of the water to the lake basin which is additionally supplied by ground water discharge at littoral zone. The discharge obtained by the ADCP instrument measurement between sub-basins shows the scale of ground water discharge to the lake in the littoral zone.

The CCHE2D flow modeling in sub-basin Northern-1 had proved a circulation of river stream leaving delta of Brda inflow. The comparison of modeling results with ADCP measurements shows similar velocities magnitude, but it was impossible to measure the velocity direction. The pattern of circular flow of Brda River inflow and additional flux of water from Central and Southern sub-basins explain high concentration of organic matter in the bottom sediments at Northern-1 sub-basin. Both fluxes bring nutrients in organic and mineral form and circulation pattern improves sedimentation conditions and increase the time of water residence.

Acknowledgements The study was a continuation of works conducted under the project EULAKES funded by the Central Europe Program (project No. 2CE243P3), and by the European Regional Development Fund (ERDF). The scientific work was financed by the funds for science in 2011–2013 granted for the implementation of an international co-funded Project (MNiSW). The authors would like to thank Anna Nadolna and Tomasz Szydeł from the Institute of Meteorology and Water Management—National Research Institute for their kind assistance in preparation of bathymetic data and help during the measurements.

References

Avalos-Cueva D, Filonov A, Tereshchenko I, Monzón CO, Velázquez-Muñoz FA (2016) Thermal structure and circulation in Lake Chapala. Mexico J Limnol 75(s1):137–143

Bajkiewicz-Grabowska E, Magnuszewski A, Mikulski Z (1993) Hydrometria. PWN, Warszawa

Barańczuk J, Borowiak D (2010) Jezioro Charzykowskie. In: Barańczuk J, Borowiak D (eds) Atlas jezior Zaborskiego Parku Krajobrazowego, Badania limnologiczne T. 6, Gdańsk, pp 32–41

Bhagowati B, Ahamad KU (2018) A review on lake eutrophication dynamics and recent developments in lake modeling. Hydrobiol, Ecohydrol. https://doi.org/10.1016/j.ecohyd.2018.03.002

Bresciani M, Giardino C (2012) Water quality of the lakes by remote sensing technique. In: Raports 3.3.2 developing remote sensing technique approach end 3.3.4.: implementing monitoring systems using remote sensing techniques. Project EULAKES European Lakes under environmental stressors (supporting lake governance to mitigate the impact of climate change) Ref. Nr. 2CE243P3

Choiński A, Gałka M, Ławniczak A (2006) Przykłady zróżnicowania gruboxci pokryw lodowych wybranych jezior. Badania Fizjograficzne Nad Polskx Zachodnią, Seria A—Geografia 57:15–20

Goszczyński J, Jutrowska E (1997) Stan czystości wód jeziora Charzykowskiego, Biblioteka Monitoringu Środowiska. Bydgoszcz Fizyczna, Tom 57:15–20

Hutter K (1986) Hydrodynamic modeling of lakes. In: Encyclopedia of fluid mechanics. Gulf Publishing Company, Houston, Texas. Chapter 22, pp 897–998

Jia Y, Wang SS, Xu Y (2002) Validation and application of a 2D model to channels with complex geometry. Int J Comput Eng Sci 3(01):57–71. https://doi.org/10.1142/S146587630200054X

Munara AM, Cavalcantia JR, Bravoa JM, Fana FM, da Motta-Marquesa D, Fragoso Jr.b CR (2018) Coupling large-scale hydrological and hydrodynamic modeling: toward a better comprehension of watershed-shallow lake processes. J Hydrol 564:424–441

Nowicka B (2012) Water—reaction to the changes, forecasts. In: Raport 4.4.1 impacts of climate change scenarios part C: Lake Charzykowskie project EULAKES European Lakes under environmental stressors (supporting lake governance to mitigate the impact of climate change) Ref. Nr. 2CE243P3, pp 14–29. http://www.eulakes.eu/servizi/Menu/dinamica.aspx?idArea=17680&idCat=17680&ID=17680&TipoElemento=area

Nowicka B, Dmitruk U, Gorzel M, Jancewicz A, Nadolna A, Tomczuk U, Pęczuła W (2013) Lake Charzykowskie: Heavy metals and pesticides. In: Raport 5.4.4 project EULAKES European Lakes under environmental stressors (supporting lake governance to mitigate the impact of climate change) Ref. Nr. 2CE243P3. http://www.eulakes.eu/servizi/Menu/dinamica.aspx?idArea=17680&idCat=17680&ID=17680&TipoElemento=area

Nowicka B, Bałandin M, Nadolna A (2017) Sensitivity of a large flow-through lake to meteorological condition and anthropogenic stress (hydromorphological assessment). Geogr Pol 90(4):401–415. https://doi.org/10.7163/GPol

Pęczuła W (2012) Badania hydrobiologiczne fitoplanktonu i makrofitów oraz ocena stanu Jeziora Charzykowskiego (w badanym zakresie) na tle danych literaturowych, project EULAKES (Ref. Nr. 2CE243P3) European Lakes under environmental stressors (supporting lake governance to mitigate the impact of climate change) (Ref. Nr. 2CE243P3). IMGW-PIB External expertise

Stangenberg M (1950) Morfometria i skład chemiczny jeziora Charzykowo. In: Marian Stangenberg (ed) Jezioro Charzykowo. Część I, PWR i L Warszawa, pp 9-42

Stangenberg M, Żemoytel K (1952) Skład chemiczny osadów głębinowych jeziora Charzykowo [chemical composition of abyssal sediments in Lake Charzykowo]. Biul Państw Inst Geol 68:139–172

Wiśniewska M, Luścińska M (2012) Long-term changes in the phytoplankton of Lake Charzykowskie. Ocean Hydrobiol Stud 41(3):90–98

LIDAR Data Application in the Process of Developing a Hydrodynamic Flow Model Exemplified by the Warta River Reach

Albert Malinger, Tomasz Kałuża and Tomasz Dysarz

Abstract The paper includes data derived from airborne laser scanning (LiDAR) to develop a 2-D hydrodynamic model of water flow in the middle reach of the Warta River between Nowe Miasto nad Wartą and Solec. Particular attention was paid to problems associated with the conversion of a digital terrain model (DTM) for the purpose of hydraulic modelling. LIDAR data were also used in the development process of a numerical roughness model by means of the so-called vegetation stratification. The results of model calculations in the form of maximum water surface elevations of the river were compared with water surface elevations derived from levelling carried out in the field. On this basis the verification of the model results was performed. The possibilities of using data from airborne laser scanning to develop a model and assess roughness coefficients of the floodplains were also examined.

Keywords LiDAR data · 2D hydrodynamic flow model · Roughness coefficients

1 Introduction

The application of hydraulic mathematical models is associated with the necessity of obtaining reliable data constituting the basis for developing a model (Mazur et al. 2016; Laks et al. 2013). With regard to calculations carried out with the use of a two-dimensional flow model, besides hydrological data on terrain characteristics and hydro-technical structure parameters, data on flow resistances are very important

---

A. Malinger (✉)
Institute of Meteorology and Water Management—National Research Institute, Warsaw, Poland
e-mail: albert.malinger@imgw.pl

T. Kałuża · T. Dysarz
Department of Hydraulic and Sanitary Engineering, Poznan University of Life Sciences, Poznań, Poland
e-mail: tomasz.kaluza99@gmail.com

T. Dysarz
e-mail: dysarz@up.poznan.pl

M. B. Kalinowska et al. (eds.), *Recent Trends in Environmental Hydraulics*, GeoPlanet: Earth and Planetary Sciences,
https://doi.org/10.1007/978-3-030-37105-0_14

(Walczak et al. 2018; Kałuża et al. 2012). In practice, even a few years ago, in order to characterise the resistances caused by the presence of vegetation in floodplains in hydraulic mathematical models (without main channel), researchers used data derived from the CORINE inventory (Eng. Coordination of Information on the Environment) containing information on land cover forms (Excimap 2007; Pestena et al. 2013) or orthophotomap interpretation. The establishment of the Topographic Data Base, currently the Data Base of Topographic Objects (BDOT10k) with the detail corresponding to a 1:10 000 topographic map has enabled the use of appropriate classes of land use. The land cover forms identified in this way can be assigned the appropriate values of Manning roughness coefficients on the basis of Chow (1959) tables.

The launch of the ISOK project, i.e.,"Informatyczny System Osłony Kraju przed nadzwyczajnymi zagrożeniami" [Eng. Information System for Protecting the Country against extraordinary threats] gave an impulse to start works related to obtaining data through airborne laser scanning. Availability, range and quality of these data mean that they can be applied in a multi-aspect manner. Apart from information about the relief, they can also be used to recognize the forms of land use and land cover as well as enable the assessment of roughness parameters of vegetation located in flood risk zones. The aim of the study was an attempt to use the information from airborne laser scanning to develop a reliable, yet simple, method of obtaining data characterising the resistances caused by the presence of vegetation during high water flows in the examined reach of the Warta River.

2 Materials

2.1 Study Site Location

The study was carried out in the reach of the middle Warta River course. Its location is shown in Fig. 1.

The section is located between Nowe Miasto nad Wartą and Solec, i.e. covering the Warta River from km 322 + 700 km to km 314 + 200. In the upper part of this reach there is a road bridge consisting of six spans with a total length of approx. 230 m; the lower part has a railway bridge in Solec (approx. km 316 + 800), consisting of 10 spans with a total length of approx. 340 m. On the right bank of the entire reach there is a flood embankment that intersects the ice-marginal valley of the Warta River and protects buildings in Solec and Krzykosy or other localities against the inflow of high water. On the left bank, in the upper part of the river reach, there is a natural high valley bank, to which the flood embankment protecting Rogusko town is connected in the middle part of the section. This section is characterised by floodplain vegetation and land use regime that are typical for the entire middle section of the Warta River.

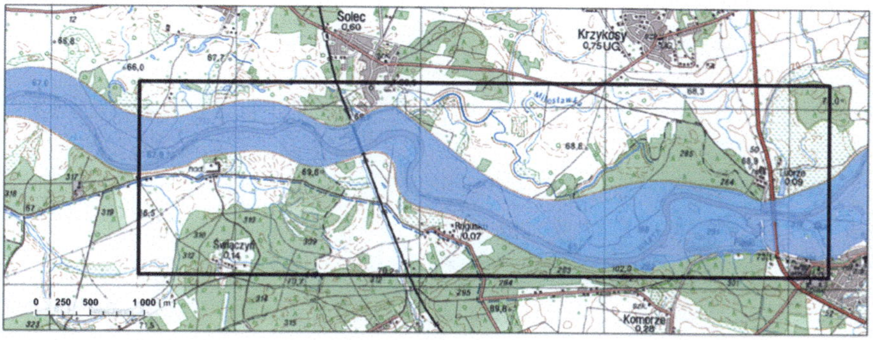

Fig. 1 Study site location of the Warta River reach

2.2 Geospatial Data

The study used geospatial data obtained with the use of various techniques. As part of the analyses, airborne LiDAR data (ALS) were used in the form of a classified cloud of points derived from the resources of Główny Urząd Geodezji i Kartografii [Eng. Central Office of Geodesy and Cartography]. Measurements related to airborne laser scanning were performed in the winter-spring period in 2011. In addition, within the aforementioned research reach, terrain data were used, the source of which was a digital terrain model in a raster format with a resolution of 1 m, developed on the basis of LiDAR data (6 points/m^2). An additional source of land cover information was the BDOT10k database with accuracy and detail corresponding to a 1:10,000 topographic map. From nine categories of topographic object classes included in BDOT10k, there was used information on land cover (PT) belonging to the third classification level in the scope of flowing water (PTWP02), stagnant water (PTWP03) and crops on agricultural land (PTTR02). In addition, the study used information obtained on the basis of the interpretation of orthophotomaps.

3 Methodology

3.1 Development of the Hybrid Model

For the purpose of the work, a hybrid model was constructed (Fig. 2a), which included the reach of the Warta river from the stream gauge in Nowa Wieś Podgórna at km 341 + 500 to km 306 + 000, i.e. 11 km below the railway bridge in Solec—called W1 area. The model was built on cross-section measurements. Geodetic cross-sections were measured in such a way that the distances between them were not greater than approx. 1000 m. In the process of model development, these data were verified and processed to eliminate errors and remove inactive parts of cross-sections. The data

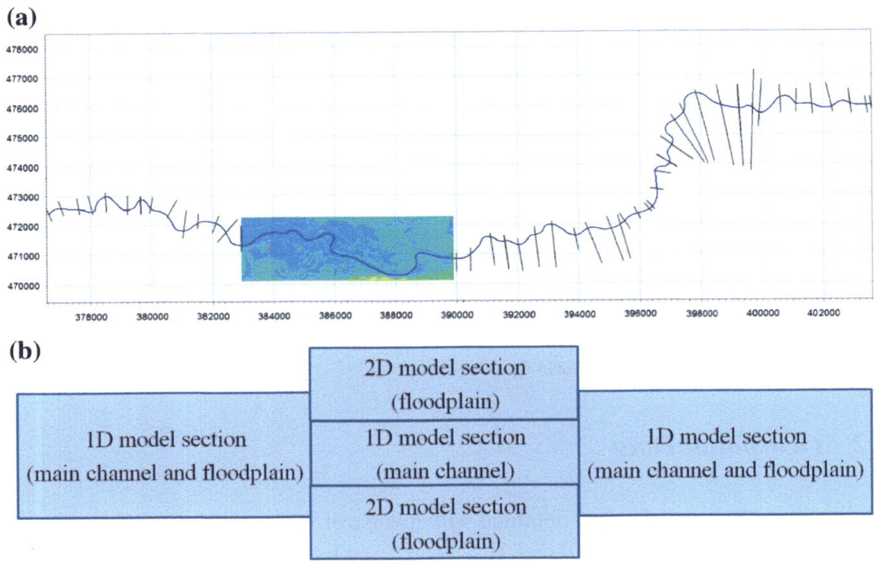

Fig. 2 1D and 2D coupling scheme modelling in the W1 reach **a**, schemes for hybrid model construction **b**

were implemented in a way that allowed for defining 1D model sections and reach with calculations performed using a two-dimensional model (Bakuła 2014). For this purpose, the range of 1D model between the area of two-dimensional sections was limited to the upper edge of scarps and the floodplains in the 1D model were removed (Fig. 2b).

Further, in accordance with the adopted principles for developing hybrid models, river-bed cross-sections in the 1D modelling section (only middle section, between 2D sections on the right and left side of floodplain) were compacted so that the distances between cross-sections were approx. 250 m. Additional cross-sections were made based on the interpolation algorithm included in MIKE11 (DHI 2011a).

3.2 Processing of the Digital Terrain Model

The applied 2D hydraulic mathematical model was constructed on a single grid. In order to define a DTM in a mathematical model, the first step was a detailed determination of spatial extent. It was emphasised that the area would cover the entire floodplain, located between the main channel and the flood embankment or the high embankment of the river valley. The W1 area was characterised with 2100 and 7500 m long sides that resulted in a field area of 15.750 km^2. Implementation of a DTM with a resolution of 1 m in a hydrodynamic mathematical model allows for including relatively detailed information about the relief; unfortunately, the calculation process itself might be quite time-consuming (Haile 2005; Haile and Rientjes 2005; Sanders 2007; Asselman 2009; Dasgupta 2015). Therefo'relatively, the search has been undertaken to find solutions allowing for reducing the DTM resolution to one that would not cause a significant quality loss, and at the same time would enable model calculations to be carried out within a reasonable time span (e.g. over a few or several days). It was planned that the conversion process of the DTM with a resolution of 1 m would be carried out using the most popular and simplest method, i.e. resampling of a regular square mesh with the use of the nearest neighbour method. The results presented in the Bakuła's work (2014) prompted a statistical analysis to determine the vertical accuracy of a converted digital terrain model. In assumptions, the results of this analysis were to be helpful when choosing the resolution of converted DTM. All analyses related to DTM processing were made using ArcGIS 10.0 by ESRI. The following indicators were used to assess the quality of converted DTMs (Bakuła 2014):

- Root Mean Square Error (*RMSE*):

$$RMSE = \sqrt{\frac{\sum_{i=1}^{n} \Delta h_i^2}{n_L}} \tag{1}$$

- correlation coefficient R_d:

$$R_d(x, y) = \frac{\sum_{i=1}^{n}\left(h_{x_i} - \bar{h}_{x_i}\right)\left(h_{y_i} - \bar{h}_{y_i}\right)}{\sqrt{\sum_{i=1}^{n}\left(h_{x_i} - \bar{h}_{x_i}\right)^2}\sqrt{\sum_{i=1}^{n}\left(h_{y_i} - \bar{h}_{y_i}\right)^2}} \tag{2}$$

where:

Δh_i —difference in height between the reduced and original model [m];
n_L —number of nodes (points) of the terrain model;
h_{x_i} —node height of the reference DTM [m];
h_{y_i} —node height of the converted DTM [m];
$\bar{h}_{x_i}, \bar{h}_{y_i}$ —average values of height [m].

In addition, the formula took into account data on extreme values, i.e. the minimum (min_h) and maximum (max_h) values of height difference and reduction factor expressed by including the percentage of observations forming part of the input data set. The original digital terrain model with a resolution of 1 m was converted into DTMs with a resolution of 2, 3, 4, 5 or 10 m. The most important result data, including the differences in the values of original digital terrain model and the values obtained after converting the DTM for the research section, are summarized in Table 1.

The analysis shows that as the DTM resolution is reduced, Root Mean Square Error (*RMSE*) increases. After converting the original DTM into a resolution of 2 m, *RMSE* was 8 cm, and in the case of the DTM with a resolution of 3 m through 9.5 cm, thus confirming Bakuła's (2014) observations. Converting the DTM into a resolution

Table 1 Comparison results of point heights for the DTM subjected to resampling by the nearest neighbour method relative to the original DTM

Parameter	GRID 2 m	GRID 3 m	GRID 4 m	GRID 5 m	GRID 10 m
Reduction coefficient r [%]	approx. 25%	approx. 11%	approx. 6%	approx. 4%	approx. 1%
$RMSE$ [m]	0.0799	0.0949	0.1259	0.1421	0.2356
min_h [m]	−3.57	−2.49	−3.09	−3.7	−3.85
max_h [m]	2.34	3.21	2.77	2.83	3.4
Correlation coefficient R_d [−]	0.987	0.994	0.978	0.974	0.943

of 10 m consequently led to the highest value of average $RMSE = 23$ cm. The effect of levelling of the terrain model was also an increase in the minimum and maximum values. It was different in the case of the DTM converted into a resolution of 3 m, for which the minimum value was lower than for the DTM with a resolution of 2 and 4 m, whereas the maximum value was higher than the maximum values for the DTMs with a resolution of 2 and 4 m.

When compared, the results of the analysis in Table 1 show that the highest correlation coefficient ($R_d \approx 0.99$) for the analysed area was achieved for the DTM with a resolution of 3 m. The lowest value of the correlation coefficient was achieved for the DTM with a resolution of 10 m. Based on the conducted analysis, the preliminary assumptions resulting from Bakuła's work (2014) regarding the propriety of selecting the DTM with a resolution of 3 m were confirmed as optimal for hydrodynamic mathematical model calculations.

3.3 Boundary Conditions of the Model

The upper boundary condition was a flow hydrograph predetermined on the basis of registered water levels at the stream gauge located in Nowa Wieś Podgórna during the flooding in 2010, whereas the lower boundary condition was the rating curve. In addition, the initial conditions of the model were determined by proper definition of the flow data and water surface elevations in the longitudinal profile. These parameters were chosen so that the model water surface at the moment t_0 was located in the main river channel. Simultaneously, the care was taken that the water surface elevations at the moment t_0 corresponded to the situation preceding the flow rate resulting from properly defined boundary conditions. To ensure the correct water exchange between 1D and 2D models in MIKE FLOOD software, standard and lateral links were applied, in accordance with the guidelines provided in the user manual (DHI 2011b).

3.4 Flow Resistances

In order to determine the values of roughness coefficients, the information on vegetation on the basis of LiDAR data was used. Conceptually, from the modelling point of view, this approach was to contribute to reliable and optimal identification of land cover forms. The source of data was LiDAR products, with LiDAR launched in mid-February 2011, i.e. less than 8 months after the flooding which occurred in 2010. The data from BDOT10k and the information obtained based on orthophotomap interpretation were used as supplement sources.

In the first stage of work, on the basis of orthophotomap analysing, several types of land cover were distinguished and verified during field visits (Tyminski and Kaluża 2012). It was assumed that the most common forms included: high forest, second-growth forest, wasteland and arable land, as well as, less frequently, small water reservoirs, most often resulting from river meandering. In the second stage of the work, a spatial analysis was carried out using GIS tools. By using DTM and DSM (digital surface model) data, there were obtained averaged heights of vegetation in the considered research section. The values were assigned to single pixels with dimensions 1 m × 1 m, creating a new thematic raster (Fig. 3a). Due to the diversity of cover found in the analysed areas, three floors of vegetation were distinguished,

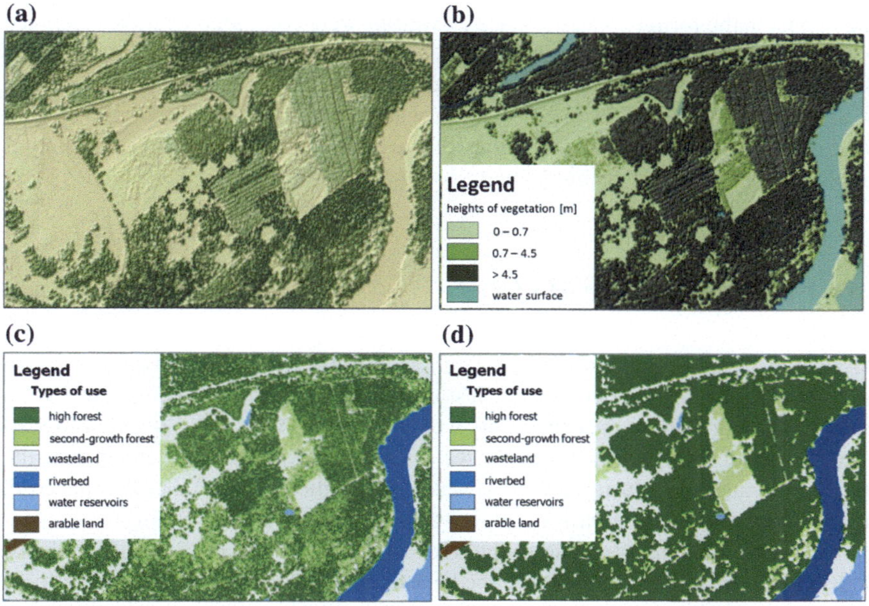

Fig. 3 Fragment showing DSM with an original resolution **a**, DSM with an original resolution including vegetation stratification **b**, DSM with an original resolution including vegetation stratification and BDOT10k data **c**, converted DSM into 3 m × 3 m resolution including vegetation stratification and BDOT10k data **d**

i.e., low, medium and high. This approach allowed for defining data (in relation to a single pixel) regarding the occurrence of "low vegetation" i.e. not exceeding 0.7 m and representing meadow or wasteland, "medium vegetation" i.e. not exceeding 4.5 m in height and representing second-growth forest, as well as "high vegetation" i.e. with a height of over 4.5 m and representing high forest (Fig. 3b).

Additionally, based on the information included in BDOT10k and on the basis of orthophotomaps, there were identified the areas corresponding to arable land, ponds/small water reservoirs and the river channel. This procedure allowed for obtaining a raster layer measuring 1 m × 1 m that defined the spatial location of three floors of vegetation, arable land, ponds/small water reservoirs, and the riverbed (Fig. 3c).

In order to determine the values of roughness coefficients, there were applied tabulated roughness coefficients published by Chow (1959). The study initially assumed the following values of roughness coefficients: high forest 0.045 ($m^{-1/3}$ s), second-growth forest 0.050 ($m^{-1/3}$ s), wasteland 0.040 ($m^{-1/3}$ s), arable land 0.038 ($m^{-1/3}$ s), water reservoirs 0.035 ($m^{-1/3}$ s), riverbed 0.030 ($m^{-1/3}$ s).

As part of the final stage of work, the data were implemented into mathematical models. The procedure consisted in transforming the raster layer to a field resolution of 3 m (Fig. 3d). This was necessary due to the use of a generalized digital terrain model with a single pixel of 3 m × 3 m.

The adopted roughness coefficients were included in preliminary calculations. The results in the form of water surface elevations were compared with the reference data. Based on the preliminary calculations, it was found out that the maximum values of water level were lower than the reference water surface elevation (data obtained on the basis of levelling of traces of high water from the 2010 flooding). Further work consisted in correcting the values of roughness coefficients and carrying out supplementary calculations until satisfactory results were obtained.

3.5 *Calibration of the Model*

Due to the fact that the analysed research section (between the railway bridge in Solec and the road bridge in Nowe Miasto nad Wartą) lacked a water gauge station and regular hydrometric measurements were not carried out, it was decided to calibrate the model taking into account the water surface elevations corresponding to the maximum water level during the flood wave culmination in 2010. For this purpose, the marks left on the tree bark by high water during the 2010 flooding were measured with special attention paid to the trees located in the Warta River floodplains. High water marks were best seen on the bark of deciduous trees. Additionally, there were measured places of high water occurrence during the culmination indicated by residents living in adjacent areas.

The operation of measuring the marks left by high water consisted in determining the elevations of several points selected over the entire length of the tested research reach. The elevation of points specifying the marks of high water was determined by means of two independent GNSS RTN receivers. Photographic documentation taken during the flight on 1 June 2010 approx. at 1:00 pm provided additional information on the floodplain area. The photos were taken less than three days after the flow culmination recorded in the water gauge located in Nowa Wieś Podgórna. During the photographic session, the water gauge in Nowa Wieś Podgórna registered the state of 611 cm, i.e., 24 cm lower than in the course of the culmination. Due to the fact that for the analysed research reach the data regarding a full course of the hydrogram with high water levels in 2010 flooding were not available, it was decided to supplement the collected information with data on water surface elevations from the period of low water levels. These measurements were carried out in five locations in August 2015.

Considering the above, the calibration process was divided into two stages. In the first stage, a series of calculations including low water flows from 2015 was carried out. Subsequent calculations were preceded by an appropriate correction of roughness coefficients specifying the flow resistances in the main channel, until the highest possible congruence of elevations measured in the field with elevations obtained as a result of model calculations was achieved. The second stage covered another series of calculations taking into account the flood wave hydrogram from 2010. Consequently, these calculations were repeated until the highest congruence of maximum elevations obtained as a result of model calculations with elevations of high water marks measured in the field was achieved. The calculations were always preceded by an appropriate correction of roughness coefficients characterising the flow resistances in the floodplain. In any case, before starting the next calculation process, the values of roughness coefficient were corrected taking into account the division due to the type of land cover.

In the process of model calibration, there were used methods of model quality classification. They allow for determining the correctness and accuracy of models with such parameters as: correlation coefficient R calculated for water table ordinates and flow rates, special correlation coefficient R_s, total square error TSE, culmination level error ΔH_{max} (DHI 2011a).

4 Calibration Results of the Hydraulic Mathematical Model

Implementation of the predefined values of roughness coefficients in the hydraulic mathematical model in accordance with the adopted principles enabled to carry out a series of model simulations. The calculations included the values of low water flows measured in August 2015. Table 2 includes the initially adopted values of Manning roughness coefficients and corrected values obtained after a series of model simulations, in the process of model calibration.

The table shows that the initially adopted values of roughness coefficients increased, falling within the range provided in the Ven Te Chow table. The biggest

Table 2 Comparison results for point heights for the DTM subject to resampling by the nearest neighbour method relative to the original DTM

Type of land cover	Preliminary values of manning roughness coefficients [$m^{-1/3}$ s]	Corrected values of manning roughness coefficients for the research section [$m^{-1/3}$ s]
High forest	0.040	0.060
Second-growth forest	0.050	0.064
Wasteland	0.040	0.056
Arable land	0.038	0.050
Water reservoirs	0.035	0.035
Riverbed	0.030	0.033

difference was in wasteland and amounted to 0.016 $m^{-1/3}$ s. Slightly smaller differences were recorded for high forest, i.e. 0.015 $m^{-1/3}$ s and for second-growth forest 0.014 $m^{-1/3}$ s as well as for arable land 0.012 $m^{-1/3}$ s. The lowest correction, however, was necessary for the value of roughness coefficient in the main channel, i.e. 0.003 $m^{-1/3}$ s.

The correction of roughness coefficients implemented in hydraulic mathematical models allowed for obtaining result elevations close to elevations obtained from measuring the water surface from the period of low water levels. These data are presented in Table 3, indicating the differences between elevations measured in the field and elevations resulting from model simulations. As can be seen, the biggest difference between water surface elevations measured in the field and water surface elevations being the result of mathematical modelling occurred in the upper part of the research reach, approaching 8 cm. The lower part of the reach was characterised by full convergence of elevations.

Finally, taking into account the elevations listed in Table 3, there were obtained the values of correlation coefficient $R = 0.998$, special correlation coefficient $R_s = 1$ and total square error $TSE = 2.78$, thus acknowledging the achieved simulation results as satisfactory. In the course of subsequent model simulations, there were obtained the data on maximum water surface elevations corresponding to elevations obtained

Table 3 Comparison of water surface elevations measured and water surface elevations resulting from mathematical modelling for the low water level in August 2015—the W1 reach

Measuring point number	Water surface elevations measured in the field [m a.s.l.]	Water surface elevations calculated [m a.s.l.]	Difference between water surface elevations [m]
1.	64.96	64.88	0.08
2.	64.78	64.74	0.04
3.	64.62	64.61	0.01
4.	64.46	64.46	0
5.	64.35	64.35	0

Table 4 Comparison of water surface elevations measured and water surface elevations resulting from mathematical modelling for the flood wave in 2010—the W1 reach

Measuring point number	Water surface elevations measured in the field [m a.s.l.]	Water surface elevations calculated [m a.s.l.]	Difference between water surface elevations [m]
1	69.82	69.82	0.01
2	69.78	69.81	0.03
3	69.76	69.80	0.03
4	69.45	69.46	0.01
5	69.40	69.37	0.03
6	69.13	69.13	0.00
7	69.03	69.00	0.03
8	68.97	68.95	0.02
9	68.96	68.91	0.05
10	68.86	68.82	0.04

from measuring the high water marks (Table 4). As can be seen, the biggest difference between water surface elevations measured in the field and water surface elevations being the result of mathematical modelling occurred in the lower part of the research reach, approaching 5 cm. These differences were slightly smaller in the upper part of the reach, with the maximum difference of 3 cm. Finally, taking into account the elevations listed in Table 4, the following values were obtained: correlation coefficient $R = 0.999$, special correlation coefficient $R_s = 1$ and total square error $TSE = 1.36$, thus considering the achieved simulation results as satisfactory.

The analysed research section lacked a water gauge station and regular hydrometric measurements were not carried out. This is why, in the proposed application the calibrated roughness could be different from the ones measured in medium flow conditions. However, this method can be especially useful for analysis of floodwater conditions.

5 Conclusions

The aim of the conducted study was to assess the possibility of using LIDAR data to construct a hydrodynamic mathematical model and to develop a reliable, yet simple method to obtain data specifying the values of resistances caused by the presence of vegetation in floodplains.

The results of the analysis allowed for concluding that due to the reduced resolution of original DTM from 1 m to 3 m, the average Root Mean Square Error (*RMSE*) was 0.08 m, i.e. just 1 cm more than when the resolution was reduced to 2 m. Simultaneously, the results achieved showed that after reducing the resolution to 3 m in the analysed research section, the correlation coefficient represented a higher value and amounted to 0.99. This aspect of the research suggested that the conversion of the DTM into a resolution of 3 m provided the acceptable parameters of converted DTM. The size of a single computing cell of 3 m × 3 m also appeared to be approvable due to the calculation time, which was approx. three weeks.

Furthermore, the results achieved on the basis of field studies and mathematical flow model calculations confirmed that DTM and DSM data obtained by means of airborne laser scanning might be used for a simplified land cover analysis that supports the assessment of floodplain capacity. Therefore, the applied method, i.e. vegetation stratification, can be an attractive alternative or complement to previously used BDOT10k or CORINE data.

References

Asselman N (2009) FloodSite. Flood inundation modelling—executive summary

Bakuła K (2014) Efektywne wykorzystanie danych LiDAR w dwuwymiarowym modelowaniu hydraulicznym [Effective use of LiDAR data in two-dimensional hydraulic modelling]. Archiwum Fotogrametrii, Kartografii i Teledetekcji 26:23–37. ISSN 2083-2214, eISSN 2391-9477. https://doi.org/10.14681/afkit.2014.002

Chow VT (1959) Open-channel hydraulics. McGraw-Hill, Inc

Dasgupta A (2015) Reduction of uncertainties in a 1D2D coupled hydrodynamic model using remote sensing data. Enschede, the Netherlands

DHI (2011a) User guide. A modelling system for rivers and channels

DHI (2011b) Reference manual. A modelling system for rivers and channels

EXCIMAP—European Exchange Circle on Flood Mapping (2007) Handbook on good practice on flood mapping in Europe. Ministry of Transport, Public Works and Water Management, The Netherlands

Haile AT (2005) Integrating hydrodynamic models and high resolution DEM (LIDAR) for flood modelling. University of Twente. www.itc.nl/library/papers_2005/msc/wrem/haile.pdf

Haile AT, Rientjes THM (2005) Effects of LIDAR DEM resolution in flood: a model sensitivity study for the city of Tegucigalpa, Honduras. In: ISPRS WG III/3, III/4, V/3 workshop "laser scanning 2005", Enschede, The Netherlands, pp 168–173

Kałuża T, Tymków P, Strzeliński P (2012) Use of remote sensing for investigating riparian shrub structures. Pol J Environ Stud 21(1):115–122

Laks I, Kałuża T, Sojka M, Walczak Z, Wróżyński R (2013) Problems with modelling water distribution in open channels with hydraulic engineering structures. Rocznik Ochrona Środowiska, 15(1):245–257

Mazur R, Kałuża T, Chmist J, Walczak N, Laks I, Strzeliński P (2016) Influence of deposition of fine plant debris in river floodplain shrubs on flood flow conditions—the Warta River case study. Phys Chem Earth, Parts A/B/C 94:106–113

Pestena R, Matias M, Canelas R, Roque D, Araujo M, Van Zeller E, Trigo-Teixeira A, Ferreira R, Proença de Oliveira R, Heleno S (2013) Calibration of 2D hydraulic inundation models in the floodplain region of the lower Tagus River. In: 2013 ESASP.722E.90P

Sanders BF (2007) Evaluation of on-line DEMs for flood inundation modeling. Adv Water Resour 30(8):1831–1843

Tyminski T, Kaluza T (2012). Investigation of mechanical properties and flow resistance of flexible riverbank vegetation. Pol J Environ Stud 21(1):201–207

Walczak N, Walczak Z, Kałuża T, Hämmerling M, Stachowski P (2018) The impact of shrubby floodplain vegetation growth on the discharge capacity of river valleys. Water 2018(10):556

On Habitat Complexity in Streams Derived from the Analysis of Tracer Data

J. Russell Manson, Steve G. Wallis, Artur Radecki-Pawlik, Karol Plesinski and Peggy Zinke

Abstract The habitat complexity of streams contributes to their biotic diversity and to their overall stream metabolism. Habitat complexity is manifested in several ways. Here we are interested specifically in physical complexity inherent in geomorphologic structures and non-uniform bed substrate at multiple spatial scales; essentially, we are considering the numerous pockets and interstices in the stream caused by both abiotic (stones, pebbles and rocks) and biotic (macro-phytes, organic debris) material. This paper explores the relationship between the dispersive fraction parameter of the Aggregated Dead Zone Model (as a surrogate for physical habitat complexity) and flow rate for streams. Results were obtained by fitting the model to a set of tracer data from several streams in Norway, the United Kingdom and Poland. The results show that while there was no apparent correlation between the dispersive fraction and flow rate, the variance in possible values for the dispersive fraction increases with decreasing flow rate. These results suggest possible estimates of this parameter but crucially also suggest uncertainties to be ascribed along with this estimate. They also suggest that as streams decrease in size the variance in physical habitat complexity increases.

Keywords Streams · Habitat · Aggregated dead zone · Tracer

J. R. Manson (✉)
Stockton University, Galloway, NJ, USA
e-mail: russell.manson@stockton.edu

S. G. Wallis
Heriot-Watt University, Edinburgh, Scotland, UK

A. Radecki-Pawlik
Cracow University of Technology, Krakow, Poland

K. Plesinski
University of Agriculture in Krakow, Krakow, Poland

P. Zinke
Norwegian University of Science and Technology, Trondheim, Norway

M. B. Kalinowska et al. (eds.), *Recent Trends in Environmental Hydraulics*, GeoPlanet: Earth and Planetary Sciences,
https://doi.org/10.1007/978-3-030-37105-0_15

1 Introduction

Stream ecologists have shown that the habitat complexity of streams contributes to their biotic diversity and consequently their whole stream metabolism (Demars et al. 2011; Besemer et al. 2009; Besemer et al. 2013). Habitat complexity is created in several ways. Here we are interested specifically in physical complexity inherent in geomorphologic structures and non-uniform bed substrate at multiple spatial scales; essentially, we are considering the numerous pockets and interstices in the stream caused by both abiotic (stones, pebbles and rocks) and biotic (macro-phytes, organic debris) material. During the 1970s such locations had been referred to as "dead zones" in the water resources literature, e.g. Valentine and Wood (1979).

Despite recent advances in terrestrial laser scanning techniques it is still impossible to completely map all these pockets and interstices and hence determine their actual volume. An alternative "holistic" way of quantifying this information is through tracer studies wherein a known mass of solute is introduced to the stream and the shapes of resulting temporal tracer concentration profiles are analysed to determine the ensemble average physical response of the stream. The analysis consists, usually, of fitting a simplified model of the stream solute transport processes to determine the model parameters which are indicative of the physical characteristics of the stream. Several models have been suggested with differing numbers of parameters and various degrees of physical realism (Rutherford 1994).

Herein we use the aggregated dead zone model with the aims of quantifying its dispersive fraction in essentially small streams over a range of flow rates and stream geomorphologies.

2 The Aggregated Dead Zone Model (ADZM)

The ADZM was developed because of reported deficiencies of the previously widely used advection-dispersion model (ADM). In a radical departure from previous work, it was postulated (Beer and Young 1983) that the dispersion occurring in dead zones dominated the shear flow dispersion. Hence a model could be constructed on the basis of a temporary trapping mechanism, known as transient storage, only or more pragmatically, the effects of all dispersive mechanisms in a river reach could be amalgamated and represented by a single effective dead zone. A further radical approach was to formulate the model only in the time domain (by spatial integration of the physical processes). This led to a much simpler mass balance equation than the ADM and access to powerful model calibration techniques via time-series analysis (Young 1984).

The model resembles simple hydrologic models for flood propagation in rivers, many of which are based on the concept of storage routing (Young and Wallis 1985; Shaw et al. 2011). Importantly, however, the ADZM includes an explicit time delay to cater for the purely advective transport processes. The transport of a conservative solute in steady flow is modeled by the following equation:

$$\frac{\partial c(t)}{\partial t} = \frac{1}{\bar{t} - \tau}[c_u(t - \tau) - c(t)] \tag{1}$$

where *c(t)* and $C_u(t - \tau)$ are the cross-sectional average solute concentrations at the downstream and upstream ends, respectively, of a river reach and t is time. The model parameters are τ, the time delay and $\bar{t}$, the total travel time. In practical terms: τ is the minimum reach travel time (time interval between the first arrival of solute at the two ends of the reach); $\bar{t} - \tau$, is the time delay caused by the presence of the dead zones, often termed the residence time. While technically this is the time the solute spends in the "aggregated dead zone" as defined in the model, one needs to remember how the ADZM is conceptualised and so it is not equivalent to the time the solute spends in the actual dead zones. None-the-less it should scale with the size of the actual dead zones. Wallis (1994) showed that the ADZM total travel time is theoretically equal to the time interval between the centroids of temporal solute concentration profiles at the two ends of a reach. The dispersive fraction, defined as $(\bar{t} - \tau)/\bar{t}$, quantifies the dispersion occurring in a reach and is defined by either the ratio of the residence time to the total travel time or as the ratio of the volume of the aggregated dead zone to the total volume of water in the reach.

Several studies have shown that the ADZM is able to reproduce solute transport in rivers very satisfactorily (Wallis et al. 1989; Green et al. 1994; Lees et al. 2000). However, the fact that the model's parameters are not related to the traditional concepts of advection and dispersion may be considered a disadvantage (Rutherford 1994). On the other hand, the time delay and the residence time have been found to vary with flow rate in a physically realistic way in single reach studies, decreasing as flow rate increases (Wallis et al. 1989; Green et al. 1994). It has also been suggested that in any given reach the dispersive fraction remains broadly constant with discharge (Wallis et al. 1989; Guymer 2002).

3 Data Collection

Data from 51 tracer studies were used in the current study. The data comes from three countries (UK, Norway and Poland) and are summarized in Table 1 which shows the ranges of flow rates and stream widths. As seen from Table 1, the streams are small (largest width is 10.5 m) and the largest flow rate is about 3 m^3s^{-1}.

Table 1 Summary of major stream characteristics

Location	Number of datasets	Flow (Ls^{-1})	Width (m)
UK	40	30–3118	0.75–10.5
Norway	9	12–1163	0.08–8.0
Poland	2	0.86–50	2–2.1

Table 2 presents the physical characteristics of the individual reaches and shows that they vary widely between man-made concrete channels (Ou Beck) and boulder strewn natural upland streams (Ousdal).

For each experiment the tracer data was obtained in essentially the same way. A known mass of conservative substance (either Rhodamine-WT or salt) was dissolved in a small volume of stream water. The tracer was then introduced to the stream either instantaneously near the stream centre or over about 30 s as a laterally distributed source. Rhodamine-WT was used for the UK streams; salt was used for the Norwegian and Polish streams. Two longitudinal stations (set apart by a short distance) were chosen at which the concentration of the tracer was sampled at a fixed time interval (typically 1–15 s). The initial mixing zone was sufficiently long (typically about 20–30 stream widths) for complete cross-sectional mixing to take place before the upper sampling station thus point measurements of concentrations

Table 2 Physical characteristics of individual reaches

River	Width (m)	Substrate	Note on geomorphology
Beinhellervatnet (Norway)	4	Gravel to cobble, in pools also sand	Meandering: riffle-pool structures
	1.2	Pebbles to cobbles, in pools also gravel	Step-pool structures
	0.08	Pebbles, cobbles	Stream in peat grassland
Kvitestakkelvi (Norway)	5	Boulders and stones	Human-affected fan stream: central thalweg
Skreåna (Norway)	8	Boulders, stones, gravel	Boulder-strewn channel: flow around boulders
Ousdal (Norway)	4	Boulders, stones	Boulder-strewn channel: flow within and around boulders
Mszana Dolna (Poland)	2	Stones, gravel: extensive macrophytes	Straight engineered channel but unmaintained
Brock (UK)	8.5	Medium cobbles	Straight with one major pool-riffle structure
Conder (UK)	5	Fine cobbles	Straight with one major pool-riffle structure
Dunsop (UK)	10.5	Coarse cobbles	Sinuous channel with one major and other minor pol-riffle structures
Ou Beck (UK)	0.75	Concrete channel	Straight engineered channel but unmaintained
Murray Burn (UK)	2.4	Cobbles	Straight channel created by landscape management

are considered representative of the cross-section. For the Rhodamine-WT experiments, either river water was pumped through a calibrated field fluorometer and logged on a microcomputer (Wallis et al. 1987) or water samples were collected and analysed using a calibrated fluorometer in the laboratory (Heron 2015). For the salt experiments, conductivity was recorded and logged in situ which was converted to a salt concentration via a calibration curve. Any naturally occurring background signal was subtracted from the observations prior to the modelling and linear scaling was used to ensure that the areas under upstream and downstream profiles were the same. This removes the effect of any lateral inflows or non-conservative behaviour of the tracer. The flow rate for each experiment was evaluated via dilution gauging, and the upstream and downstream temporal tracer profiles were analysed to elucidate stream transport parameters, as described below.

4 Application of the Model

Equation (1) was solved using a two-level, time-centered finite difference scheme. The rationale behind this choice is that the method is unconditionally stable and robust, and second order accurate in time. Wallis (2007) found that time centered numerical integration for such an equation, albeit formulated differently using linear dynamical systems theory, was a satisfactory approach. The model was fitted to the observed downstream tracer curves by parameter optimization, minimising the sum of squared residuals, SSR, by a modified Levenberg-Marquardt algorithm (Press et al. 1992). A Python script was used to undertake the minimisation by calling a C code implementation of the model. Python and C were linked using the BOOST library. A normalised fitting parameter (NRMSE) was defined for comparing the performance of the model to different data sets, defined by:

$$\mathrm{NRMSE} = \frac{1}{\max(C_{\mathrm{OBS}})}\left(\frac{\mathrm{SSR}}{N}\right)^{0.5} \tag{2}$$

where N is the number of data points and C_{OBS} refers to the observed concentration profile. Since our minimum concentration is always zero (background removed) this is equivalent to a standard practice of normalizing with respect to the range of values.

5 Results and Discussion

The model was successfully optimized to the 51 sets of tracer data. Results are presented below. Figure 1 shows the typical fit of the successfully optimized model to one of the tracer studies (Mszana Dolna, Poland).

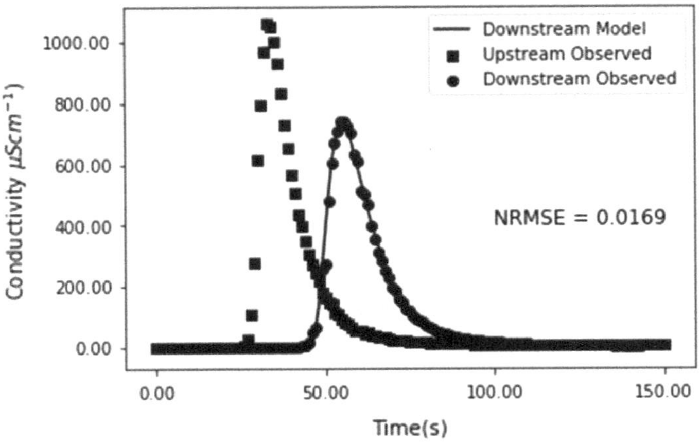

Fig. 1 A typical tracer modeling result showing goodness of fit

Figure 2 shows the dispersive fraction computed for all tracer studies plotted against flow rate. It shows no clear relationship but there is a greater variance in dispersive fraction values in the smaller streams, i.e. there is a large scatter in the data, particularly at flows <0.2 m^3s^{-1}. We postulate that this large variance is due to the influence of wide variations in the channel geomorphology having great influence on the dispersive fraction in low flow regimes; at larger flows such variations in geomorphological detail would tend to be "drowned" out. To further explore this, we have separated the results (albeit arbitrarily) into those for flows less and greater than 1 m^3s^{-1}. These results are plotted as boxplots in Fig. 3 and indicate the much

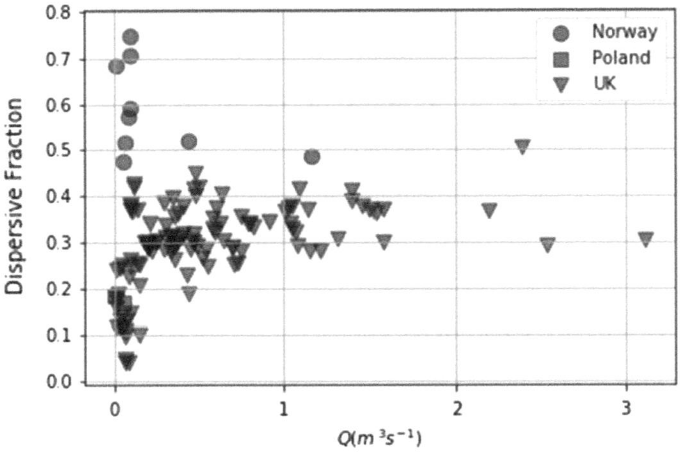

Fig. 2 Dispersive fraction versus flow rate

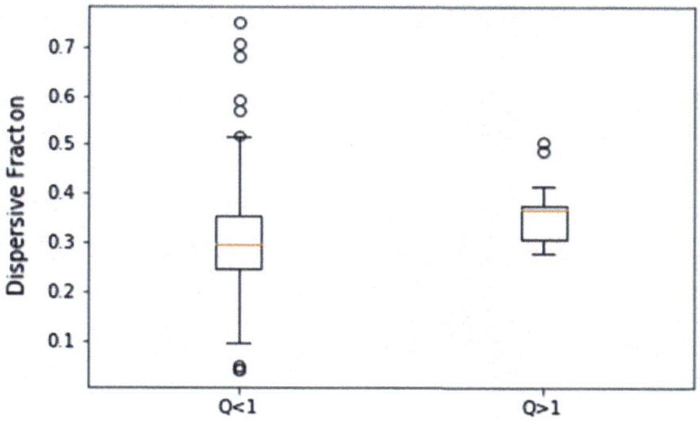

Fig. 3 Boxplots for dispersive fraction. **a** flows less than 1.0 m^3s^{-1}, **b** flows greater than 1.0 m^3s^{-1}

Table 3 Statistics for dispersive fraction for flows less and greater than 1 m^3s^{-1}

Statistic	Q < 1 (m^3s^{-1})	Q > 1 (m^3s^{-1})
Mean	0.302	0.358
Standard deviation	0.127	0.057
Maximum	0.750	0.485
Minimum	0.043	0.279
Median	0.298	0.367

larger variance in the set of lower flows. In Table 3 we further quantify this by reporting statistical parameters for the lower and higher flows indicating approximately equivalent means but standard deviations that differ by a factor of approximately 2.

The difference between mean and median indicates that the distribution of dispersive fraction values is not normal. This is illustrated in Fig. 4.

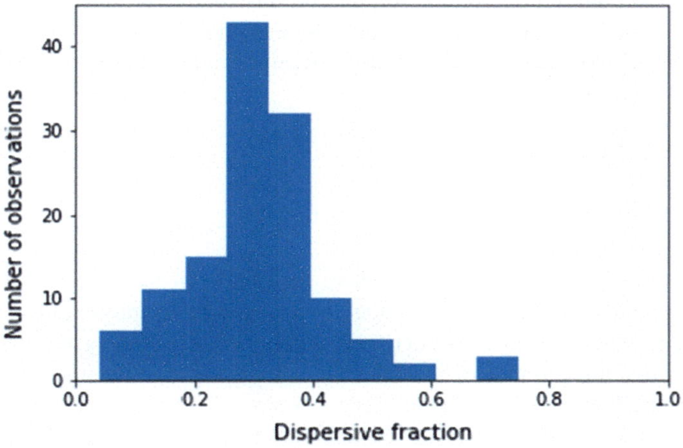

Fig. 4 Histogram of dispersive fraction values from all tracer data sets

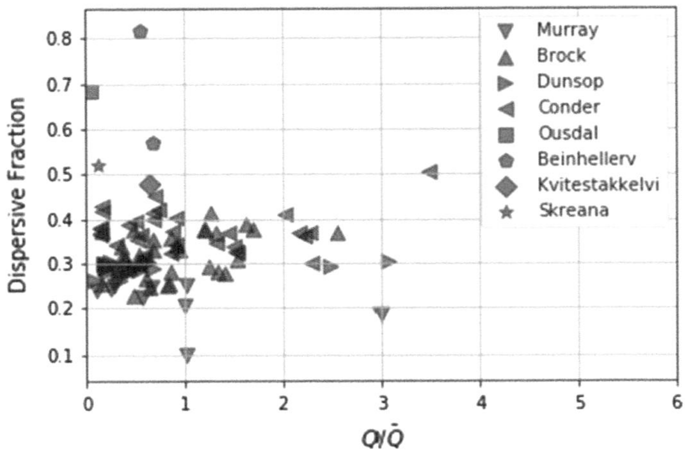

Fig. 5 Dispersive fraction versus dimensionless flow

There is likely a gradient of dispersive fraction values between simple and complex geomorphologies. Table 2 describes the different stream substrates and geomorphologies, and Fig. 5 shows the dispersive fraction plotted against dimensionless flow (flow divided by mean annual flow) for steams for which data was available. Murray Burn which is a relatively uniform modified natural channel has low dispersive fraction values (~0.2); Brock, Conder and Dunsop which are natural channels containing one or more geomorphological structures have medium values (0.3–0.4); all the Norwegian streams have high values (0.5–0.7) reflecting steps, pools, and flow within and around large boulders. Mszana Dolna and Ou Beck are man-made concrete channels with no geomorphological structures and have very low dispersive fractions (~0.1) but are not plotted on Fig. 5 because the mean annual flow was not available. Figure 5 also suggests that a non-dimensional flow of 1 would be a more rigorous way of separating small and large streams than the previous and arbitrary suggestion of a flow rate of 1 m^3/s.

6 Conclusions

A solute transport model was optimized to many sets of tracer data collected in streams to quantify the "transient storage" as a surrogate for physical habitat complexity. The results show no apparent correlation between the dispersive fraction and flow rate but that the variance in possible values for the dispersive fraction increases with decreasing flow rate. These results suggest possible estimates of this parameter for small flows but crucially also suggest uncertainties to be ascribed along with this estimate. They also suggest that as streams decrease in flow rate the range in physical habitat complexity increases.

Acknowledgements The field investigations in Norway were conducted as part of the project SusWater funded by the Norwegian Research Council (NFR code 244505). Special thanks to Muriel Z. Brückner, Philipp Neubauer, Jie Qin and Hans-Petter Fjeldstad for the technical assistance in the field.

References

Beer T, Young PC (1983) Longitudinal dispersion in natural streams. J Env Eng Am Soc Civ Eng 109:1049–1067

Besemer K, Singer G, Hödl I, Battin TJ (2009) Bacterial community composition of stream biofilms in spatially variable-flow environments. Appl Environ Microb 75(22):7189–7195

Besemer K, Singer G, Quince C, Bertuzzo E, Sloan W, Battin TJ (2013) Headwaters are critical reservoirs of microbial diversity for fluvial networks. Proc Biol Sci 280(1771):20131760. https://doi.org/10.1098/rspb.2013.1760

Demars BOL, Manson JR, Olafsson JS, Gislason GM, Friberg N (2011) Stream hydraulics and temperature determine the metabolism of geothermal Icelandic streams. Knowl Manag Aquat Ec 402:05

Green HM, Beven KJ, Buckley K, Young PC (1994) Pollution prediction with uncertainty. In: Beven K, Chatwin P, Millbank J (eds) Mixing and transport in the environment, Wiley, pp 113–137

Guymer I (2002) A national database of travel time, dispersion and methodologies for the protection of river abstractions. Research and Development Technical Report P346, Environmental Agency of England and Wales

Heron AJ (2015) Pollutant transport in rivers: estimating dispersion coefficients from tracer experiments. MPhil Thesis, Heriot-Watt University, UK

Lees MJ, Camacho LA, Chapra S (2000) On the relationship of transient storage and aggregated dead zone models of longitudinal solute transport in streams. Water Resour Res 36:213–224

Press WH, Teukolsky SA, Vetterling WT, Flannery BP (1992) Numerical recipes in C: the art of scientific computing, 2nd edn. Cambridge University Press

Rutherford JC (1994) River mixing. Wiley

Shaw EM, Beven KJ, Chappell NA, Lamb R (2011) Hydrology in practice, 4th edn. Spon Press

Valentine EM, Wood IR (1979) Experiments in longitudinal dispersion with dead zones. J Hyd Div Am Soc Civ Eng 105:999–1016

Wallis SG (1994) Simulation of solute transport in open channel flow. In: Beven K, Chatwin P, Millbank J (eds) Mixing and transport in the environment. Wiley, pp 89–111

Wallis SG (2007) On the numerical solution of the ADZ model. In: Rowinski PM (ed) Transport phenomena in hydraulics, monographic, vol E-7(401). Institute of Geophysics, Polish Academy of Sciences, pp 263–269

Wallis SG, Blakeley C, Young PC (1987) A microcomputer based fluorometric data logging and analysis system. J Inst Water Eng Sci 41:122–134

Wallis SG, Young PC, Beven KJ (1989) Experimental investigation of the aggregated dead zone model for longitudinal solute transport in stream channels. Proc Inst Civ Eng 87, Part 2, 1–22

Young PC (1984) Recursive estimation and time series analysis: an introduction. Springer Verlag

Young PC, Wallis SG (1985) Recursive estimation: a unified approach to the identification, estimation and forecasting of hydrological systems. Appl Math Comp 17:299–334

Quantification of Flood Hazards Due to Assumed Breaching of Attabad Landslide Dam, Pakistan

Arham Mansoor, Noor Muhammad Khan, Aziz Akbar, Yasir Abbas and Muhammad Umar Farooq

Abstract A massive landslide of 210 m height occurred in year 2010 at Hunza River near Attabad village, about 100 km upstream of the confluence of Hunza River with Gilgit River. Keeping in view the high storage capacity of this landslide, overflow or piping can washout this landslide mass within a few hours, once the collapse start, with disastrous impacts downstream. Dam Break Analysis has been carried out, using hydrodynamic module of MIKE 11, to predict flood wave arrival time, peak discharge and wave height at Daniyor Bridge upstream of the confluence of River Hunza and River Gilgit. The water levels observed at Daniyor Bridge for years 2011 and 2012 are used for calibration and validation of the routing model, and the results are found to be very satisfactory. Sensitivity analysis is also performed on the model's main input parameters and their effect on the peak outflow, flood wave arrival time and wave height are studied. The study incorporates computations under six different breach triggering and breach shape scenarios, which include trapezoidal and triangular breach shapes with failure due to either piping phenomenon or overtopping of the dam crest or user-defined breach of the dam. The results of the dam break analysis can be used for the preparation of an Emergency Action Plan for the affected area. The paper presents the details of the project, the model set up and details of the studies carried out.

Keywords Dam break modelling · Attabad Landslide Dam · River Hunza · Daniyor Bridge · MIKE 11

1 Introduction

A massive landslide occurred in Hunza River basin on January 04, 2010, near Attabad village, which is located about 100 km upstream of the confluence of Hunza and Gilgit Rivers. Due to this landslide, a natural dam of 126–210 m height has been created

A. Mansoor (✉) · N. M. Khan · A. Akbar · Y. Abbas · M. U. Farooq
Department of Civil Engineering, University of Engineering and Technology Lahore, Lahore, Pakistan
e-mail: arham_mkhan@yahoo.com

M. B. Kalinowska et al. (eds.), *Recent Trends in Environmental Hydraulics*, GeoPlanet: Earth and Planetary Sciences,
https://doi.org/10.1007/978-3-030-37105-0_16

across Hunza River which resulted in the formation of a lake on the upstream side which gradually rose in time to a height of about 120 meters above river bed level, inundating about 19 km of the Karakoram Highway (KKH) and the Old Shishkat Bridge, and submerging upstream villages. The width of dam crest is approximated as 350 m whereas its length along the river is 2 km. The volume of the water that would be stored in the lake up to the top of the landslide has been estimated as 305 Mm^3.

The landslide location is right on a known local fault. This is the main reason why the rocks of the area, which are otherwise quite hard, are highly jointed and shattered at this location. This shattered state of rocks and the tectonic activity associated with the faults are the main reasons of occurrence of such massive landslides in the area.

The composition and behavior of dam at Attabad is very different from similar man-made structures. This landslide dam contains huge rock blocks (weighing up to hundreds of tons), stones and fine soil. Composition of landslide material visible on the surface is mostly blackish organic soil with sand and gravel and some small boulders. As there is no proper spillway in the landslide mass, it would eventually overtop after being water-filled to its top level. The overtopping of the landslide mass would cause erosion of the soil, which is expected to progress rapidly as the downstream slope of the landslide dam is quite steep (roughly 0.7 V:1H). This rapid erosion of the soil can progress so fast that it may washout most of the landslide mass within a few hours; which in the case of a dam break event is very high.

There are several settlements (villages) along the Hunza River from landslide location to confluence with Gilgit River. These villages are very close to river bank and under direct threat of inundation due to dam break flood.

2 Study Area

For Attabad Landslide Dam, the coordinates of the upstream end are 74° 49′ 50″ E, 36° 18′ 10″ N and it extends downstream up to 74° 48′ 11″ E, 36° 18′ 26″ N. Water accumulation due to river blockage starts at 74° 49′ 50″ E, 36° 18′ 10″ N and extends in the upstream direction. The length of reservoir has been reported as 15.5 km on 7th May 2010.

Location of the study area on map of Pakistan is shown in Fig. 1.

3 Dam Break Modelling

The dam break study can be divided into two parts:

1. Dynamics of breach section development
2. Routing of resulting flood wave.

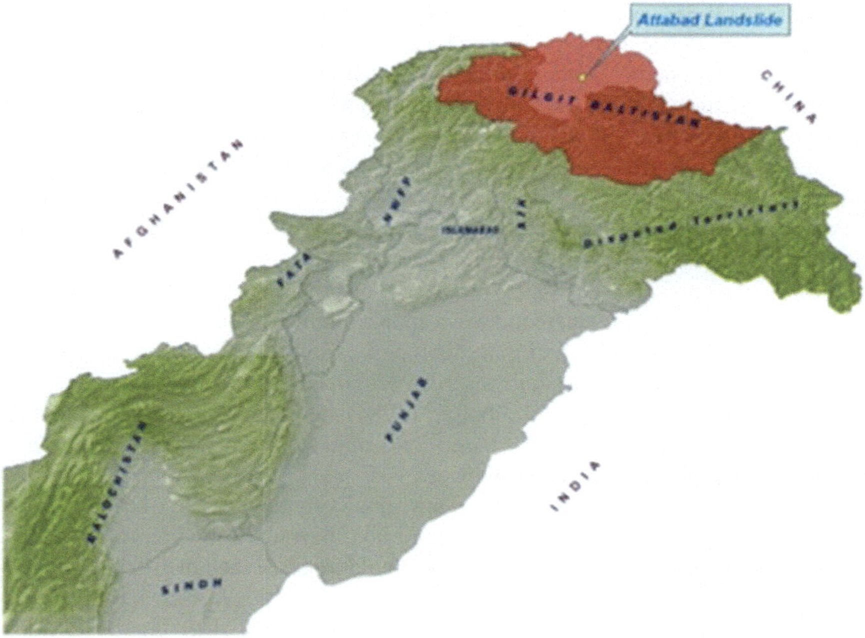

Fig. 1 Location of study area on a map of Pakistan. *Source* National Engineering Services of Pakistan (Pvt.) Limited

The first part deals with the characteristics of breach triggering phenomenon, progress of the breach, and computations of the outflow hydrograph. The second part (flood routing) deals with the determination of the change in flow characteristics of the flood wave during to its propagation in a channel, flood plains, or a network of channels, as per the situation in hand.

The mechanism of breach development is largely dependent on the type of dam and its construction material. Review of the historical breaches depicts that concrete arch and gravity dams breach by sudden collapse, overturning or sliding away of the structure, whereas dam breach of earth fill dam is triggered either by the erosion or overtopping of embankment material and seepage/piping through the embankment, foundation, and/or abutments of the dam.

The dam breach phenomenon is a time dependent, multi-phase and non-homogeneous in nature. Once a breach has developed, the discharging water erodes the vicinities of the breach, until either the reservoir water is depleted or the breach resists further erosion. The requirement to simulate these processes is essential for:

- Flood forecasting
- Crises planning and management
- Decision making for taking safety measures in time.

Cristofano (1965) was the first to simulate the growth of breach in an earthen dam (as described by Ponce and Tsivoglon 1981). Using the geo-technical principles, he equated forces on flowing water through the breach to resistive shear strength acting on the bottom surface of the overflow channel. Thus, he was able to relate rate of change of erosion to rate of change of water through the overflow channel.

Chow and Zvi (1973) for the first time addressed hydrodynamic modeling of two dimensional watershed flow.

Ponce and Tsivoglon (1981) presented a comprehensive literature review regarding "Sudden Dam Breach" and "Gradual Dam Breach", with major emphasis on the latter case.

Hunt (1982) worked on kinematic wave approximation of St. Venant equation for dam-break flood wave and found that this type of solution becomes valid after the flood wave has advanced about four reservoir lengths downstream of the dam.

MacDonald and Langridge-Monoplois (1984) developed, a parameter called "Breach Formation Factor (BFF)", based on a number of dam failure data. Then they found a graphical relationship between BFF and volume of material removed and peak rate of outflow from breach. They also correlated, with a best fit curve, Breach development time and volume of material removed during breach, no doubt, a remarkable work which can be utilized for simulation of breach in different models as input data.

MacDonald and Langridge-Monoplois (1984) classified dam breach forming mechanisms into two general categories:

1. Sudden removal of a portion or the entire embankment structure (bursting), as a result of overstressing forces.
2. Progressive erosion of the embankment material.

Basco (1989) after Brutsaert (1971) again verified the applicability of Saint Venant equation and found that the equation can be applied to dam break wave with sufficient accuracy but cannot be applied to the waves having time-period less than 100 s. He also added some terms to St. Venant equation to account for stream line curvature effect and made a general form of Boussinesq equation.

Gozali and Hunt (1993) obtained numerical and experimental data for flow within reservoir after a partial dam breach.

Froehlich (1995) described an empirical method for quickly estimating the peak outflow from a breached embankment dam.

Recent research at the Natural Resources Conservation Service (NRCS) laboratories and sites in Norway (Hoeng et al. 2004) have helped further to understand breach process.

The National Interagency Committee on Dam Safety, in 2009, measured the extent or magnitude of a dam failure event in terms of the classification of the dam. According to them, high hazard dams as those where failure or mis-operation will cause loss of human life, whereas dams classified as significant are those at which failure or mis-operation probably would not result in loss of human life but could cause economic loss, environmental damage, disruption of lifeline facilities, or other significant damage. Low hazard potential dams are those at which failure or mis-operation probably would not result in a loss of human life but would cause limited economic and/or environmental losses. Losses would be limited mainly to the owner's property.

Dam break studies can be carried out by either:

- Scaled physical hydraulic models
- Mathematical simulation using computers.

A modern tool for the dam break analysis is the mathematical model which is most cost effective, and approximately solves the governing flow equations of continuity and momentum by computer simulation. Sophisticated computer programs, such as MIKE 11 and DAMBRK, developed in 1994 and 1982, respectively, are dependent on certain inputs regarding the geometric and temporal characteristics of the dam breach. The techniques employed for estimating these breach characteristics are not as advanced as that of the computer programs, and therefore these are the limiting factors in the dam break analysis.

4 Mike 11 Dam Break Module Description

MIKE 11 is a commercial software package that is developed at Danish Hydraulic Institute (DHI) and the Water Quality Institute for the simulation of fluid flows, sediment transport and water quality in estuaries, rivers, irrigation systems, and similar water bodies.

St. Venant's continuity and momentum equations are used for performing flood routing which is the evaluation of an unsteady flow through rivers, flood plains and network of channels. Modeling equations of MIKE 11, i.e., continuity and momentum equations are as follows:

$$\frac{\partial Q}{\partial x} + \frac{\partial A}{\partial t} = q \quad (1)$$

$$\frac{1}{A}\frac{\partial Q}{\partial t} + \frac{1}{A}\frac{\partial}{\partial x}\left(\frac{Q^2}{A}\right) + g\frac{\partial h}{\partial x} - \frac{gQ|Q|}{ARC^2} = 0 \quad (2)$$

where A is the flow area (m^2), R is the resistance radius (m), C is the Chezy's resistance coefficient (m$^{1/2}$ s^{-1}), g is the gravitational acceleration (m s^{-2}), h is stage above horizontal reference level (m), Q is discharge (m^3 s^{-1}), A is momentum distribution coefficient and q is the lateral inflow (m^2 s^{-1}).

MIKE 11 can accommodate any system of channels, taking into account:

- Lateral discharges
- Sub-critical and super critical flow conditions
- Free overflow and submerged flow at weirs (internal boundary condition)
- Different reservoir operation policies
- Flow related resistance and local head losses
- Flooding and drying of low level areas

- Two-dimensional flow conditions on flood plains
- Wind stresses, etc.

MIKE 11 can also simulate flooding associated with the failure of dams. Dam failure may be simulated as a sudden or a gradual process, depending on the type of dam construction. The probable changes in channel and flood plain geometry, caused by sediment erosion and deposition, can also be modeled/simulated.

5 Methodology

In this study, the core system is the dam break (DB) module which is used to model the dam breach process. The hydrodynamic (HD) module which represents the heart of MIKE 11 was used to calculate the outflow, discharge and water level in the channel. To setup the model for this study, parameters such as river network, cross sections, boundary data, time series and hydrodynamic were created. The network for this study was setup using the network editor which is a central unit of MIKE 11 model. Branches and points were inserted indicating the river network and dam structures. In this study, three branches were created to represent the reservoir formed upstream of the landslide mass and the downstream channels i.e. River Hunza and River Gilgit, respectively. Dam break structure was described based on crest level and crest length, whilst breaching of the structure was described using the breach calculation methods, i.e., Mike 11 energy equation or NWS dam break equation, failure moment, i.e., hours after start, reservoir water level and date and time, and mode of failure, i.e., time dependent and erosion based. Points were inserted along the channels and defined as chainage, by assigning specific cross-section which was described in the cross-section file. Figure 2 shows the entire network of River Hunza from Attabad Landslide Dam to Alam Bridge.

The river cross-sections from chainage 0.00 to chainage 32,822.76 are from topographic survey, while cross-sections from chainage 33,486.69 to chainage 96,859.69 are extracted from SRTM data. These cross sections were inserted in the model to join the entire network while cross-sections from bathymetric survey were used to represent the reservoir. In the boundary section, the breach parameters and the inflow hydrograph in time series were inserted into the models. The upstream of the dam was specified as 'Inflow' boundary type and downstream as 'Water level'. Dam is specified as 'Dam break' where the dam breach parameters in time series were inserted. For Probable Maximum Flood (PMF) scenarios, PMF inflow hydrograph (Fig. 3) was inserted in the model. The Probable Maximum Flood (PMF) is the flood that may be expected from the most severe combination of critical meteorological and hydrologic conditions. PMF also has physical meanings which provide an upper limit of the interval within which the engineer must operate and design. The PMF is generated using the input values of Probable Maximum Precipitation (PMP) in this study area.

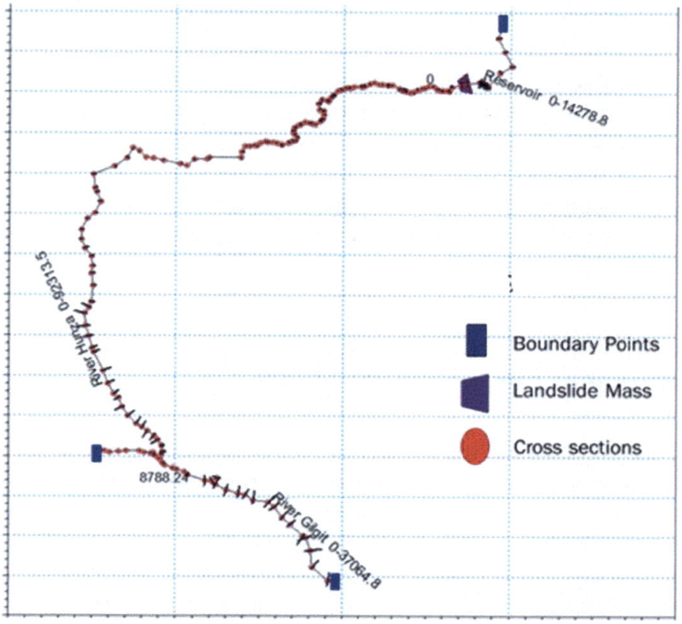

Fig. 2 Network branch of Hunza River from Attabad Landslide Dam to Alam Bridge

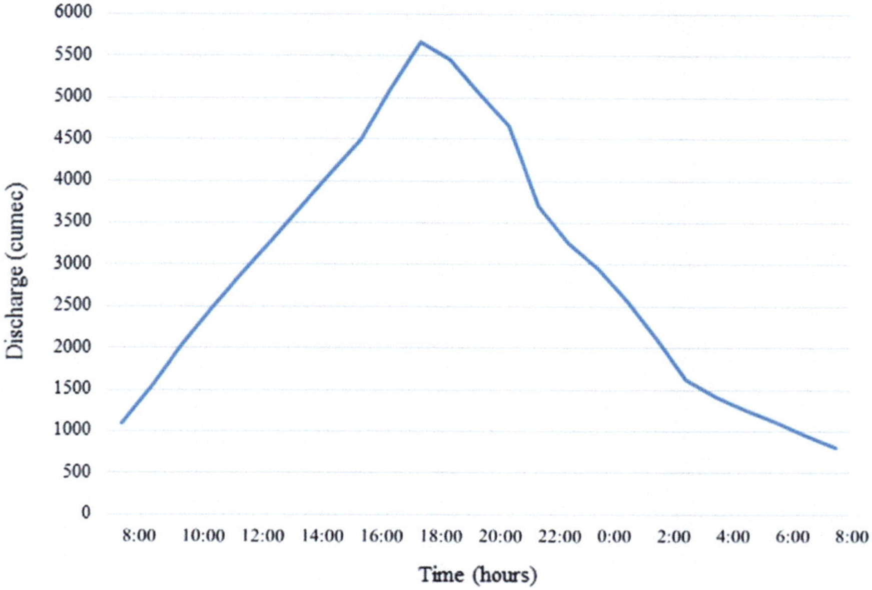

Fig. 3 Maximum Flow Hydrograph entering into the reservoir upstream of the landslide dam

For hydrodynamic part, initial water level is required to be justified in order to run a hydrodynamic computation. The river bed was examined to be composed of different sizes of particles, i.e., from silt and clay to boulders and cobbles hence, the recommended Manning's (n) for mountain streams, no vegetation in channel, banks usually steep, trees and brush along banks submerged at high stages, bottom gravels, cobbles, and few boulders equals to 0.03 for main channel and 0.035 for overbank.

For Attabad Landslide Dam, one of the worst cases is taken to be the failure moment, i.e., when 100-year flood peak reaches the reservoir, while it is already at its maximum level (2427 m).

6 Calibration of Flood Routing Model

Calibration and validation of the model has been done by using the depth discharge time series data of two years, i.e., 2011 and 2012, respectively.

Trial and error method has been adopted to calibrate the Manning's roughness coefficient, n, in this study. Calibration/validation of 'n' is done with the help of the water depths and discharges obtained at Daniyor Bridge.

For good calibration, resistance factor equivalent to 0.033 Manning's n (M = 30) is taken (Chow 1973) which shows that the channel is very rough, having boulders and rocks in it. So, the calibration is made selecting "M" value of 30 for the whole channel. The calibration results can be seen in Figs. 4 and 5 whereas validation results are presented in Figs. 6 and 7, respectively.

Coefficient of determination has been used to correlate the water depths gauged at site and those obtained from the model for calibration and validation purpose.

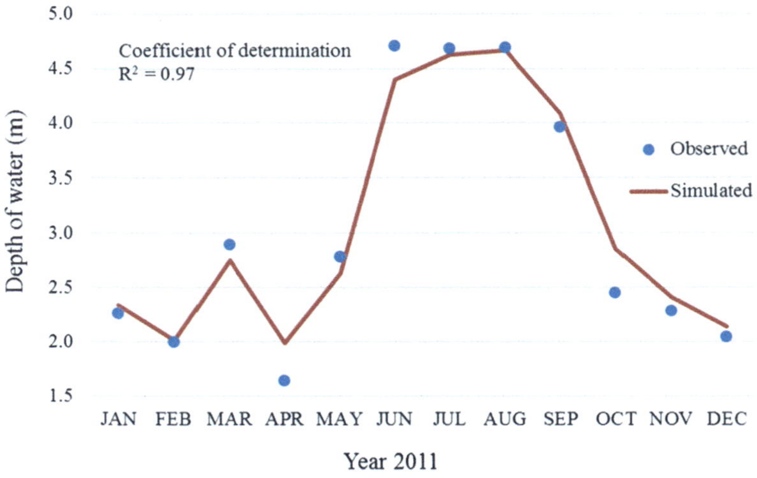

Fig. 4 Calibration of water depths at Daniyor Bridge

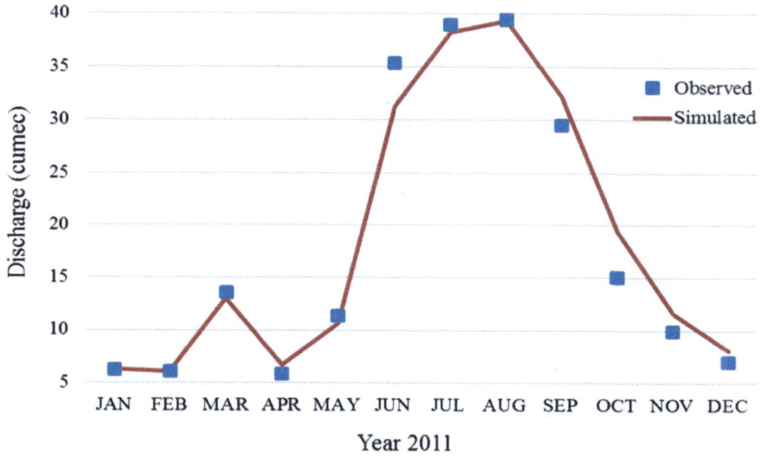

Fig. 5 Calibration of discharge at Daniyor Bridge

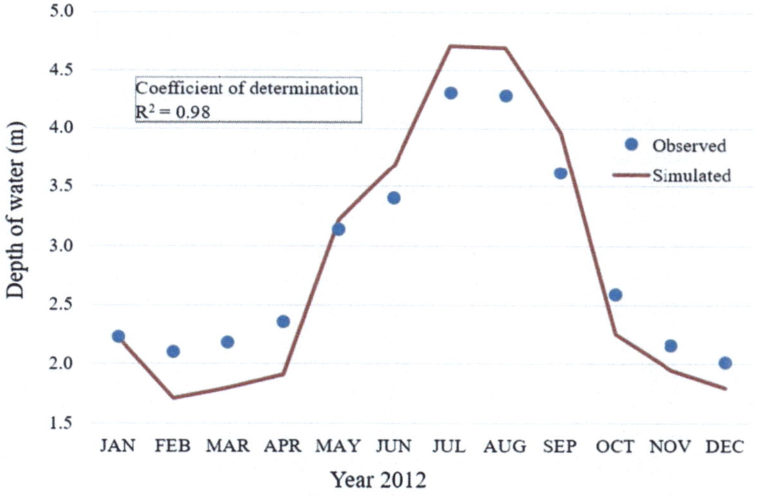

Fig. 6 Validation of water depths at Daniyor Bridge

7 Results of Dam Breach

Breaking of Attabad Landslide Dam has been studied by both sediment erosion method and by user-specified method. Erosion based dam-break is further studied with both options of triggering due to piping and triggering due to overflow, and each case is then further studied by both options of trapezoidal and triangular breach shape. Hence, six cases of dam-break have been studied in total.

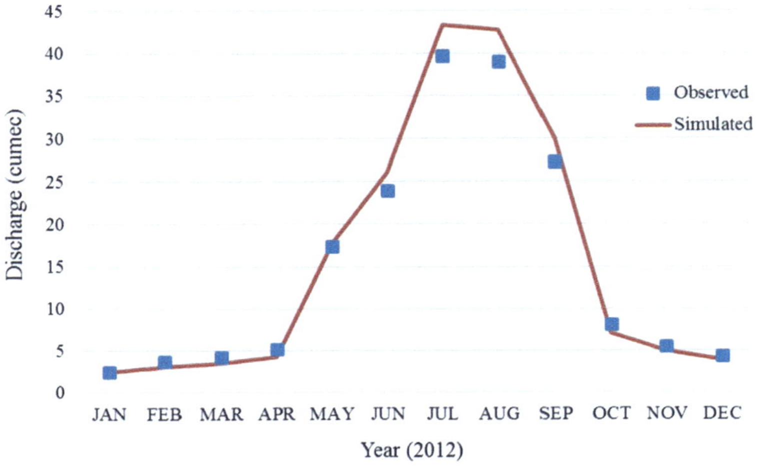

Fig. 7 Validation of discharge at Daniyor Bridge

For carrying out user-specified method of dam break, area of breach, i.e., the bottom level, side slope and bottom width of the breach has to be specified for a fixed time interval. The model uses this given information for the initial time step, whereas dimensions of breach are dependent on the sediment erosion in the succeeding time steps. To measure the sediment erosion, MIKE 11 calculates upstream and downstream water level and calculates flow velocity over the breach. This velocity is used to measure the shear stresses. These are compared with shear stress limit (shield's parameter) and if found greater, the rate of erosion is calculated. This rate of erosion in m^3/sec per unit width is then utilized to measure the new breach dimensions using upstream, downstream slopes of the dam, and width of the crest.

Sediment erosion based dam breach case shows that triggering of dam breach due to piping with the final shape of the breaching section to be trapezoidal causes a peak outflow of 10,528 m^3/s, whereas, when the final breach is triangular in shape, a peak discharge of 4862 m^3/s occurs. Arrival time of peak discharge at Daniyor Bridge for trapezoidal shape breach is 8 h, whereas for triangular shape breach it is 11 h after the start of breach. Table 1 shows that time of peak for triggering of dam breach due to overtopping and trapezoidal breach shape is 9 h, whereas for triangular breach shape the peak discharge occurs 12 h following the breach initiation. Figure 8 presents discharge hydrographs for various dam break scenarios.

8 Results of Flood Routing

As a result of dam-break, the water level and discharge at each node along with discharge in between the two nodes are calculated with the help of hydro-dynamic

Table 1 Comparison of discharge and time to peak at Daniyor Bridge for different cases of dam break

Breach growth	Triggering	Shape of breach	Discharge at Daniyor Bridge (cumec)	Time to peak (h)
Erosion	Piping	Trapezoidal	10,528	8.0
		Triangular	4862	11.0
	Over-topping	Trapezoidal	11,545	9.0
		Triangular	9017	12.0
User specified	–	Trapezoidal	4205	13.0
	–	Triangular	4069	13.0

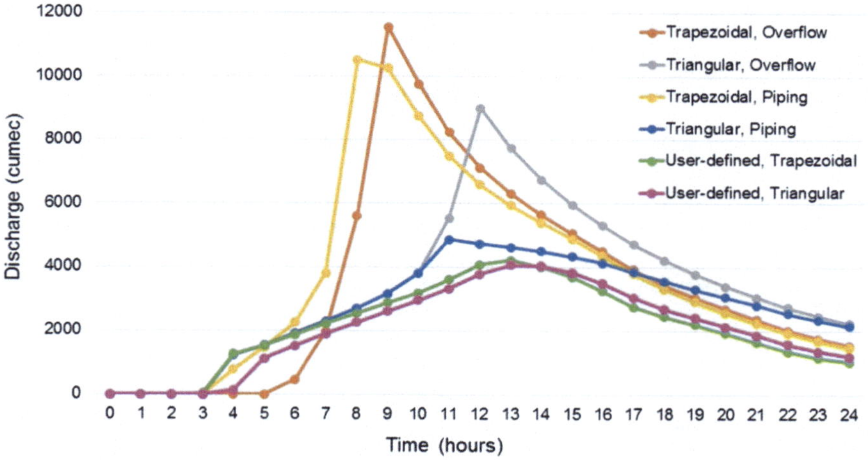

Fig. 8 Discharge hydrograph at Daniyor Bridge as a result of dam break

module of MIKE 11. The time step for storing of the result is taken 1 h, while that for calculation is one second. Water levels of the six dam breach cases under study at Daniyor Bridge are presented in Fig. 9.

Table 2 shows that the maximum water level at Daniyor Bridge is higher for trapezoidal case as compared to that of triangular case. The maximum computed water level of 1420.55 m, i.e., 13.2 m depth of water is observed at Daniyor Bridge caused by dam break due to overtopping and final breach shape to be trapezoidal.

9 Sensitivity Analysis

Sensitivity analysis of dam breach parameters is done by changing different input parameters on both sides so that its effect on peak outflow and time of peak can be

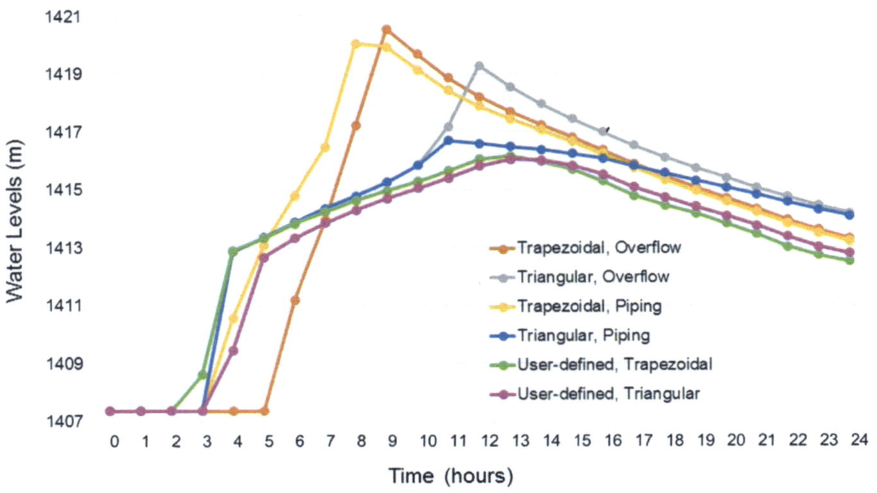

Fig. 9 Water levels at Daniyor Bridge as a result of dam break

Table 2 Comparison of discharge and time to peak at Daniyor Bridge for different cases of dam break

Breach growth	Triggering	Shape of breach	Peak water level at Daniyor Bridge (m, amsl)	Water depth at Daniyor Bridge (m)	Time to peak (h)
Erosion	Piping	Trapezoidal	1420.06	12.67	8.0
		Triangular	1416.69	9.30	11.0
	Over-topping	Trapezoidal	1420.55	13.16	9.0
		Triangular	1419.27	11.88	12.0
User specified	–	Trapezoidal	1416.15	8.76	13.0
	–	Triangular	1416.03	8.64	13.0

seen. Results are summarized in Table 3. It shows that peak discharge is directly proportional to the final width (b_f) and the side slopes of the breaching section and in case of failure due to piping peak discharge varies directly with the change in the initial diameter of pipe (d_o), whereas it varies inversely with variation of the initial breach level (Z_o).

Roughness factor 'M' (=l/n, reciprocal of Manning's n) is changed for carrying out sensitivity analysis for the flood routing model. A 20% increase and 20% decrease is introduced to see the effect on water level and discharge. Figure 10 represents the effect on water level at Daniyor Bridge with ±20% variation of M. It is evident that

Table 3 Summary of sensitivity analysis effect of parameters on peak outflow and time of peak outflow

Parameter	Case	Q_{max}	T_{Qmax}
Initial Breach Level	↗	↘↘	↗↗
	↘	↘↘	↘↘
Initial Pipe Diameter	↗	↘↘	↘↘
	↘	↘↘	↘↘
Breach Side Slope	↗	↗	↘↘
	↘	↘↘	↗
Breach Final Width	↗	↘	↗
	↘	-	-
Roughness Factor	↗	↘	↘
	↘	↘	↘

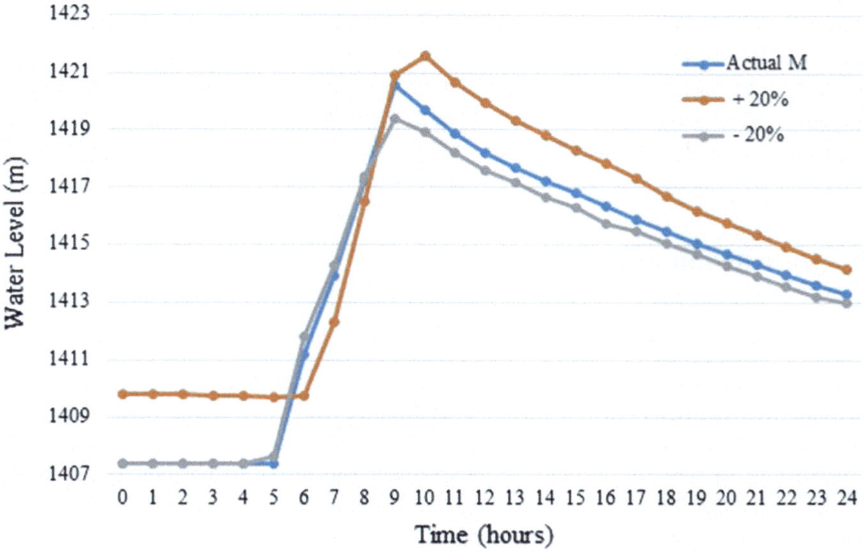

Fig. 10 Effect of change of roughness factor 'M' on computed water level hydrograph at Daniyor Bridge (erosion, overflow, trapezoidal)

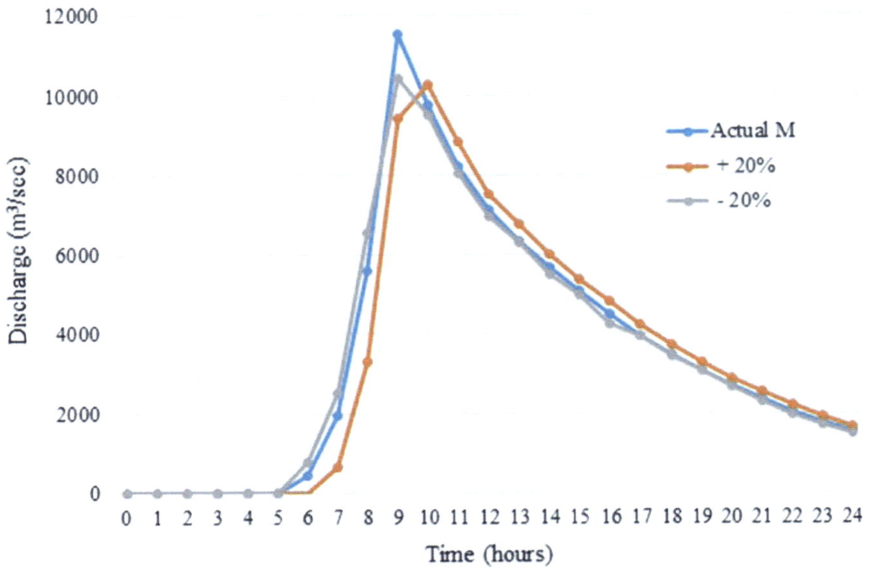

Fig. 11 Effect of change of roughness factor 'M' on computed discharge hydrograph at Daniyor Bridge (erosion, overflow, trapezoidal)

the water level and M are inversely proportional because of the fact that channel conveyance is reduced with the reduction in the roughness factor; hence, the water levels increase. The inversely proportional relationship between discharge and roughness coefficient 'M' can be seen in Fig. 11. It shows that the increase of M introduces a leading effect (occurs earlier) on discharge peak and reduction of M produces a lagging effect on the discharge peak.

10 Conclusion and Recommendations

The results and experiences gained from this study can be concluded as follows:

- The peak outflow through the dam in case of trapezoidal shape is about three times greater than the peak outflow through the dam for triangular shape, despite the fact that all parameters of the two cases were same except for the shape of the breach.
- According to sensitivity analysis, peak discharge through the dam is inversely proportional to the initial breach level and is directly proportional to the initial diameter of the pipe, final width of breaching section and breach side slopes.
- Peak flow reaching Daniyor Bridge due to trapezoidal breach is 2.3 times the historic maximum flood, whereas for triangular breach it is 1.5 times the flood magnitude (i.e. 5000 m^3/s) which Daniyor Bridge has faced in 1967.

- There are still a number of important dam breach parameters that researchers have not yet properly investigated. Significantly more physical modelling experiments need to be carried out so that numerical modelers can better understand the dam breaching phenomena.
- Multiple materials are often used to construct earthen embankment dams. Composite structures breach much differently than a homogenous dam. However, MIKE 11 only allows the user to use one type of sediment for the entire dam. This limits the model to breaching only homogeneous dams or causes the user to make possibly unrealistic assumptions.
- Present numerical models do not take into account the location, shape, or imperfection of the initial breach of real dams. For example, MIKE 11 assumes the initial breach notch to be rectangular in shape and located at the midpoint of the embankment crest. However, in reality, it is almost unlikely for a real dam to start breaching from the mid-point of the crest of the dam. Therefore, researchers should try to incorporate the randomness of breaching initiation in future numerical models and investigate characteristics on the initial breach in order to determine the worst case scenario of a dam breach.

Acknowledgements Authors are highly thankful to Civil Engineering Department of University of Engineering and Technology Lahore for providing conducive environment for research, Danish Hydraulic Institute (DHI) for rendering the student package of Mike 11 and Surface Water Hydrology Project (SWHP) as well as National Engineering Services of Pakistan (Pvt.) Limited (NESPAK) for sharing the relevant data.

References

Basco DR (1989) Limitation of de Saint Venant equations in dam-break analysis. J Hydraul Eng 115(7):950–965

Brutsaert W (1971) De Saint Venant Eq. experimentally verified. J Hydraul Eng 97(HY 9):1387–1401

Chow VT (1973) Open channel hydraulics. McGraw Hill Book Co., Chap. 20

Chow VT, Zvi AB (1973) Hydrodynamic Modelling of two-dimensional watershed flow. J Hydraul Div ASCE 99(HY11):2023–2039

Froehlich DC (1995) Peak outflow from breached embankment dam. J Water Resour Plann Manage 121(1):90–97

Gozali S, Hunt B (1993) Dam-break solution for a partial breach. J Hydraul Res 31(2):205–214

Hoeng K, Lovoll A, Vaskinn KA (2004) Stability and breaching of embankment dams: Field tests on 6 m high dams. Int J Hydropower Dams 11(1):88–92. http://www.mikebydhi.com

Hunt B (1982) Asymptotic solution for dam-break problem. J Hydraul Div ASCE 108(HY1):115–126

MacDonald TC, Langridge-Monopolis J (1984) Breaching characteristics of dam failures. J Hydraul Div ASCE 110(5):567–586

Ponce VM, Tsivoglon AJ (1981) Modelling gradual dam breaches. J Hydraul Div ASCE 107(HY7):829–838

Monitoring of Riparian Vegetation Growth on Fluvial Sandbars

Michael Nones, Massimo Guerrero and Renata Archetti

Abstract The paper proposes a simplified methodology to track the evolution of vegetation patterns over a central sandbar of the Po River, Italy, by means of a fixed video camera installed on the top of a bridge pier. Looking downstream, the camera acquires images every twelve hours while hourly water levels are derived from a radar hydrometer located 300 m upstream of the study area. The vegetation growth rate is computed analysing several images covering the period July–December 2017, characterized by a dry period during the summer/autumn and a flood at the end of the year. The tracking of the vegetation patterns bounds provides some general indications on the role of a transient hydrology on the plants' dynamics.

Keywords Po River · Riparian vegetation · Time-lapse photography · Video monitoring

1 Introduction

Aside from the hydrological drivers, the management of vegetated sandbars in fluvial environments requires the combined knowledge of morphological and vegetational dynamics. Even if the correlation between a changing hydrology and the morphological evolution of fluvial bars is well understood (Luchi et al. 2010), a knowledge gap is still evident, especially in recognizing how riparian plants respond to subsequent

M. Nones (✉)
Institute of Geophysics, Polish Academy of Sciences, Ksiecia Janusza 64, 01-452 Warsaw, Poland
e-mail: mnones@igf.edu.pl

M. Guerrero · R. Archetti
Department of Civil Chemical, Environmental and Materials Engineering, University of Bologna, viale Risorgimento 2, 40136 Bologna, Italy
e-mail: massimo.guerrero@unibo.it

R. Archetti
e-mail: renata.archetti@unibo.it

M. B. Kalinowska et al. (eds.), *Recent Trends in Environmental Hydraulics*, GeoPlanet: Earth and Planetary Sciences,
https://doi.org/10.1007/978-3-030-37105-0_17

floods and droughts, and how they develop interacting with river morphology, sediment, and hydro-climatic dynamics (Vicente-Serrano et al. 2013; Serlet et al. 2018) at the local scale.

Time is a key element that regulates the plant establishment in river corridors and permits a comparison with the recurrence interval of extreme events affecting the fluvial morphology and the ecological succession (Räpple et al. 2017). Therefore, it is necessary to develop low-cost techniques for long-term monitoring of the vegetation growth on river sandbars. One solution can involve the use of remote methods to create consistent dataset without affecting the environment with invasive techniques.

As already demonstrated, a reliable real-time monitoring of shoreline changes in coastal environments (Aarninkhof et al. 2003; Archetti and Romagnoli 2011), or of woody debris (Kramer and Wohl 2014), biota (Deacy et al. 2016) and hydro-morphodynamics (Nichols et al. 2016; Nones et al. 2018b; Schoener 2018) in river contexts can be performed by means of time-lapse photography. This technique, in fact, represents a major tool in monitoring landform evolution with nonintrusive methods, thanks to the reduced costs and good image resolution in comparison with others such as video monitoring (Chandler 1999; MacVicar and Piégay 2012; Kramer and Wohl 2014). In addition, freeware software devoted to image analysis is spreading worldwide, encouraging also non-experts to move towards photogrammetry and time-lapse photography to remotely monitor fluvial dynamics at manifold spatiotemporal scales (Pyle et al. 1997; Stojic et al. 1998; Schoener 2018). As recently discussed (Kyes et al. 2016), indeed, the application of cost-effective, user-friendly techniques for examining the eco-hydrology of streams and river corridors through digital imagery is not yet completely exploited, in particular regarding the vegetation dynamics of seasonal plants growing on river sandbars.

Using a small reach of the Po River in Italy as a case study, this research will provide additional details in this direction. After discussing the adopted method, the paper shows a preliminary analysis of the images captured with a fixed camera installed on the top of a bridge pier. After processing the images with Matlab®, the vegetation patterns over the sandbar between July and December 2017 are derived, pointing out a time lag between the maximum of the vegetation growth and the maximum of the measured water elevation.

2 Materials and Methods

To reduce costs and measurement-related disturbances associated with field campaigns, the monitoring of sandbar dynamics and herbaceous vegetation encroachment over a central sandbar on the Po River in Italy (Fig. 1a) was performed using a fixed camera installed on a railway bridge pier (Fig. 1b). By means of Matlab® and AutoCAD®, the obtained images were rectified, the areas covered by vegetation computed and these data coupled with hydrographic information measured in the same reach to evaluate the correlation between the local hydrology and the growth

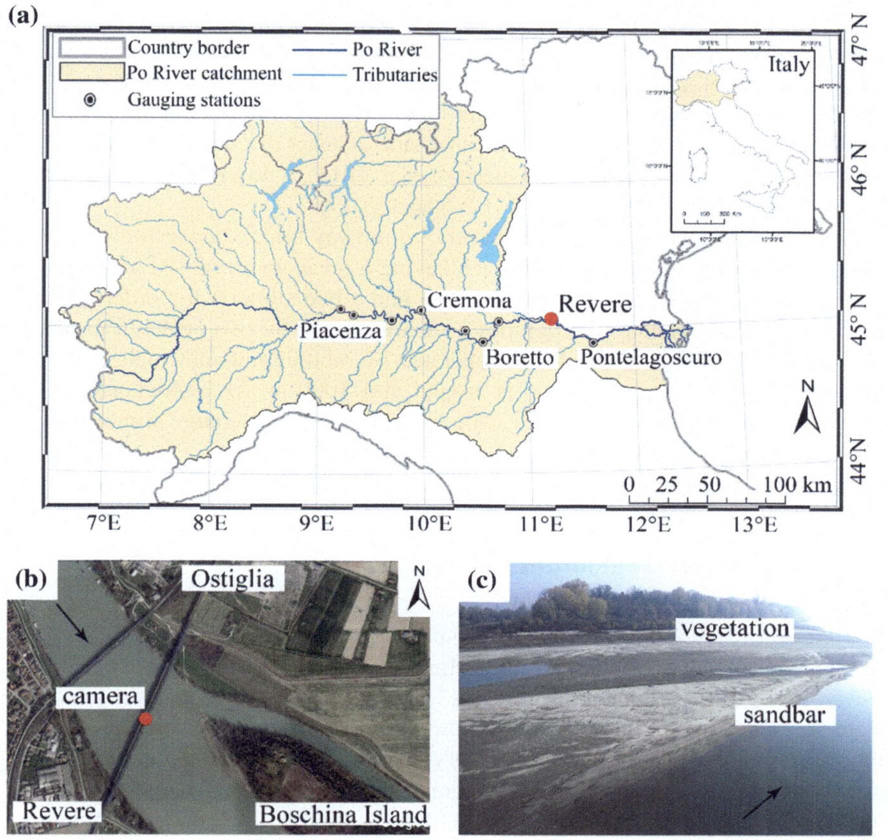

Fig. 1 **a** Po River catchment; **b** satellite image of the studied area and location of the camera; **c** example of an image acquired with the fixed camera: vegetation growth over the sandbar in the near field, and the Boschina Island in the far field

rate of the herbaceous vegetation in the period July–December 2017. Summer and autumn of 2017 were very dry and prolonged, and the first relevant flood happened in mid-December.

2.1 Case Study

The Po River (Fig. 1a) has a total length of around 660 km and drains a basin of around 74,500 km^2, highly anthropized especially in the middle and lower reaches (Domeneghetti et al. 2015). In fact, the present river bathymetry results from the interaction between natural processes and anthropic pressures affecting the reach since centuries, like damming, water exploitation for several uses, land use change for agriculture, river channel embanking and sediment mining, which drove to a significant degradation of the river bed (Surian and Rinaldi 2003; Lanzoni et al. 2015; Nones et al. 2018a) with consequent negative effects on local fauna and flora. On the one side, a reduction of the water flow is detrimental for riparian vegetation and biota (Nagler et al. 2018). On the other side, a water discharge decrease means a reduction in sediment transport (Guerrero et al. 2013), with a detrimental effect on the ecological status (Vesipa et al. 2017: Nones 2019).

The study area is located downstream of a railway bridge between the cities of Revere and Ostiglia (Fig. 1b), in a reach having a mean width of 400 m, mean bed elevation of few meters above the sea level and a bed slope in the order of 10^{-5}. Usually, the local annual hydrography is characterized by two peaks in discharge in autumn and spring, respectively, generated by rainfall and snowmelt, while the long-term mean annual discharge is around 800–1000 m^3 s^{-1} (Montanari 2012). However, 2017 was a very dry year, with an annual mean discharge around 700 m^3 s^{-1} and a few flooding events only during the winter.

2.2 *Tracking of Vegetation Patterns*

Starting from July 2017, the site was monitored with a fixed camera Mobotix MX-M15D-SEC having a resolution of 12.5 Mpix installed on a pier of the railway bridge Revere-Ostiglia (coordinates 45° 03′ 13.4″ N, 11° 08′ 16.2″ E), just on the left of the waterway (Fig. 1b). To monitor both the morphological dynamics of the sandbar and the vegetation, images were acquired every 12 h (midday and midnight). Because of light conditions, only daily images are here analysed and compared to recognize the herbaceous vegetation growing on the sand deposit created downstream of the pier (Fig. 1c).

By means of a chessboard and the Image Processing Toolbox of Matlab®, the image distortion due to the camera lens was corrected, while the vegetation patterns were extracted using an own-developed semi-automatic Matlab® procedure already applied for the tracking of the edge-of-water line displacements (Archetti et al. 2016; Nones et al. 2018b). In addition, linear homographic equations (Guerrero and Lamberti 2007) were applied to transform image plane coordinates to real plane coordinates based on seven ground control points acquired during a topographic survey made in July 2017 and combined with information retrieved from satellite images of the same period (Nones et al. 2018b). Following the homographic procedure, the camera position corresponds to real plane coordinates 0.0 m (Fig. 2c).

As visible from Fig. 2a, the vegetation patterns were monitored in the near-field, involving positioning errors of a few meters, as discussed by Nones et al. (2018b). The homographic transformation of the images was applied, and the obtained bounds of the vegetation patterns were analyzed with AutoCAD® to compute the area of each pattern, assuming the continuity between upper and lower bounds and a complete vegetation cover within the bounds.

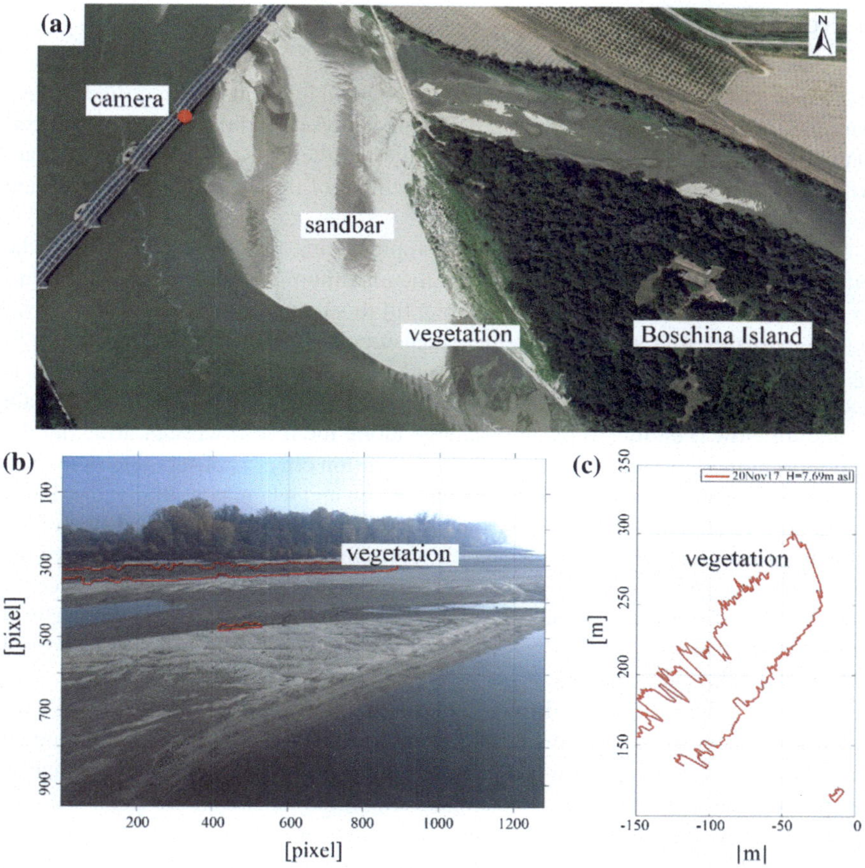

Fig. 2 **a** Aerial view of the studied area, Google Earth image acquired on June 21, 2017. **b** Reconstruction of the vegetation bounds on the raw image taken from the camera on November 20, 2017; **c** vegetation pattern after the homographic transformation for the same image

The local hydrology was continuously measured by means of a hydrometer located around 300 m upstream of the railway bridge. The acquired data were averaged every hour and adjusted accounting for the water slope. The resulting water levels corresponding to the acquisition time of the camera was adopted as an indication for the hydrological state.

3 Results and Discussion

To compare the images and estimate the areas covered by plants, ten rectified images spanning the period July–December 2017 were analysed and the respective vegetation patterns extracted (Fig. 3a). During this period the Po River basin was characterized by an important drought, with an increment of the water levels only during the last days of observation in December.

Aiming to obtain a dimensionless ratio representing the vegetation growth, the computed vegetated areas were divided by the maximum one, measured on November 20 (Fig. 3b, orange line). As observable, till this maximum the vegetation cover slightly increases after a strong increment measured between July and August, reaching a quasi-stable state that lasts for around three months. Such state, in fact, is practically unaffected by hydrological variations (Fig. 3b, blue line), if they remain comparably low (i.e., low to mean discharge along the reach) without affecting the local morphology. A sudden change of the vegetation cover is observable in the last days of November, at the end of the growing season and when, potentially, the floods become more frequent thus changing the sedimentary pattern with the erosion of the sandbar (Nones et al. 2018b). However, the correlation between an increment of the water levels (i.e., flow discharge) and a decrement of the herbaceous vegetation cover is not so clear from the present application, pointing out the necessity to perform additional studies spanning a longer temporal horizon and based on an improved calibration of the video camera.

Notwithstanding the short time period studied, the preliminary outcomes presented here seem to confirm the ones obtained by Xia et al. (2018), who studied herbaceous vegetation in a semi-humid river in China. By means of field information they pointed out that, in the case of sandbars and floodplains with low bank stability typical of sandy rivers like the Po, erosive phenomena associated with floods disturb or interrupt the process of vegetation succession, reducing the growing period and the plant diversity of the vegetation covering these areas. However, an in-depth analysis of the adopted methodology is necessary to confirm the preliminary outcomes and to show the potential of time-lapse photography in describing the dynamics of the riparian vegetation forced by the local morphology. In addition, for the future, a better rectification of the images is forecasted, having acquired additional control points both along the banks and within the main channel.

4 Conclusions

The paper shows a preliminary estimate of the vegetation growth on a sandbar of the Po River, in Italy, during the period July–December 2017, which was very dry with only a relevant flood at the end of the year. The analysis was performed by means of the time-lapse photography, analyzing diurnal images acquired from a fixed camera located on the top of a bridge pier. Based on these first set of images, it is observable

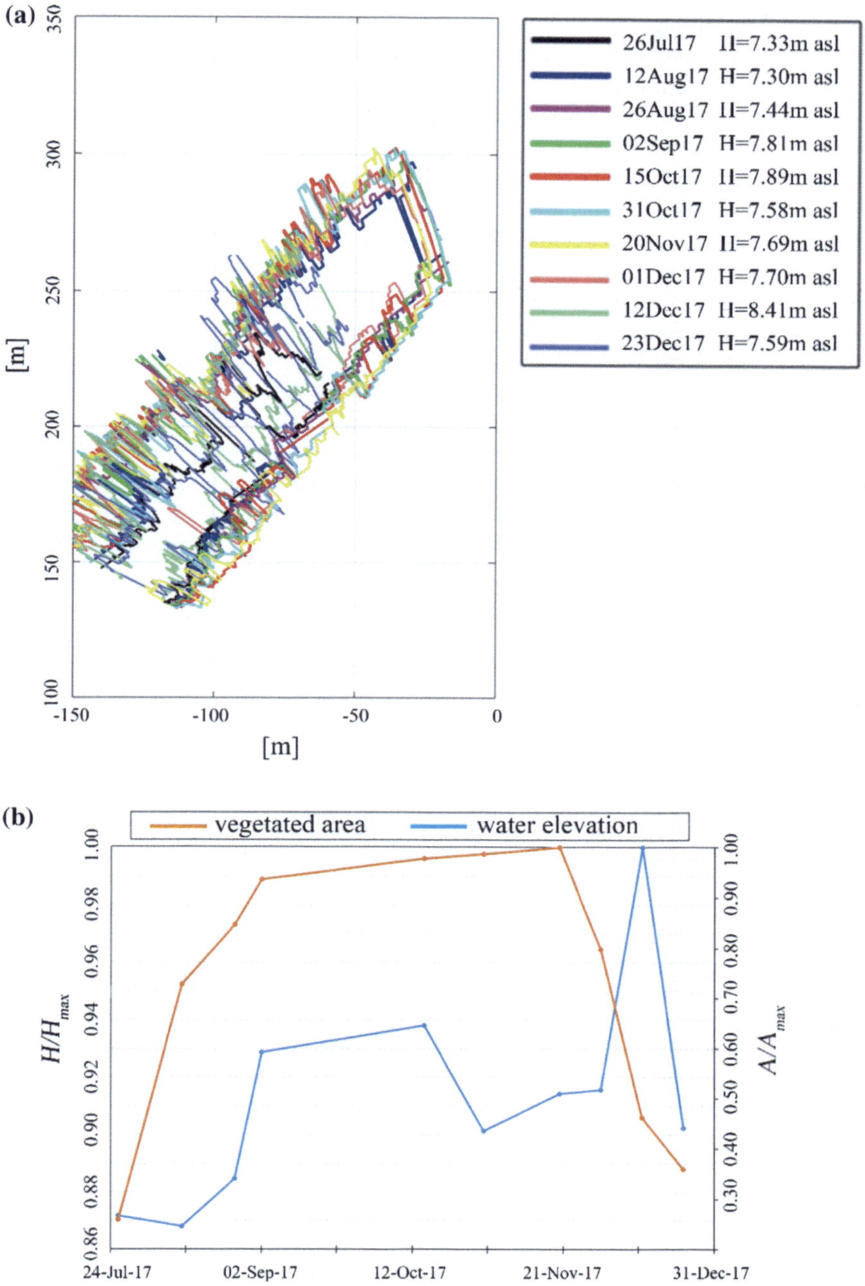

Fig. 3 **a** Monitored vegetation bounds, with indications of measured water elevations at the acquisition date; **b** computed vegetated areas covering the sandbar and measured water elevations at the gauging station of Revere

that a prolonged drought phase can enhance the encroachment of seasonal plants until the season is favourable, fixing the sandbar and changing the natural morphological pattern towards a single-thread one (Bertoldi et al. 2014). Indeed, if the vegetation colonizes the sandbar in a stable manner, the active channel, where water and sediments are conveyed, becomes narrow and deeper, with possible consequences on infrastructures and local habitat. However, a longer period of monitoring is necessary to adequately characterize the reciprocal feedbacks between the local hydrology, the bar morphology and the vegetative dynamics.

Although correlated with several uncertainties, this method can represent a good way to estimate the vegetation growth rate on a fluvial sandbar. On the one hand, there are many strengths in using time-lapse photography to monitor fluvial vegetation: (i) it is a low-cost, low-maintenance, power-efficient method; (ii) it can be used to detect trends at different temporal scales, from days to years; (iii) it provides rough estimates of vegetation growth rates allowing for a comparison with the water levels. On the other hand, several limitations are embedded in the method: (i) estimates are imprecise, depending both on the number and positioning of the ground control points and the camera calibration parameters; (ii) being the results affected by the camera characteristics, its choice is paramount in obtaining images with an adequate resolution and covering the desired area; (iii) the computed vegetated areas rely on many assumptions on the vegetation dynamics (density, type of vegetation, seasonality, etc.).

For the future, therefore, additional works should be focused on the improvement of the precision of the methodology, analyzing new ground control points better distributed to reduce the positioning errors (Rangel et al. 2018). Moreover, field samples of the local vegetation are necessary to corroborate the assumptions made in the analysis, as well as to recognize the growing stage. Given that the time range is paramount in the biological dynamics, a necessary and obvious next step is the evaluation of a longer period, covering alternating dry and wet conditions, and not just a single flood event, as well as several seasons, allowing for an analysis of the vegetation response to significant and frequent changes of the fluvial hydrology.

Acknowledgements This research has been partially developed in the framework of the project INFRASAFE—Monitoraggio intelligente per infrastrutture sicure, April 2016–March 2018, founded by the Emilia-Romagna Region of Italy, through the POR FESR 2014–2020.

Part of the work performed by Michael Nones was supported by the statutory activities No. 3841/E-41/S/2018 of the Ministry of Science and Higher Education of Poland.

References

Aarninkhof SGJ, Turner IL, Dronkers TDT, Caljouw M, Nipius L (2003) A video-based technique for mapping intertidal beach bathymetry. Coast Eng 49(4):275–289

Archetti R, Romagnoli C (2011) Analysis of the effects of different storm events on shoreline dynamics of an artificially embayed beach. Earth Surf Proc Land 36(11):1449–1463

Archetti R, Paci A, Carniel S, Bonaldo D (2016) Optimal index related to the shoreline dynamics during a storm: the case of Jesolo beach. Nat Hazards Earth Syst Sci 16(5):1107–1122

Bertoldi W, Siviglia A, Tettamanti S, Toffolon M, Vetsch D, Francalanci S (2014) Modeling vegetation controls on fluvial morphological trajectories. Geophys Res Lett 41(20):7167–7175

Chandler JH (1999) Effective application of automated digital photogrammetry for geomorphological research. Earth Surf Proc Land 24(1):51–63

Deacy WW, Leacock WB, Eby LA, Stanford JA (2016) A time-lapse photography method for monitoring salmon (*Oncorhynchus* spp.) passage and abundance in streams. PeerJ 4:e2120

Domeneghetti A, Carisi F, Castellarin A, Brath A (2015) Evolution of flood risk over large areas: quantitative assessment for the Po River. J Hydrol 527:809–823

Guerrero M, Lamberti A (2007) Clouds image processing for velocity acquisition. In: Proceedings of the 32nd IAHR world congress, Venice, Italy

Guerrero M, Di Federico V, Lamberti A (2013) Calibration of a 2-D morphodynamic model using water-sediment flux maps derived from an ADCP recording. J Hydroinformatics 15(3):813–828

Kramer N, Wohl E (2014) Estimating fluvial wood discharge using time-lapse photography with varying sampling intervals. Earth Surf Proc Land 39(6):844–852

Kyes TA, Jones CN, Scott DT, Chuquin D (2016) A cost-effective image processing approach for analyzing the ecohydrology of river corridors. Limnol Oceanogr Methods 14(6):359–369

Lanzoni S, Luchi R, Bolla Pittaluga M (2015) Modeling the morphodynamic equilibrium of an intermediate reach of the Po River (Italy). Adv Water Resour 81:92–102

Luchi R, Zolezzi G, Tubino M (2010) Modelling mid-channel bars in meandering channels. Earth Surf Proc Land 35(8):902–917

MacVicar B, Piégay H (2012) Implementation and validation of video monitoring for wood budgeting in a wandering piedmont river, the Ain River (France). Earth Surf Proc Land 37(12):1272–1289

Montanari A (2012) Looking for changing patterns in river discharge. Hydrol Earth Syst Sci 16:3739–3747

Nagler PL, Jarchow CJ, Glenn EP (2018) Remote sensing vegetation index methods to evaluate changes in greenness and evapotranspiration in riparian vegetation in response to the Minute 319 environmental pulse flow to Mexico. Proc IAHS 380:45–54

Nichols MH, Nearing M, Hernandez M, Polyakov VO (2016) Monitoring channel head erosion processes in response to an artificially induced abrupt base level change using time-lapse photography. Geomorphology 265:107–116

Nones M, Pugliese A, Domeneghetti A, Guerrero M (2018a) Po River morphodynamics modelled with the open-source code iRIC. In: Kalinowska M, Mrokowska M, Rowiński P (eds) Free surface flows and transport processes. GeoPlanet: earth and planetary sciences. Springer, Cham, Switzerland

Nones M, Archetti R, Guerrero M (2018b) Time-lapse photography of the edge-of-water line displacement of a sandbar as a proxy of riverine morphodynamics. Water 10(5):617

Nones M (2019) Numerical modelling as a support tool for river habitat studies: an Italian case study. Water 11(3):482

Pyle CJ, Richards KS, Chandler JH (1997) Digital photogrammetric monitoring of river bank erosion. Photogram Rec 15(89):753–764

Rangel JMG, Gonçalves GR, Pérez JA (2018) The impact of number and spatial distribution of GCPs on the positional accuracy of geospatial products derived from low-cost UASs. Int J Remote Sens 39(21):7154–7171

Räpple B, Piégay H, Stella JC, Mercier D (2017) What drives riparian vegetation encroachment in braided river channels at patch to reach scales? Insights from annual airborne surveys (Drôme River, SE France, 2005–2011). Ecohydrology 10(8):e1886

Schoener G (2018) Time-lapse photography: low-cost, low-tech alternative for monitoring flow depth. J Hydrol Eng 23(2):06017007

Serlet AJ, Gurnell AM, Zolezzi G, Wharton G, Belleudy P, Jourdain C (2018) Biomorphodynamics of alternate bars in a channelized, regulated river: An integrated historical and modelling analysis. Earth Surf Proc Land 43(9):1739–1756
Stojic M, Chandler JH, Ashmore P, Luce J (1998) The assessment of sediment transport rates by automated digital photogrammetry. Photogramm Eng Remote Sens 64(5):387–395
Surian N, Rinaldi M (2003) Morphological response to river engineering and management in alluvial channels in Italy. Geomorphology 50(4):307–326
Vesipa R, Camporeale C, Ridolfi L (2017) Effect of river flow fluctuations on riparian vegetation dynamics: Processes and models. Adv Water Resour 110:29–50
Vicente-Serrano SM, Gouveia C, Camarero JJ, Beguería S, Trigo R, López-Moreno JI, Azorin-Molina C, Pasho E, Lorenzo-Lacruz J, Revuelto J, Moran-Tejeda E, Sanchez-Lorenzo A (2013) Response of vegetation to drought time-scales across global land biomes. PNAS 110:52–57
Xia H, Kong W, Li X, Zhang Y, Guo F, Sun OJ (2018) Variations in herbaceous vegetation structures and vegetation-environment relationships from floodplain to terrace along a large semi-humid river. Ecol Res 33(5):1049–1058

Hydrodynamics of Water-Worked and Screeded Gravel-Bed Flows

Ellora Padhi, Nadia Penna, Subhasish Dey and Roberto Gaudio

Abstract The turbulence parameters over a water-worked gravel-bed (WGB) and a screeded gravel-bed (SGB) are studied, measuring the flow by a particle image velocimetry system in a rectangular flume keeping the flow conditions identical in both the beds. Owing to the water action, the orientation and the alignment of the surface gravels in the WGB were entirely different from those in the SGB. As a result, the WGB possesses higher roughness than the SGB. Analysis of vorticity contours infers that the gravels cause the fluid streaks to have both clockwise and counterclockwise motions in the near-bed flow zone. Besides, the double-averaged streamwise velocity, spatially averaged (SA) Reynolds and dispersive shear stress, SA and dispersive turbulent kinetic energy (TKE) fluxes, and SA TKE budget were analyzed in both the beds. Owing to the higher roughness in the WGB than in the SGB, the turbulence parameters, especially in the near-bed flow zone, remain higher in the former than in the latter.

Keywords Water-worked beds · Turbulent flow · Open-channel flow · Gravel-bed flow

1 Introduction

Laboratory experiments to study the turbulence phenomena in open-channel flows are usually carried out under idealized gravel beds, called the 'screeded gravel beds' (SGBs), where the bed roughness structures created by the gravels are randomly poised. By contrast, in natural gravel-bed rivers, the bed roughness is formed by the water action on surface gravels. As a result, the natural riverbeds possess roughness structures having quasi-organized orientation, alignment, and spacing. Such

E. Padhi (✉) · S. Dey
Department of Civil Engineering, Indian Institute of Technology, Kharagpur, West Bengal 721302, India
e-mail: ellora@iitkgp.ac.in

N. Penna · R. Gaudio
Dipartimento di Ingegneria Civile, Università della Calabria, 87036 Rende, CS, Italy

M. B. Kalinowska et al. (eds.), *Recent Trends in Environmental Hydraulics*, GeoPlanet: Earth and Planetary Sciences,
https://doi.org/10.1007/978-3-030-37105-0_18

beds are termed 'water-worked gravel-beds' (WGBs). To resolve the spatial flow heterogeneity in a gravel bed flow, the Double Averaging Methodology (DAM) was implemented in macro-rough flows (Dey and Das 2012; Dey 2014). In a gravel bed flow, the near-bed flow zone is called 'roughness layer' and is influenced by the individual gravels. The roughness layer is divided into the form-induced and the interfacial sublayers. The form-induced sublayer is the flow layer that occupies the volume of the fluid above the crest level up to the top edge of the roughness layer. On the other hand, the interfacial sublayer is the layer that occupies the volume of the fluid below the crest level.

Nikora et al. (1998) conducted an experiment in the WGB to study the effects of water work on bed topography. Barison et al. (2003) claimed that owing to the water action, the near-bed spatial distribution of time-averaged velocity is different in the WGBs from the SGBs. The presence of spatially coherent flow structures was discovered in the near-bed flows in the WGBs (Cooper and Tait 2008; Hardy et al. 2009). Cooper and Tait (2010) observed the spatial pattern of time-averaged velocity in the near-bed flows in both the WGBs and SGBs. However, in recent years, to explore the effects of water work on the DA streamwise velocity, Reynolds shear and normal stresses, dispersive shear stress, etc., some researchers were interested in conducting experimental studies on the WGBs (Cooper et al. 2013; Padhi et al. 2018a). Despite these experimental studies on the WGBs, a detailed analysis of turbulence characteristics, such as SA and dispersive TKE fluxes, and SA TKE budget for a WGB is still to be performed.

Therefore, this study aims to examine the DA streamwise velocity and SA turbulent flow parameters, e.g. SA Reynolds and dispersive shear stress, SA and dispersive streamwise and vertical TKE fluxes, and SA TKE budget, in a WGB with respect to an SGB keeping the flow conditions identical in both the beds. Further, the robustness of Kolmogorov's two-thirds law in the WGB and SGB is verified.

2 Experimental Setup

Experiments were conducted in a rectangular tilting flume of length 9.6 [m], width 0.485 [m], and depth 0.50 [m] for both, WGB and SGB. Uniform coarse gravel with median size $d_{50} = 4.81$ [mm] was used to create the SGB. The flow measurements were taken at a distance of 7.18 [m] from the flume inlet, where the fully developed turbulent flow prevailed.

Initially, the gravels were placed in the flume and screeded manually to form an SGB, whose surface fluctuations were measured by a 3D terrestrial laser scanner. Next, a WGB was created transporting surface gravels with a higher flow over the SGB. The bed slopes in the WGB and SGB were found to be 0.4% and 0.7%, respectively. From the bed topography, the crests of the beds, z_c, were found to be at $z = 1.96$ [mm] and 2.83 [mm] from the virtual bed level in the WGB and SGB, respectively, with average roughness heights of $\Delta_k = 1.25$ [mm] and 1.04 [mm]. The details of the WGB formation were reported in Padhi et al. (2018a).

The flow measurements were carried out for a flow depth of $h = 0.10$ [m] and average flow velocity of 0.43 [m s^{-1}] in a clear-water condition, using a two-dimensional (2D) particle image velocimetry (PIV) system (see Padhi et al. 2018a). From the Reynolds shear stress (RSS) distributions, the shear velocities $u_*\left[= \left(-\langle\overline{u'w'}\rangle\right)^{0.5}\right.$, where u' and w' are the fluctuations of streamwise and vertical velocity components, respectively] were obtained as 0.068 [m s^{-1}] and 0.077 [m s^{-1}] in the WGB and SGB, respectively. The shear Reynolds number R_* ($= u_*\Delta_k/\nu$, where ν is the kinematic viscosity, assumed as 10^{-6} [m^2 s^{-1}]) was obtained as 85 and 81 in the WGB and SGB, respectively, confirming the rough-turbulent flow.

3 Results and Discussion

3.1 *Time-Averaged Vorticity, Flow Field and DA Streamwise Velocity*

Dimensionless time-averaged vorticity $\bar{\omega}h/u_*$ contours and flow velocity vectors [magnitude $\left(\bar{u}^2 + \bar{w}^2\right)^{0.5}$ and direction $\tan^{-1}(\bar{w}/\bar{u})$] on a dimensionless central vertical plane (x/d_{50} by z/h) in the WGB and SGB are shown in Fig. 1. Here, $\bar{u}$ and $\bar{w}$ are the time-averaged streamwise and vertical velocity components, respectively, and $\bar{\omega}$ is the time-averaged vorticity, whose negative and positive signs indicate the

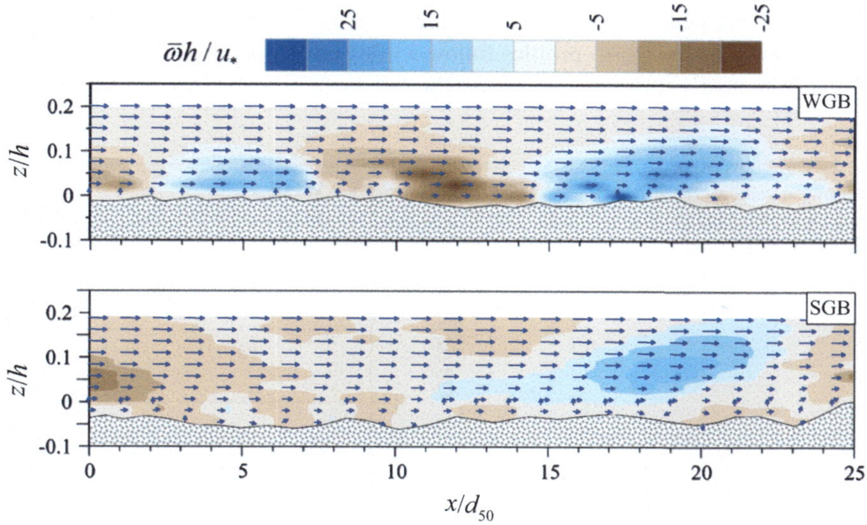

Fig. 1 Time-averaged vorticity contours $\bar{\omega}h/u_*$ and velocity vectors on a central vertical plane in the WGB and SGB. The vector → 0.25 [m s^{-1}]

anticlockwise and clockwise fluid motions, respectively. In fact, the anticlockwise fluid motion is to accelerate the flow, causing a downward momentum transport in the streamwise direction. By contrast, the clockwise fluid motion is to decelerate the flow, resulting in an upward momentum transport in the upstream direction. Interestingly, in the near-bed flow, $\bar{\omega}$ changes its sign alternatively throughout the measuring section. It implies that the gravels incite the flow to be heterogeneous. As a result, flow perturbation occurs. Further, the change of sense of fluid motions is more frequent in the WGB than in the SGB. It is attributed to the higher WGB roughness than the SGB. To substantiate the near-bed flow heterogeneity, the velocity vectors are closely examined. In the near-bed flow, the effects of spatial heterogeneity are apparent from the scattered velocity vectors owing to the flow mixing induced by the gravels. In essence, the near-bed velocity vectors in the WGB are less scattered than the SGB, because the orientations of the gravels were more organized in the former than in the latter.

Figure 2 represents the variations of dimensionless DA streamwise velocity $\langle \bar{u} \rangle / u_*$ with $(z + \Delta z)/\Delta z$ in the WGB and SGB, where Δz is the zero-plane displacement. To fit the data to a logarithmic law within the extremity of the wall shear layer from the crest, the DA streamwise velocity $\langle \bar{u} \rangle$ and z are scaled by u_* and Δz, respectively. The logarithmic law is expressed as $\langle \bar{u} \rangle / u_* = \kappa^{-1} \ln[(z + \Delta z)/z_0]$ where κ is the von Kármán coefficient (=0.41) and z_0 is the zero velocity level. Using the method proposed by Dey and Das (2012), in the WGB, Δz and z_0 were obtained as 2.3×10^{-3} [m] and 1.7×10^{-3} [m], respectively. Similarly, in the SGB, the Δz and z_0 were obtained as 4.81×10^{-3} [m] and 3.59×10^{-3} [m], respectively.

Referring to Fig. 2, in the near-bed flow, the $\langle \bar{u} \rangle / u_*$ profiles have an inflectional shape owing to the effects of flow mixing within the interstices of gravels and a momentum sink within the interfacial sublayer. Owing to the subsurface flow underneath the gravels, the $\langle \bar{u} \rangle / u_*$ profiles follow a third-order polynomial law below the crest and are nearly invariant to z toward the roughness trough (Dey and Das 2012).

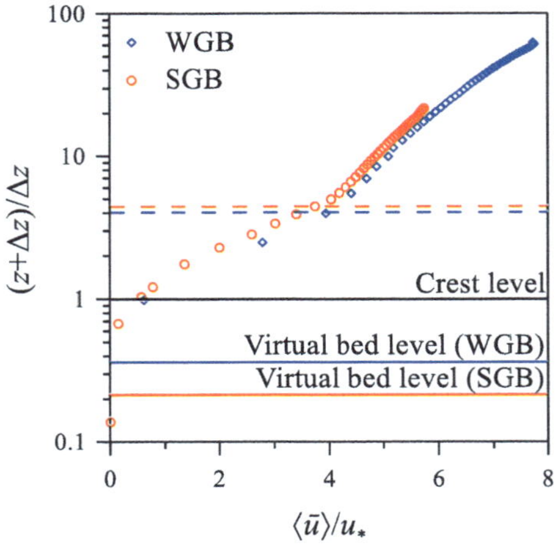

Fig. 2 Variations of DA streamwise velocity $\langle \bar{u} \rangle / u_*$ with $(z + \Delta z)/\Delta z$ in the WGB and SGB

Interestingly, the $\langle\bar{u}\rangle/u_*$ profiles follow a linear law up to the top edge of the form-induced sublayer. It happens because of the extended effect of the bed roughness. Above the form-induced sublayer, the profiles exhibit a logarithmic variation with z toward the free surface. The reason is accredited to the gravel roughness, being in conformity with Pokrajac et al. (2008) and Mignot et al. (2009). It is apparent that in the near-bed flow zone, the $\langle\bar{u}\rangle/u_*$ is greater in the WGB than in the SGB. The reason is ascribed to the well-organized roughness structure in the WGB. The water-work action orients the longest axis of the surface gravels to be streamwise, which glides the WGB flow to be streamlined relative to the SGB flow.

3.2 SA Reynolds Shear Stress and Dispersive Shear Stress

For a steady, uniform flow over a macro-rough bed, the SA total shear stress $\langle\bar{\tau}\rangle$ can be expressed as $\langle\bar{\tau}\rangle = \langle -\overline{u'w'}\rangle + \langle -\tilde{u}\tilde{w}\rangle + \mu \mathrm{d}\langle\bar{u}\rangle/\mathrm{d}z$, where $\langle -\overline{u'w'}\rangle$ is the SA RSS, $\langle -\tilde{u}\tilde{w}\rangle$ is the dispersive shear stress, $\tilde{u}$ and $\tilde{w}$ are the spatial fluctuations of $\bar{u}$ and $\bar{w}$, $\mu \mathrm{d}\langle\bar{u}\rangle/\mathrm{d}z$ is the DA viscous shear stress and μ is the dynamic viscosity of water. Further, $\langle\bar{\tau}\rangle$ follows the linear profile $\langle\bar{\tau}(z/h \geq 1)\rangle \times u_*^{-2} = 1 - (z/h)$. Importantly, the DA viscous shear stress is vanishingly small, because the flow is rough. Hence, $\langle\bar{\tau}\rangle$ approximately equals $\langle -\overline{u'w'}\rangle + \langle -\tilde{u}\tilde{w}\rangle$, at a given z.

In the WGB and SGB, the variations of the dimensionless SA RSS $\langle -\overline{u'w'}\rangle/u_*^2$ with z/h are depicted in Fig. 3. In both the beds, the $\langle -\overline{u'w'}\rangle$ profiles start developing from the interfacial sublayer and then grow gradually with an increase in z until they

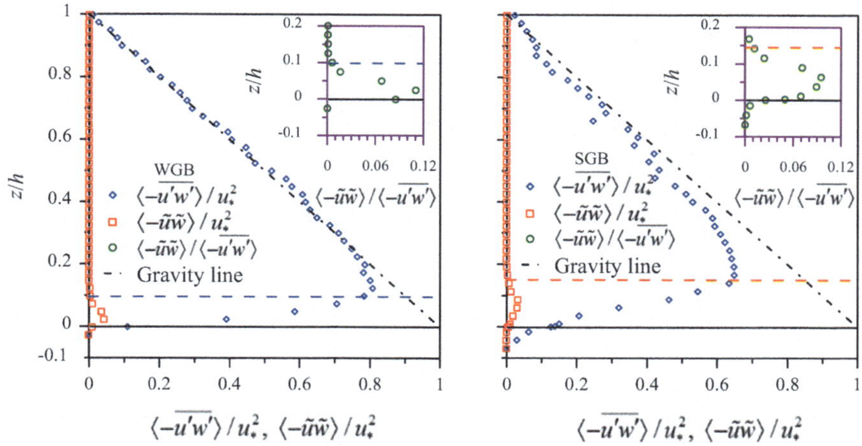

Fig. 3 Variations of SA RSS $\langle -\overline{u'w'}\rangle/u_*^2$ and dispersive shear stress $\langle -\tilde{u}\tilde{w}\rangle/u_*^2$ with z/h in the WGB and SGB. In the respective insets, variations of the ratio of dispersive shear stress and SA RSS $\langle -\tilde{u}\tilde{w}\rangle/\langle -\overline{u'w'}\rangle$ with z/h in the WGB and SGB are shown

reach their peak values. The peaks of $\langle -\overline{u'w'} \rangle$ appear at $z/h = 0.125$ and 0.145 in the WGB and SGB, respectively. It can be noted that in both the beds, after reaching the peak values, the $\langle -\overline{u'w'} \rangle$ profiles follow the linear law (linear gravity line). However, owing to the effects of near-bed flow heterogeneity, the $\langle -\overline{u'w'} \rangle$ profiles diminish. As a result, within the roughness layer, the $\langle -\overline{u'w'} \rangle$ profiles depart from the linear gravity line. Interestingly, in the near-bed flow, the reduction in $\langle -\overline{u'w'} \rangle$ is compensated for by the appearance of dispersive shear stress $\langle -\tilde{u}\tilde{w} \rangle$, as shown in Fig. 3. It is pertinent to mention that the term $\langle -\tilde{u}\tilde{w} \rangle$ results from the form-induced fluctuations owing to the spatial flow heterogeneity (Manes et al. 2007; Dey and Das 2012). Further, from Fig. 3, one may notice that the SA RSS peaks appear at the top edge of the form-induced sublayer. According to Nezu and Nakagawa (1993), the gravels cause the flow perturbations to enhance. It implies that u' and w' reach their maxima at this level owing to an intense flow mixing process in the near-bed flow zone. This corroborates the observations of Manes et al. (2007) and Ferraro et al. (2016). Moreover, comparison of $\langle -\tilde{u}\tilde{w} \rangle$ profiles in both the beds show that in the near-bed flow, the higher roughness in the WGB induces a greater value of $\langle -\overline{u'w'} \rangle$ than that in the SGB. This indicates that the bed roughness is directly associated with the u' and w' and, in turn, with the SA RSS.

Reverting to the $\langle -\tilde{u}\tilde{w} \rangle$ profiles, they start developing from $z/h = 0.10$ and 0.14 in the WGB and SGB, respectively. These levels represent the lower extremities of the wall shear layer and are considered to be the threshold of form-induced sublayer development (Dey and Das 2012). It is apparent that in both, WGB and SGB, the $\langle -\tilde{u}\tilde{w} \rangle$ profiles grow gradually as z decreases within the form-induced sublayer. The $\langle -\tilde{u}\tilde{w} \rangle$ profiles attain their peaks at $z/h = 0.025$ in the WGB and 0.08 in the SGB, and then decline with a further decrease in z in both the beds. After Dey and Das (2012), within the form-induced sublayer, the $\tilde{u}$ and $\tilde{w}$ are larger and thus they give rise to higher $\langle -\tilde{u}\tilde{w} \rangle$ values. By contrast, within the interfacial sublayer ($z < 0$), the $\tilde{u}$ and $\tilde{w}$ are considerably small and thus the $\langle -\tilde{u}\tilde{w} \rangle$ profiles reduce considerably. Comparing the results, it is found that at a given z, the $\langle -\tilde{u}\tilde{w} \rangle$ profile is greater in the WGB than in the SGB. It implies that the higher roughness in the WGB than in the SGB causes to enhance the $\tilde{u}$ and $\tilde{w}$.

Besides, insets of Fig. 3 illustrate the variations of the ratio of dispersive shear stress to SA RSS $\langle -\tilde{u}\tilde{w} \rangle / \langle -\overline{u'w'} \rangle$ with z/h in the WGB and SGB. Since the $\langle -\tilde{u}\tilde{w} \rangle$ profiles are quite small for $z/h > 0.20$, the $\langle -\tilde{u}\tilde{w} \rangle / \langle -\overline{u'w'} \rangle$ ratios in both the beds almost vanish. Thus, the ordinates are shown up to $z/h = 0.20$. The $\langle -\tilde{u}\tilde{w} \rangle / \langle -\overline{u'w'} \rangle$ profiles start growing from the top edge of the form-induced sublayer and then gradually increase with a decrease in z. The $\langle -\tilde{u}\tilde{w} \rangle / \langle -\overline{u'w'} \rangle$ ratios attain their peaks at $z/h = 0.025$ in the WGB and 0.06 in the SGB and then they decrease rapidly with a further decrease in z. The damping in $\langle -\tilde{u}\tilde{w} \rangle / \langle -\overline{u'w'} \rangle$ is associated with the reduction in $\langle -\tilde{u}\tilde{w} \rangle$ relative to $\langle -\overline{u'w'} \rangle$ (Dey and Das 2012). Hence, it implies that the spatial velocity fluctuations are responsible, rather than the temporal velocity fluctuations, to reduce $\langle -\tilde{u}\tilde{w} \rangle / \langle -\overline{u'w'} \rangle$.

3.3 SA and Dispersive Turbulent Kinetic Energy Fluxes

In a two dimensional turbulent flow, the SA streamwise and vertical TKE fluxes ($\langle f_{kw}\rangle$ and $\langle f_{kw}\rangle$) are given by $0.75\left(\langle\overline{u'u'u'}\rangle + \langle\overline{u'w'w'}\rangle\right)$ and $0.75\left(\langle\overline{u'u'w'}\rangle + \langle\overline{w'w'w'}\rangle\right)$, respectively (Dey 2014). Similarly, the dispersive streamwise and vertical TKE fluxes ($\langle f_{fu}\rangle$ and $\langle f_{fw}\rangle$) are expressed as $0.75\left(\langle\tilde{u}\tilde{u}\tilde{u}\rangle + \left\langle\overline{\tilde{u}\tilde{w}\tilde{w}}\right\rangle\right)$ and $0.75\left(\langle\tilde{u}\tilde{u}\tilde{w}\rangle + \left\langle\overline{\tilde{w}\tilde{w}\tilde{w}}\right\rangle\right)$, respectively. Figure 4 represents the variations of dimensionless SA streamwise and vertical TKE fluxes $\left(\langle f_{ku}\rangle/u_*^3 \text{ and } \langle f_{kw}\rangle/u_*^3\right)$ with z/h in the WGB and SGB. In both the beds, the $\langle f_{ku}\rangle$ profiles start with positive values, attaining their peaks above the crest, and then diminish with a further increase in z. At $z/h = 0.10$ in the WGB and 0.089 in the SGB, the $\langle f_{ku}\rangle$ profiles become negative and are continued for remaining portion of z. The positive and negative signs infer a downstream and an upstream transport of $\langle f_{ku}\rangle$, respectively. Moreover, for a given z, the $\langle f_{ku}\rangle$ in the WGB is greater than that in the SGB, which is attributed to high temporal velocity fluctuations induced by the higher WGB roughness.

Examining the $\langle f_{kw}\rangle$ profiles, it is evident that they initiate with small negative values, attaining their negative peaks at $z/h = 0.05$ and 0.025 in the WGB and SGB, respectively. Thereafter, they diminish with an increase in z, becoming positive at $z/h = 0.10$ in the WGB and 0.089 in the SGB. The negative sign signifies a downward transport of $\langle f_{kw}\rangle$ and the positive sign suggests an upward transport. However, for a given z, the $\langle f_{kw}\rangle$ in the WGB is larger than that in the SGB and the reason is the same as before.

Interestingly, the $\langle f_{ku}\rangle$ and $\langle f_{kw}\rangle$ profiles provide information about the bursting events. The combination of a positive $\langle f_{ku}\rangle$ and a negative $\langle f_{kw}\rangle$ gives rise to the sweep

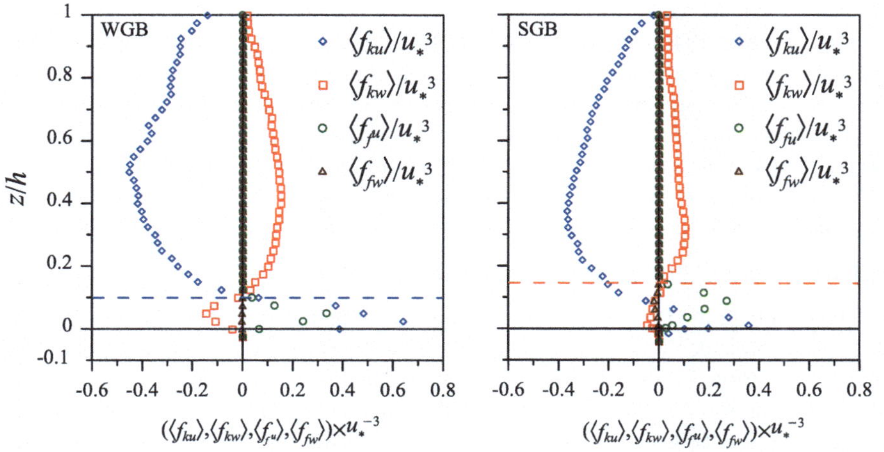

Fig. 4 Variations of SA streamwise TKE flux $\langle f_{kw}\rangle/u_*^3$, SA vertical TKE flux $\langle f_{kw}\rangle/u_*^3$, dispersive streamwise TKE flux $\langle f_{kw}\rangle/u_*^3$, and dispersive vertical TKE flux $\langle f_{kw}\rangle/u_*^3$ with z/h in the WGB and SGB

events. On the other hand, the combination of a negative $\langle f_{ku}\rangle$ and a positive $\langle f_{kw}\rangle$ gives rise to the ejection events. The results infer that the sweeps are the dominating events for $z/h \leq 0.10$ in the WGB and $z/h \leq 0.089$ in the SGB, while the ejections become the governing mechanism for $z/h > 0.10$ in the WGB and $z/h > 0.089$ in the SGB (Dey and Das 2012).

In addition, the variations of the dimensionless dispersive streamwise and vertical fluxes ($\langle f_{fu}\rangle$ and $\langle f_{fw}\rangle$) with z/h in the WGB and SGB are depicted in Fig. 4. In both the beds, the $\langle f_{fu}\rangle$ profiles start with the positive values, increasing gradually as one moves toward the top edge of the roughness layer. They attain their positive peaks above the crest, and thereafter tend to decrease with a further increase in z. The $\langle f_{fu}\rangle$ profiles almost vanish above the roughness layer. Unlike the $\langle f_{fu}\rangle$ profiles, the $\langle f_{fw}\rangle$ profiles start growing with the negative values, as z increases. They attain their negative peaks above the crest in both the beds. After achieving their peaks, they diminish with an increase in z, becoming negligible above the roughness layer. It is inferable that $\tilde{w}$ is lesser than $\tilde{u}$, resulting in a smaller $\langle f_{fw}\rangle$ than $\langle f_{fu}\rangle$. Comparative study shows that owing to higher $\tilde{u}$ and $\tilde{w}$ values in the WGB than in the SGB, both the $\langle f_{fu}\rangle$ and $\langle f_{fw}\rangle$ profiles attain higher magnitude in the former than in the latter.

3.4 SA Turbulent Kinetic Energy Dissipation and Budget

The structure function fitting method is one of the most useful tools to estimate the TKE dissipation ε. Following Kolmogorov's second similarity hypothesis, one can establish the relationship of the second-order velocity increment structure function with the streamwise separation distance r and the ε within the inertial subrange (Frisch 1995). According to this hypothesis, the second-order streamwise velocity increment structure function is defined as $\langle \Delta u\rangle^2 = \left\langle [u'(x+r) - u'(x)]^2 \right\rangle$ where Δu is the streamwise velocity increment along a spatial distance in a streamwise direction [m s^{-1}], r is the separation distance between two measuring locations [m], x is the measuring distance in the streamwise direction from a suitable starting location [m]. Following the said hypothesis, $\langle \Delta u\rangle^2$ within the inertial subrange is $\langle \Delta u\rangle^2 = C_2\langle \varepsilon r\rangle^{2/3}$, where C_2 is a universal constant equaling 2.12 (Sreenivasan 1995). It is called the two-thirds law and is valid only for a steady, homogeneous, isotropic turbulence (Frisch 1995). Thus, the SA TKE dissipation is $\langle \varepsilon\rangle = r^{-1}\left(\left\langle \overline{\Delta u}\right\rangle^2 / C_2\right)^{3/2}$.

According to the two-thirds law, $\overline{\Delta u^2}$ roughly equals the two-thirds power of the separation distance r within the inertial subrange. Therefore, it is important to validate the said criterion. To do this, the variations of dimensionless second-order velocity structure function $\overline{\Delta u^2}/u_*^2$ with dimensionless separation distance r/d_{50} for different z/h in the WGB and SGB are plotted in Fig. 5. For the validation of the two-thirds law, a 2/3 slope line is also shown. Having a close examination of Fig. 5, in both the beds, one can find a specific trend of $\overline{\Delta u^2}/u_*^2$ with the 2/3 slope at small values of r/d_{50}. It is evident that the $\overline{\Delta u^2}$ increases with a decrease in z value. In the near-bed flow zone at $z/h = 0.05$, the $\overline{\Delta u^2}$ is maximum for both the beds. It signifies that the

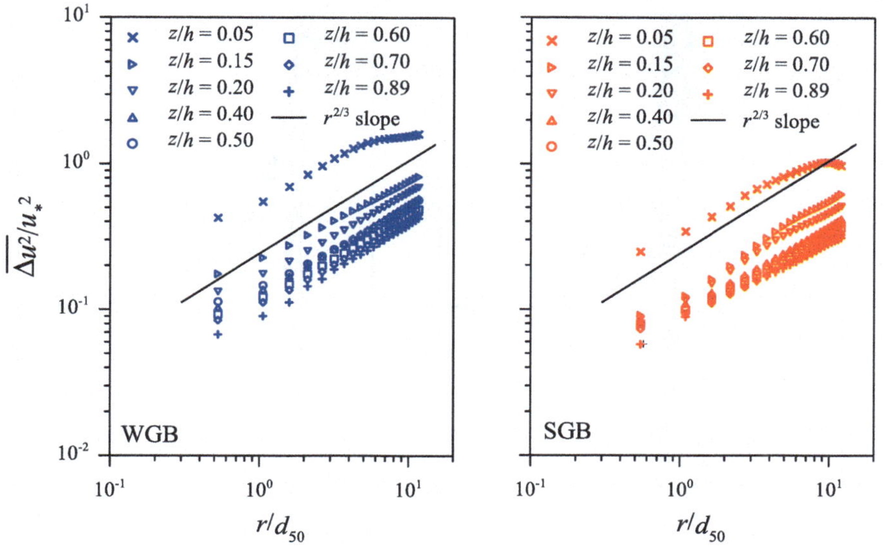

Fig. 5 Variations of second-order velocity structure function $\overline{\Delta u^2}/u_*^2$ with r/d_{50} for different z/h in the WGB and SGB

gravels cause an enhancement of u' in the near-bed flow. In addition, in the WGB, the $\overline{\Delta u^2}$ profiles for all z/h are almost parallel to the 2/3 slope for $r/d_{50} \approx 1.2$–4.9, while in the SGB, they are parallel for $r/d_{50} \approx 1.2$–4.3. The aforementioned ranges are deemed to be the inertial subrange in the respective beds. Therefore, Kolmogorov's two-thirds law is preserved within the inertial subranges in both, WGB and SGB.

Figure 6 shows the variations of the two-thirds law (made dimensionless multiplying by h/u_*^3) with the dimensionless separation distance r/d_{50} in the WGB and SGB. In both the beds, for all z/h values, black horizontal broken lines indicate the plateaus of the two-thirds law profiles. The plateau indicates the SA TKE dissipations

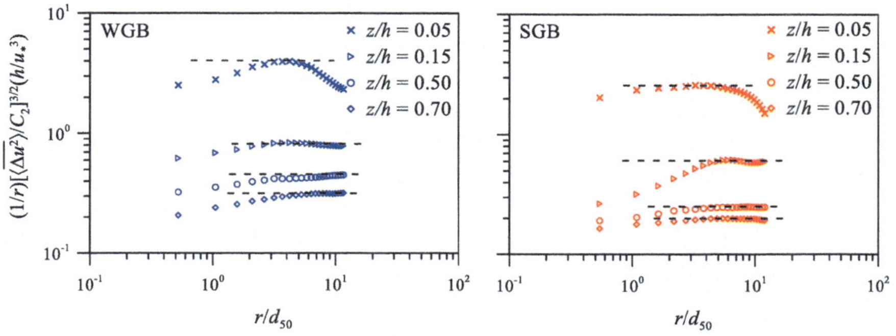

Fig. 6 Variations of Kolmogorov's SA two-thirds law with r/d_{50} for different z/h in the WGB and SGB

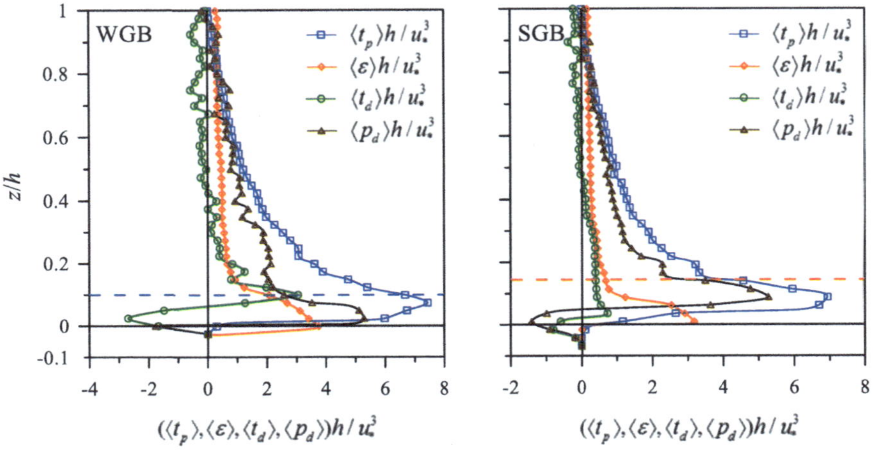

Fig. 7 Variations of SA TKE production $\langle t_p\rangle h/u_*^3$, SA TKE dissipation $\langle \varepsilon\rangle h/u_*^3$, SA TKE diffusion $\langle t_d\rangle h/u_*^3$ and SA pressure energy diffusion $\langle p_d\rangle h/u_*^3$ with *z*/*h* in the WGB and SGB

$\langle \varepsilon\rangle$ corresponding to a given *z*/*h* (Padhi et al. 2018b). It is apparent that the $\langle \varepsilon\rangle$ values increase with a decrease in *z*. For *z*/*h* = 0.70 and 0.50, the two-thirds law profiles are quite consistent. However, for *z*/*h* = 0.15 and 0.05 (in the near-bed flow zone), the two-thirds law profiles do not remain the same. They initially increase with an increase in separation distance *r*, reaching their respective peaks and then decrease, as *r* increases. The plausible reason behind this is attributed to the flow mixing in the near-bed flow zone induced by the gravels. It can be noted that using the two-thirds law, the $\langle \varepsilon\rangle$ profiles can be obtained over the entire flow depth, as shown in Fig. 7 in both, WGB and SGB.

The SA TKE budget equation for a flow over a gravel bed is given by

$$\underbrace{\langle\overline{u'w'}\rangle\frac{\partial\langle\bar{u}\rangle}{\partial z}}_{\langle t_p\rangle} = \langle\varepsilon\rangle + \underbrace{\frac{\partial\langle f_{kw}\rangle}{\partial z}}_{\langle t_d\rangle} + \underbrace{\frac{1}{\rho}\frac{\partial\langle\overline{p'w'}\rangle}{\partial z}}_{\langle p_d\rangle} - \underbrace{\nu\frac{\partial^2\langle k\rangle}{\partial z^2}}_{\langle \nu_d\rangle} \tag{1}$$

where $\langle t_p\rangle$ is the SA TKE production, $\langle t_d\rangle$ is the SA TKE diffusion, $\langle p_d\rangle$ is the SA pressure energy diffusion, $\langle \nu_d\rangle$ is the viscous diffusion and p' indicates the pressure fluctuations. For a rough turbulent flow, the effect of viscous diffusion $\langle \nu_d\rangle$ is negligible (Dey 2014). Further, $\langle p_d\rangle$ is estimated as $\langle p_d\rangle = \langle t_p\rangle - \langle\varepsilon\rangle - \langle t_d\rangle$.

Figure 7 illustrates the variations of dimensionless SA TKE budget parameters, $\langle t_p\rangle h/u_*^3$, $\langle \varepsilon\rangle h/u_*^3$, $\langle t_d\rangle h/u_*^3$, and $\langle p_d\rangle h/u_*^3$, with *z*/*h* in the WGB and SGB. In both the beds, the $\langle t_p\rangle$ profiles develop below the crest, increasing rapidly with *z* within the form-induced sublayer. They attain their peaks near the top edge of the form-induced sublayer and then start diminishing gradually as one goes toward the free surface. However, at a given *z*, $\langle t_p\rangle$ is higher in the WGB than in the SGB. The reason is attributed to the difference in roughness in both the beds.

Further, from Fig. 7, it is apparent that in both the beds, the $\langle \varepsilon \rangle$ profiles are almost negligible below the crest; however, they have their peaks at the crest. Subsequently, they decrease, as z increases. Note that for $z \leq 0.20\ h$, the reduction rate of $\langle \varepsilon \rangle$ with z is higher than that for $z > 0.20\ h$, owing to the intense fluid mixing in the presence of gravels. Importantly, for a given z, $\langle \varepsilon \rangle$ is found to be larger in the WGB than in the SGB, which is accredited to the higher WGB roughness.

Furthermore, the $\langle t_d \rangle$ profiles in both, WGB and SGB, develop with the negative values, increasing rapidly with z. In the WGB, its peak value is apparent just above the crest, while in the SGB, it appears below the crest. Having reached their peaks, the $\langle t_d \rangle$ profiles reduce with an increase in z, changing their signs at $z/h = 0.05$ and 0.02 and attaining the positive peaks at $z/h = 0.1$ and 0.04 in the WGB and SGB, respectively. Then, they decrease and again change their signs, becoming negative at $z/h = 0.40$ in the WGB and 0.45 in the SGB and remaining same for the rest portion of z. It is evident that the WGB possesses higher $\langle t_d \rangle$ than the SGB. The underlying reason is same as before.

Finally, the $\langle p_d \rangle$ profiles in both the beds begin with negative values, attaining their negative peaks at the crest. Then, they decrease (in absolute magnitude) sharply, becoming positive above the crest. In the WGB, the $\langle p_d \rangle$ profile reaches the positive peak at $z/h = 0.025$, while in the SGB, it attains at $z/h \approx 0.09$. However, they decline with a further increase in z. The negative sign of $\langle p_d \rangle$ indicates a gain in TKE production (Dey et al. 2011). However, for a given z, the $\langle p_d \rangle$ attains a greater value in the WGB than in the SGB and the reason is same as before.

4 Conclusions

The water action changes the randomly poised SGB roughness structure to the organized WGB roughness structure with a higher roughness. The gravels cause the near-bed flow perturbation, resulting in a number of clockwise and anticlockwise motions of fluid streaks. The organized roughness structure in the WGB causes the near-bed flow to be more streamlined than in the SGB. As a result, for a given vertical distance, the DA velocity is greater in the WGB than in the SGB.

In the near-bed flow, for a given vertical distance, the SA RSS is greater in the WGB than in the SGB owing to the larger temporal velocity fluctuations in the former than in the latter. In addition, the dispersive shear stress is greater in the WGB than in the SGB, because the higher WGB roughness induces larger spatial velocity fluctuations. The SA TKE fluxes infer that in the near-bed flow, the sweeps are the governing events, while in the main flow, the ejections dominate. For small values of separation distance, the second-order velocity structure function follows the 2/3 slope, indicating the presence of inertial subrange in both the beds. The SA TKE budget analysis reveals that at a given vertical distance, all the DA TKE budget parameters are higher in the WGB than in the SGB.

References

Barison S, Chegini A, Marion A, Tait SJ (2003) Modifications in near bed flow over sediment beds and the implications for grain entrainment. In: Proceedings of XXX international association of hydraulic research congress, Thessalonki, Greece, pp 509–516
Cooper JR, Aberle J, Koll K, Tait SJ (2013) Influence of relative submergence on spatial variance and form-induced stress of gravel-bed flows. Water Resour Res 49(9):5765–5777
Cooper JR, Tait SJ (2008) The spatial organisation of time-averaged streamwise velocity and its correlation with the surface topography of water-worked gravel beds. Acta Geophys 56(3):614–641
Cooper JR, Tait SJ (2010) Spatially representative velocity measurement over water-worked gravel beds. Water Resour Res 46(11):W11559
Dey S (2014) Fluvial hydrodynamics: Hydrodynamic and sediment transport phenomena. Springer, Berlin
Dey S, Das R (2012) Gravel-bed hydrodynamics: Double-averaging approach. J Hydraul Eng 138(8):707–725
Dey S, Sarkar S, Ballio F (2011) Double-averaging turbulence characteristics in seeping rough-bed streams. J Geophys Res Earth Surf 116(F03020):1–16
Ferraro D, Servidio S, Carbone V, Dey S, Gaudio R (2016) Turbulence laws in natural bed flows. J Fluid Mech 798:540–571
Frisch U (1995) Turbulence: the legacy of A. N. Kolmogorov. Cambridge University Press, Cambridge
Hardy RJ, Best JL, Lane SN, Carbonneau PE (2009) Coherent flow structures in a depth-limited flow over a gravel surface: The role of near-bed turbulence and influence of Reynolds number. J Geophys Res 114(1):F01003
Manes C, Pokrajac D, McEwan I (2007) Double-averaged open-channel flows with small relative submergence. J Hydraul Eng 133(8):896–904
Mignot E, Barthelemy E, Hurther D (2009) Double-averaging analysis and local flow characterization of near-bed turbulence in gravel-bed channel flows. J Fluid Mech 618:279–303
Nezu I, Nakagawa H (1993) Turbulence in open-channel flows. Balkema, Rotterdam
Nikora VI, Goring DG, Biggs BJF (1998) On gravel-bed roughness characterization. Water Resour Res 34(3):517–527
Padhi E, Penna N, Dey S, Gaudio R (2018a) Hydrodynamics of water-worked and screeded gravel beds: a comparative study. Phys Fluids 30(8):085105
Padhi E, Penna N, Dey S, Gaudio R (2018b) Spatially averaged dissipation rate in flows over water-worked and screeded gravel beds. Phys Fluids 30(12):125106
Pokrajac D, McEwan I, Nikora V (2008) Spatially averaged turbulent stress and its partitioning. Exp Fluids 45(1):73–83
Sreenivasan KR (1995) On the universality of the Kolmogorov constant. Phys Fluids 7(11):2778–2784

Quantitative Characterization of the Roughness of Four Artificially Prepared Gravel Surfaces

Jie Qin, Teng Wu and Deyu Zhong

Abstract Two-dimensional second-order structure functions (2DSSFs) have been widely used to investigate the roughness characteristics of gravel surfaces. Because of the complex patterns of the 2DSSFs, the interpretation of the 2DSSFs results remains challenging, and the explanations are mostly qualitative. Recently, a novel quantitative method for the description of the roughness of gravel-beds was proposed by Qin et al. (J Hydraul Res 57:90–106, 2019), which determines the statistical significance of structure functions based on Monte Carlo simulations. This new method is evaluated by four artificially prepared gravel surfaces to test the capability of the method in quantitatively differentiating predefined gravel structures. The four surfaces were prepared according to dissimilar arrangements of gravels and tried to represent gravel structures of different scales. The results successfully differentiated the roughness characteristics of the four surfaces and demonstrated the promising capability of the new method.

Keywords Gravel surfaces · Roughness scales · Structure functions

J. Qin · T. Wu
College of Harbour, Coastal and Offshore Engineering,
Hohai University, Nanjing, China
e-mail: wuteng@hhu.edu.cn

J. Qin (✉)
Key Laboratory of Port, Waterway and Sedimentation Engineering
of Ministry of Transport, Nanjing, China
e-mail: qinjie@alumni.cuhk.net

D. Zhong
State Key Laboratory of Hydroscience and Engineering,
Tsinghua University, Beijing, China
e-mail: zhongdy@tsinghua.edu.cn

M. B. Kalinowska et al. (eds.), *Recent Trends in Environmental Hydraulics*, GeoPlanet: Earth and Planetary Sciences,
https://doi.org/10.1007/978-3-030-37105-0_19

1 Introduction

Recent developments in photogrammetry and laser scanning allow for the determination of high precision digital elevation models (DEMs) of gravel surfaces. Based on the DEMs, various random-field approaches, such as spectral analysis, wavelet analysis, correlation functions, structure functions, as well as fractal and multi-fractal analysis, have been used to investigate the roughness characteristics of gravel surfaces (Millane et al. 2006; Qin and Ng 2011; Guala et al. 2014; Bertin and Friedrich 2016; Powell et al. 2016).

Second-order structure functions (SSFs) are commonly used random-field approaches to investigate the spatial correlation of roughness (Smart et al. 2002; Aberle et al. 2010; Bertin and Friedrich 2014; Powell et al. 2016) and the scaling behavior of gravel beds (Nikora et al. 2004; Qin and Ng 2012). The SSFs can be calculated based on a 1D profile (1D second-order structure functions, 1DSSFs) or over an area (2D second-order structure functions, 2DSSFs). By analyzing the SSFs of gravel surfaces, a lot of roughness characteristics of gravel surfaces have been revealed, such as the grain and bedform scales (Clifford et al. 1992), the universal scaling range (Nikora et al. 1998), the relationship between the orientation of contours on 2DSSFs and historical flow direction (Nikora et al. 2004), and topographic evolution processes (Qin et al. 2017). Butler et al. (2001) summarized previous studies related to the SSFs analysis of gravel surfaces and proposed three scaling regions corresponding to: subgrain, grain, and bedform scales. As the subgrain-scale is associated with small spatial lags most of which belong to discrete particle surfaces and some correspond with breaks between particles, this scale characterizes the smooth surface of individual grains. This scale is consistent with the universal scaling region defined by Nikora et al. (1998). The subsequent grain-scale is related to the distribution of larger grains on the surface and hence the development of gravel structures (Strom and Papanicolaou 2008; L'Amoreaux and Gibson 2013). This scaling region is characterized by anisotropic roughness and a length scale proportional to the largest grain size. The bedform scale is associated with large-scale roughness, like undulations of the surface (Robert 1988; Clifford et al. 1992).

In spite of the considerable progress being made based on the 2DSSFs, the interpretation of the results remains challenging due to the complex patterns on the 2DSSFs. The challenge is particularly significant in terms of the description of gravel structures. Different surface structures have been observed in gravel-bed rivers, such as gravel clusters (Brayshaw 1985; Curran and Waters 2014), stone cells (Church et al. 1998), and transverse ribs (Richards and Clifford 2016). However, quantitative geometrical descriptions of the gravel structures are rare, and this may be associated with insufficient methods in quantifying the gravel structures. Since the 2DSSFs reflect the spatial correlation of roughness, this method is commonly used to investigate the structures on gravel surfaces. However, the explanation based on the 2DSSFs are mostly qualitative.

Qin et al. (2019) proposed a novel quantitative method to characterize the roughness of gravel-beds by determining the statistical significance of the 2DSSFs. Such an approach allows for the identification of characteristic length scales of roughness based on statistical significance testing and hence for the simplification of the interpretation of correlation analyses. In the current study, this new method is applied to four artificially prepared gravel surfaces to test the capability of the method in quantifying the roughness of different scales.

2 Data and Methods

2.1 Data

Four artificially prepared gravel surfaces are investigated in this study, and the DEMs of the surfaces are shown in Fig. 1. The four surfaces (S-R, S-X, S-F, and S-FG) were prepared using the same gravels but were arranged into different configurations on the surfaces. The grain size distribution of the gravels is shown in Fig. 2, with $d_{50} = 24.6$ mm and $d_{90} = 35.7$ mm. For the first configuration (Fig. 1a), the gravels

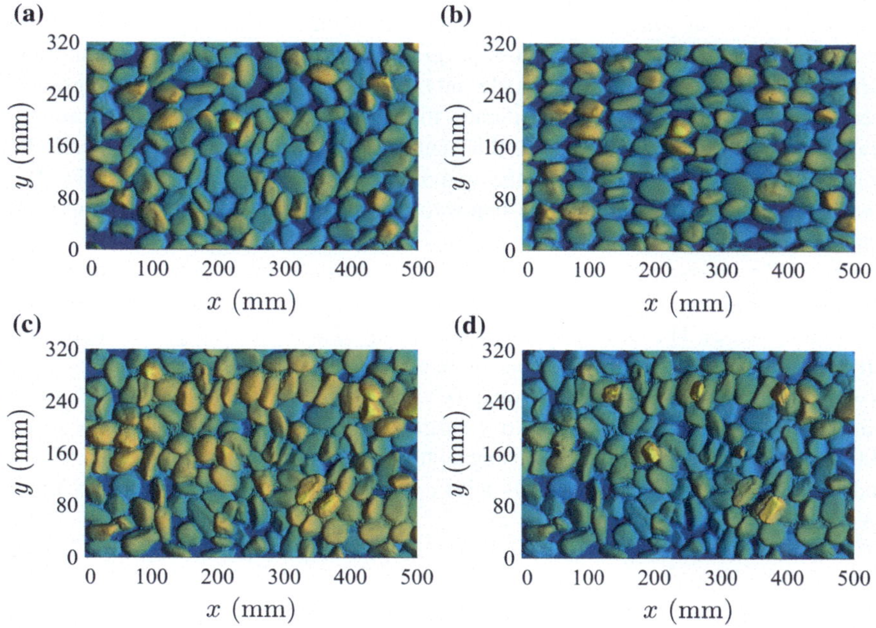

Fig. 1 DEMs of the four surfaces prepared by artificially placed gravels. The four surfaces were built by the same gravels but were rearranged into different configurations: **a** random arrangement (S-R), **b** with long axes being parallel to the x direction (S-X), **c** random arrangement of gravels with longitudinal bedforms underneath (S-F), and **d** some gravels being superposed to the third surface (S-FG)

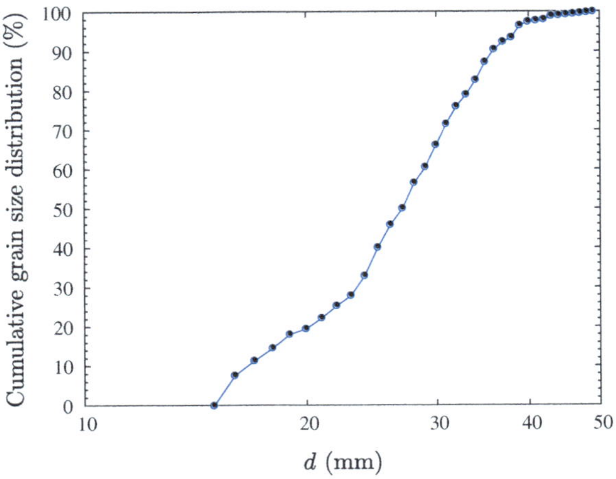

Fig. 2 Grain size distribution of the gravels used to prepare the four gravel surfaces

were randomly distributed on a flat surface. In Fig. 1b, the gravels were rotated to align with the x direction (the longitudinal direction). The third configuration is similar to the first one, except for bedforms being added under the gravels. In the fourth configuration, some gravels superposed on the third surface. The first surface (S-R) characterized by a random distribution of gravels and thus the spatial correlation is supposed to exist at the subgrain scale only. For the second surface (S-X), gravels were repeatedly presented which indicates that the spatial correlation existed among gravels, e.g. the grain scale. On the third surface, bedforms existed underneath the gravels indicating the presence of the spatial correlation at the bedform scale. The fourth surface reflects a small disturbance to the third one, which is used to test the sensitivity of the new method.

The DEMs were collected using a close-range photogrammetry method. The photos were sequentially acquired in the x direction with an overlap of 60% between sequential images. The photos were captured using a Nikon D800 digital camera with a fixed-focus lens of 50 mm. The maximum resolution of the camera is 7360×4912 pixels, and each pixel has a dimension of 0.0049 mm. The flying height of the camera was around 1 m that corresponds to a vertical precision of ± 0.1 mm (Lane et al. 2001). The photos were processed using a photogrammetric soft-ware (VisualSFM), and produced high-precision DEMs with a horizontal resolution of 1 mm × 1 mm (Wu 2013).

2.2 *Methods*

This study applied a new method proposed by Qin et al. (2019) for the analysis of the gravel roughness based on the 2DSSFs. In comparison to the traditional 2DSSFs, a statistical significance test is used in the new method to obtain significant roughness characteristics. This method is briefly introduced in this paper, and detailed information can be found in Qin et al. (2019).

2.2.1 Second-Order Structure Functions (SSFs)

The one dimensional SSFs is defined as

$$D_1(l) = \frac{1}{N-n}\sum_{i=1}^{N-n}[z(x_i + n\delta x) - z(x_i)]^2 \tag{1}$$

where $l = n\delta x$ denotes the spatial lag, n the control variable for l, δx the sampling interval, N the number of observed elevation points along the profile, and z the bed-elevation at the longitudinal coordinate x. The roughness pattern of gravel-bed rivers is generally anisotropic due to the influence of both grain shape (typically elliptic), grain clusters, or other bedforms, and anisotropy of roughness may become important if the flow direction changes with variations in discharge (Butler et al. 2001; Cooper et al. 2008). In order to investigate the anisotropic characteristics of roughness, the 2D-structure functions can be used

$$D_2(l_x, l_y) = \frac{1}{(N-n)(M-m)}\sum_{i=1}^{N-n}\sum_{j=1}^{M-m}\left[z(x_i + n\delta x, y_j + m\delta y) - z(x_i, y_i)\right]^2 \tag{2}$$

where $l_x = n\delta x$, and $l_y = m\delta y$ denote the spatial lags in streamwise (x) and transverse direction (y) with n and m being the control variables, respectively, δx and δy are the sampling intervals usually equal to the resolution of models, and N and M are the total numbers of measured bed elevations in x and y directions, respectively (Aberle and Nikora 2006). In order to analyze granular surface structures, large scale morphological features and trends (i.e., bed slopes) should be filtered from the topographic data. Detrending the surfaces can be achieved by linearly detrending the surface (Aberle and Nikora 2006) or a spline interpolation of the surface (Smart et al. 2002; Powell et al. 2016; Bertin et al. 2017; Qin et al. 2019). In this study, the grains were placed on a flat and nearly horizontal surface, and thus the linear detrending method was used to remove the trends.

2.2.2 Surrogate Data Analysis

To determine the statistical significance of the SSFs, an empirical distribution of the structure function is needed which was obtained using the surrogate method proposed by Theiler et al. (1992). Surrogate data analysis is based on statistical hypothesis testing, and a set of surfaces needs to be generated preserving certain properties of the original surface defined by a null hypothesis. If a significant difference exists between the generated and the original surface, the null hypothesis will be rejected.

Please note that the original surface refers to the DEM which was obtained by a photogrammetric method and was post-processed by de-noising and interpolation.

In the present study, the null hypothesis assumes that the surface grains are randomly distributed, and no spatial correlation persists at scales larger than the subgrain-scale. The surrogate surfaces are generated by subdividing the original surface into squared tiles with side lengths in the order of magnitude of d_{84}. The tiles of each DEM were then reshuffled, i.e. randomly placed and oriented (rotated through 0°, 90°, 180°, 270°) to form a new surface with the same spatial extent as the original surface. The reshuffling of the tiles results in an increase of the variance at small scales as discontinuities in the surface are introduced along the tile-boundaries. This issue as well as the influence of the tile size on the reshuffling were discussed in Qin et al. (2019). The Monte Carlo method was applied to generate surrogate surfaces. The required number of simulated surfaces can be determined by investigating the convergence of the sample mean value and the variance of SSFs at given lags. As an example, Fig. 3 depicts the sample mean and variance of the second-order structure function values computed for the arbitrarily spatial lags of $l_x = 50$ mm and $l_y = 50$ mm for surface S-X. The figure indicates that both the sample mean and variance reach stable values for around $n_x = 100$ surrogate surfaces. The convergence can be quantitatively evaluated using the coefficients of variation for the sample mean and variance (Ballio and Guadagnini 2004).

For the surrogate analysis in this study, only one observation is available (here the DEM of the gravel surface). The significance can be defined in terms of p-values from the distribution of the surrogate surfaces (Theiler et al. 1992). Using $n_s = 100$ surrogate surfaces of the surface S-X, the critical values of the 1% significance level are exemplarily shown in Fig. 4 (0.5% for the blue mesh and 99.5% for the yellow mesh) together with the function values of the undisturbed DEM. The difference between the blue and yellow meshes for each lag defines the non-rejection region, the 2DSSFs being insignificant. Areas the 2DSSFs protruding through the blue mesh indicate 2DSSFs values are significantly different from the surrogate surfaces.

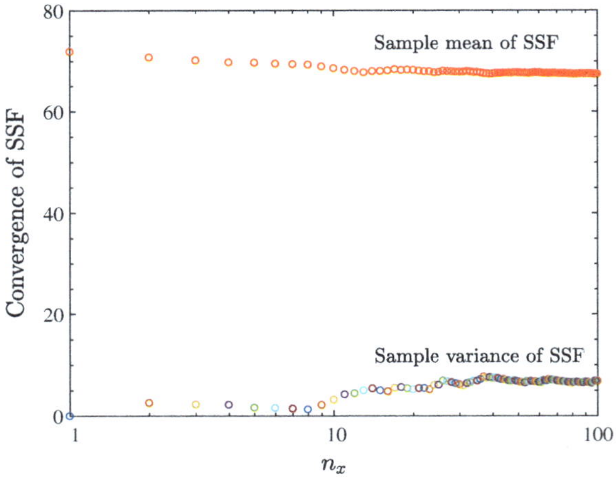

Fig. 3 Convergence of the sample mean and variance of the SSFs for the surface S-X at the monitoring lag ($l_x = 50$ mm and $l_y = 50$ mm) as a function of the number of Monte Carlo simulations

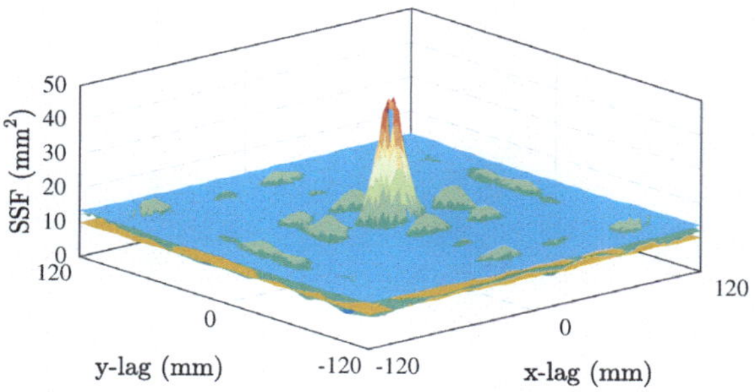

Fig. 4 Critical values (0.5% for the blue and 99.5% for the yellow meshes) of significance level 1% for the surface S-X. The regions emerging from the blue mesh indicate significant spatial correlations at the corresponding lags

3 Results and Discussion

The 2DSSFs of the four surfaces are shown in Fig. 5. In these figures, some qualitative characteristics can be observed. The SSFs in the figures were normalized by σ^2, and the thick lines in these plots represent $0.7\sigma^2$. The value of 0.7 is arbitrarily chosen to represent spatial lags exhibiting weak spatial correlations, and values of 0.8 or 0.9 may be used also.

In Fig. 5a, the thick line is a circle which reflects the isotropic characteristic of the random arrangement of gravels, e.g. no preferential orientation of gravels on this surface. In Fig. 5b, the gravels were aligned with the x direction, the thick line becomes an ellipse in this plot with the long axis being parallel to the x direction. In addition, because the gravels are well sorted and thus more or less uniform in size, the regular distribution of the gravels creates a duplicate pattern in Fig. 5b. In Fig. 5c, the presence of bedforms on this surface creates parallel contours at larger scales. At smaller scales, the contours showing circular shapes which is similar to the patterns in Fig. 5a. The contours at larger scales represent strong anisotropic behavior introduced by the bedform, whereas the circular contours near the center reflect isotropic features indicating the random orientations of gravels. The thick line in this plot represents a transition from the isotropic roughness to anisotropic roughness. As a consequence, the thick line in this plot poorly reflected the roughness for both smaller and larger scales. When placing some gravels on the S-F surface, the DEM looks rather different, as shown in Fig. 1, whereas the pattern of SSFs is more or less the same.

Figure 6 shows the filtered 2DSSFs by the 1% confidence limit. In this figure, the insignificant areas are filtered out. In Fig. 6a, the filtered SSFs of the randomly distributed gravels show a circular area reflecting the isotropic characteristics at the subgrain scale. The diameter of the circular area is 32 mm which corresponds to d_{76} of the surface. This length scale is somewhat larger than that obtained in Qin et al. (2019). In Fig. 6a, rather weak spots around the central circular area can be observed. These spots, although significant in the statistic test, represent very weak spatial correlations that any explanation related to these weak correlations should be careful.

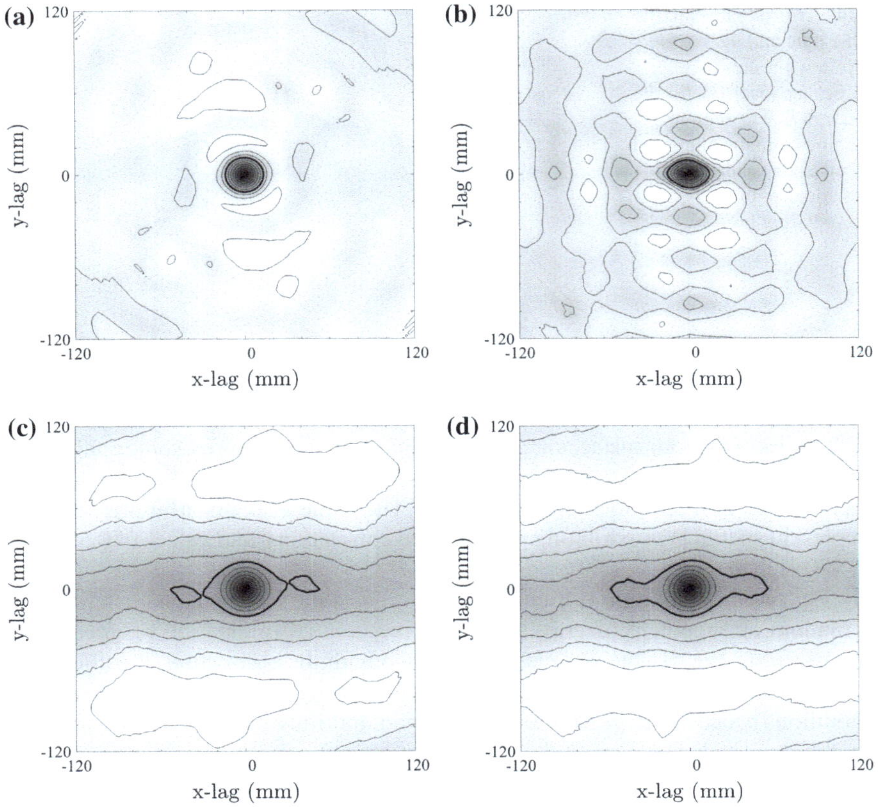

Fig. 5 The 2D second-order structure functions of the four surfaces. Plots (**a**), (**b**), (**c**), and (**d**) correspond to surfaces S-R, S-X, S-F, and S-FG, respectively. The values in these plots are normalized by σ^2, and the thick black lines represent $0.7\sigma^2$

In Fig. 6b, the duplicate patterns can be easily identified. An elliptic spot exists at the central area in comparison to the circular spot in Fig. 6a. This elliptic spot reflects the ellipsoid shape of gravels. The long (A_l) and short (A_s) axes of the ellipse area are 46 and 28 mm, respectively, which indicates an elliptic rate of 1.64. Including the central spot, there are five large spots exist in the x direction at $y = 0$, and the intervals between neighboring spots, I_x, are 54, 42, 42, and 54 mm. Normalizing the intervals by A_l, I_x/A_l equals 1.17, 0.91, 0.91, and 1.17. In the y direction at $x = 0$, there are also seven spots, although two of them are rather small in size in comparison to the other five spots. The intervals between neighboring spots, I_y, are 32, 34, 30, 30, 34, and 32 mm. Normalizing I_y by A_s, the intervals (I_y/A_s) are 1.14, 1.21, 1.07, 1.07, 1.21, and 1.14. The mean I_x and I_y are $1.04A_l$ and $1.14A_s$, respectively, which well reflects the regular arrangement of the gravels.

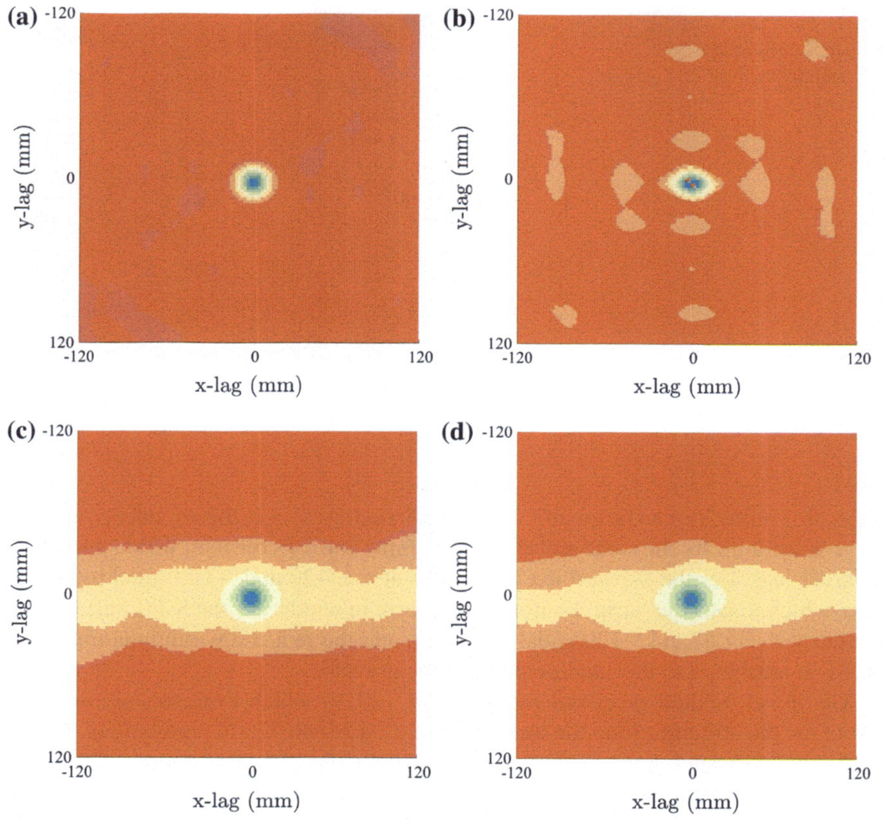

Fig. 6 The 2DSSFs of the four surfaces filtered by the 1% confidence limit, i.e. the values lower than the 1% confidence limit are preserved. Plots (**a**), (**b**), (**c**), and (**d**) correspond to surfaces S-R, S-X, S-F, and S-FG, respectively

In Fig. 6c and d, since both surfaces have strip bedforms underneath the gravels, the filtered 2DSSFs readily exhibit the existence of the bedforms. Comparing Fig. 6c and d, the areas showing significant spatial correlations are similar in both size and shape. This is attributed to the fact that superposing several gravels on a gravel surface has a negligible effect on the overall pattern of 2DSSFs, although the DEMs look different because of the high elevations of the added gravels. In this figure, the difference between the 2DSSFs can still be observed. Superposing gravels to the surface S-F increases the variance for all spatial lags; however, this effect is diluted by the numerous pairs of data at a certain lag. As a result, the increasing variation effect introduced by superposing gravels increases with the decrease of the number of samples in Eqs. (1) and (2). On the other hand, the number of samples decreases with increasing spatial lags, which means that the effect of increasing variance is more evident at larger lags than smaller lags. By comparing the width of the 'strip'

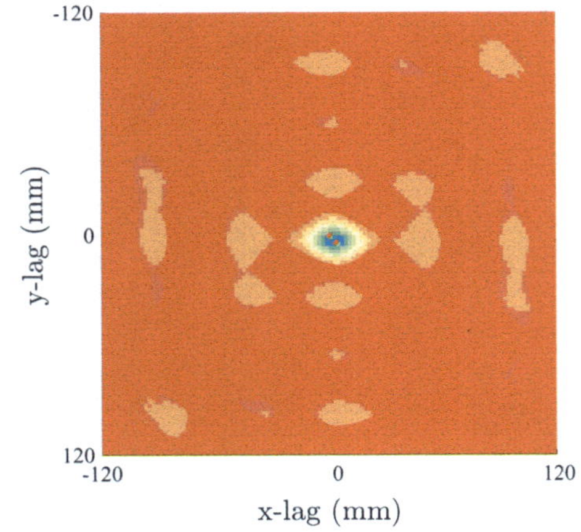

Fig. 7 The 2DSSFs of the surface S-X filtered by the 5% confidence limit

areas, this influence can be quantitatively evaluated. At $x = 0$, the widths of 'strip' of the two surfaces are the same and equal to 80 mm. The difference in widths increases at larger lags. For the S-F surface, the widths at $x = 120$ and -120 equal to 80 mm also, which means the 'strip' bedform is uniform in width. For the S-FG surface, the widths at $x = 120$ and -120 equal to 62 mm, and the reduced width in comparison to S-F is attributed to the superposition of the gravels.

The novel method proposed by Qin et al. (2019) which evaluates significance levels for the structure function analysis based on Monte Carlo simulations. In this study, a significance level of 99% is used to filter the 2DSSFs, whereas Qin et al. (2019) used a 95% level to investigate the roughness characteristics. To evaluate the influence of different significance levels, the filtered 2DSSFs based on the 95% significance level is presented in Fig. 7. Individual spots obtained by the 99% significance level (Fig. 6b) show somewhat smaller sizes in comparison to those of the 95% level, whereas the overall patterns are the same. As a result, both 95% and 99% significance levels can be employed to study the roughness characteristics.

4 Summary

This study tested the capability of a new method in differentiating the predefined structures on four artificially prepared gravel surfaces. For the randomly distributed gravel surface, an isotropic area with a diameter of d_{76} characterizing the subgrain feature of gravels was observed. For the regularly aligned gravel surface, the duplicate spot pattern reflects the regular arrangement of gravels and the distances between neighboring spots in the x and y directions correspond to the long and short axes of the central elliptic area. The filtered 2DSSFs were similar for the third and the fourth surfaces that both of them were characterized by a 'strip' pattern reflecting the presence of the bedforms. Because of the superposing gravels, the spatial variation increases in the fourth surface which is confirmed by the relatively narrow 'strip' at larger spatial lags. The results produced by the new method demonstrated the capability of the method proposed by Qin et al. (2019) which successfully differentiated the subgrain-, grain- and bedform-scale roughness. Different significance levels, 95 and 99%, considered in this study, produced similar results that both can be used to filter the structure functions.

Acknowledgements This work is supported by the National Key R&D Program of China (2016YFC0402506), the CRSRI Open Research Program (CKWV2019724/KY), and the Open Funding of the Key Laboratory of Port, Waterway and Sedimentation Engineering of Ministry of Transport (Yn918004).

References

Aberle J, Nikora VI (2006) Statistical properties of armored gravel bed surfaces. Water Resour Res 42:W11414

Aberle J, Nikora VI, Henning M, Ettmer B, Hentschel B (2010) Statistical characterization of bed roughness due to bed forms: a field study in the Elbe River at Aken, Germany. Water Resour Res 46:W03521

Ballio F, Guadagnini A (2004) Convergence assessment of numerical Monte Carlo simulations in groundwater hydrology. Water Resour Res 40:W04603

Bertin S, Friedrich H (2014) Measurement of gravel-bed topography: evaluation study applying statistical roughness analysis. J Hydraul Eng ASCE 140:269–279

Bertin S, Friedrich H (2016) Field application of close-range digital photogrammetry (CRDP) for grain-scale fluvial morphology studies. Earth Surf Process Landf 41:1358–1369

Bertin S, Groom J, Friedrich H (2017) Isolating roughness scales of gravel-bed patches. Water Resour Res 53:6841–6856

Brayshaw AC (1985) Bed microtopography and entrainment thresholds in gravel-bed rivers. Bull Geol Soc Am 96:218–223

Butler JB, Lane SN, Chandler JH (2001) Characterization of the structure of river-bed gravels using two-dimensional fractal analysis. Math Geol 33:301–330

Church M, Hassan MA, Wolcott JF (1998) Stabilizing self-organized structures in gravel-bed stream channels: field and experimental observations. Water Resour Res 34:3169–3179

Clifford NJ, Robert A, Richards KS (1992) Estimation of flow resistance in gravel-bedded rivers: a physical explanation of the multiplier of roughness length. Earth Surf Process Landf 17:111–126

Cooper JR, Aberle J, Koll K, McLelland SJ, Murphy BM, Tait SJ, Marion A (2008) Observation of the near-bed flow field over gravel bed surfaces with different roughness length scales. In: Altinakar M, Kokpinar MA, Aydin I, Cokgor S, Kubaba SK (eds) Proceedings of the international conference on fluvial hydraulics river flow 2008, Cesme, Turkey, pp 739–746

Curran JC, Waters KA (2014) The importance of bed sediment sand content for the structure of a static armor layer in a gravel bed river. J Geophys Res 119:1484–1497

Guala M, Singh A, BadHeartBull N, Foufoula-Georgiou E (2014) Spectral description of migrating bed forms and sediment transport. J Geophys Res 119:123–137

L'Amoreaux P, Gibson S (2013) Quantifying the scale of gravel-bed clusters with spatial statistics. Geomorphology 197:56–63

Lane SN, Chandler JH, Porfiri K (2001) Monitoring river channel and flume surfaces with digital photogrammetry. J Hydraul Eng ASCE 127:871–877

Millane RP, Weir MI, Smart GM (2006) Automated analysis of imbrication and flow direction in alluvial sediments using laser-scan data. J Sediment Res 76:1049–1055
Nikora VI, Goring DG, Biggs BJF (1998) On gravel-bed roughness characterization. Water Resour Res 34:517–527
Nikora VI, Koll K, McEwan IK, McLean S, Dittrich A (2004) Velocity distribution in the roughness layer of rough-bed flows. J Hydraul Eng ASCE 130:1036–1042
Powell DM, Ockelford AM, Rice SP, Hillier JK, Nguyen T, Reid I, Tate NJ, Ackerley D (2016) Structural properties of mobile armors formed at different flow strengths in gravel-bed rivers. J Geophys Res 121:1–21
Qin J, Ng SL (2011) Multifractal characterization of water-worked gravel surfaces. J Hydraul Res 49:345–351
Qin J, Ng SL (2012) Estimation of effective roughness for water-worked gravel surfaces. J Hydraul Eng ASCE 138:923–934
Qin J, Zhong D, Wu T, Wu L (2017) Evolution of gravel surfaces in a sediment-recirculating flume. Earth Surf Process Landf 42:1397–1407
Qin J, Aberle J, Henry PY, Wu T, Zhong D (2019) Statistical significance of spatial correlation patterns in armoured gravel beds. J Hydraul Res 57:90–106
Richards K, Clifford N (2016) Fluvial geomorphology: structured beds in gravelly rivers. Prog Phys Geogr 15:407–422
Robert A (1988) Statistical properties of sediment bed profiles in alluvial channels. Math Geol 20:205–224
Smart GM, Duncan MJ, Walsh JM (2002) Relatively rough flow resistance equations. J Hydraul Eng ASCE 128:568–578
Strom KB, Papanicolaou AN (2008) Morphological characterization of cluster microforms. Sedimentology 55:137–153
Theiler J, Eubank S, Longtin A, Galdrikian B, Doyne Farmer J (1992) Testing for nonlinearity in time series: the method of surrogate data. Physica D 58:77–94
Wu C (2013) Towards linear-time incremental structure from motion. In: 2013 international conference on 3D vision (3DV). IEEE, pp 127–134

Rosette Diffuser for Dense Effluent—Puck Bay Case Study

Małgorzata Robakiewicz

Abstract The construction of underground gas stores, on-going since October 2010, results in the production of brine discharged into the Puck Bay (south Baltic Sea). A rosette diffuser located 2300 m off-shore at a depth of 8 m, discharges at maximum of 300 $m^3\ h^{-1}$, with saturation not exceeding 250 kg m^{-3}. It consists of 16 risers, with 3 nozzles, 0.009 m in diameter each. Due to ecological restrictions, maximum salinity cannot exceed 9.2 PSU, while the excess salinity is limited to 0.5 PSU in the near field of the installation. Assessment of excess salinity can be done using two alternative approaches: analysis of in situ measurements, or analysis of brine parameters and discharge conditions. Continuous salinity measurements in a vertical located at the centre of the installation enable simple operational assessment of excess salinity; however, the velocity measured in situ can support interpretation of registered data.

Keywords Brine discharge · Mixing · Salinity · Puck Bay

1 Introduction

Discharging wastes into natural water bodies is a common practice. The discharge systems are constructed to enhance mixing with ambient to fulfil environmental restrictions. Up till now more attention was paid to effluents characterized by density lower than ambient, e.g. discharges from treatment plants. However, diffusers discharging dense effluents require similar attention.

In Poland, brine production and discharge were associated with mine industry. Recently, due to construction of underground gas stores in salt deposits, brine discharge into the Puck Bay (the inner part of the Gulf of Gdańsk) is on-going. The construction site is located about 4 km away from the coast, while the discharge system 2300 m off-shore (Fig. 1). Purified water from the waste water treatment plant

M. Robakiewicz (✉)
Institute of Hydro-Engineering, Polish Academy of Sciences, Kościerska 7, 80-328 Gdańsk, Poland
e-mail: marob@ibwpan.gda.pl

M. B. Kalinowska et al. (eds.), *Recent Trends in Environmental Hydraulics*, GeoPlanet: Earth and Planetary Sciences,
https://doi.org/10.1007/978-3-030-37105-0_20

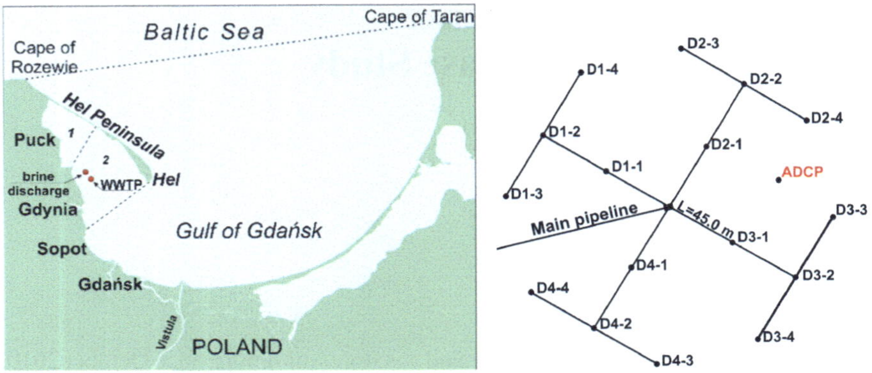

Fig. 1 Location of brine discharge in the Puck Bay (left; 1—Puck Lagoon, 2—Outer Puck Bay) and a general scheme spatial distribution of a rosette type diffuser system (right)

(WWTP) Dębogórze, located in the vicinity, is used for diluting caves. For ecological reasons, the total volume of brine is limited to 300 m^3 h^{-1}, while its saturation cannot exceed 250 kg m^{-3}. The water permit limits the maximum salinity to 9.2 PSU, and the excess value in the vicinity of installation to 0.5 PSU. The restrictions imposed resulted from data analysis carried out by Nowacki and Jarosz (1998) for the proposed discharge region, where the maximum measured value reached 8.7 PSU. The natural salinity variation of 0.5 PSU was recognized as tolerable by local organisms.

The aim of a discharge system is to intensify process of effluent mixing with the ambient. In general there are two solutions: (1) discharge by a single jet outfall, (2) discharge by multiple jets diffuser outfall (in the form of: a single pipeline, system of pipelines or rosette with a number of risers). Several elements are responsible for effluent dilution: (1) density difference between effluent and ambient, (2) nozzle parameters, (3) configuration of a discharge system, and (4) hydrodynamic conditions in the ambient. The detailed analysis carried out in the pre-investment study for brine discharge to the Puck Bay (Robakiewicz and Robakiewicz 2008) resulted in choice of the diffuser system located 2300 m off-shore at a depth of 8 m consisting of 4 arms. Each of them has 4 risers of 3 m height, spaced every 45 m (denoted D, see Fig. 1). From each riser 3 nozzles, distributed every 120° of circumference, discharge brine with an angle of incidence $\Theta = 45°$ into the sea to assure negligible interaction between brine jets. Based on the pre-investment analysis, nozzles of 0.008 m diameter were proposed. Hydraulic tests carried out in the first months of system operation, resulted in enlargement of nozzles' diameter to 0.009 m.

2 Monitoring of Salinity Changes Based on in Situ Measurements

The Puck Bay, a western sub-region of the Gulf of Gdańsk, is separated from the open sea by the Hel Peninsula. It is divided into two parts with different circulation patterns: the eastern part, called the Outer Puck Bay (av. depth ~20.5 m), and the western part, called the Puck Lagoon (av. depth ~3 m). The Gulf of Gdańsk is a non-tidal region where currents are generated mainly by spatially and temporarily changeable wind, and the accompanying water level variations. Coastline and bottom configuration significantly influence velocity patterns in the region. The spatial and temporal variability of salinity in the Gulf of Gdańsk depends on two main factors: the inflows of fresh water from the Vistula river, and of saline water originating from the south Baltic Sea and modified by the currents.

Monitoring of salinity variations due to brine discharge is carried out continuously since starting dilution of underground stores (X 2010). It covers operational monitoring carried out by Gas Storage Poland sp. z o. o., and detailed monitoring of brine mixing in the Puck Bay carried out by the Institute of Hydro-Engineering PAS.

2.1 Continuous Operational Monitoring

Continuous operational monitoring covers parameters of discharge (saturation, brine discharge in division into 4 arms), salinity measurements on 4 depths (1.5, 3.5, 5.5, 7.5 m below MSL) in the centre of installation (location S, Fig. 2), and hydro-

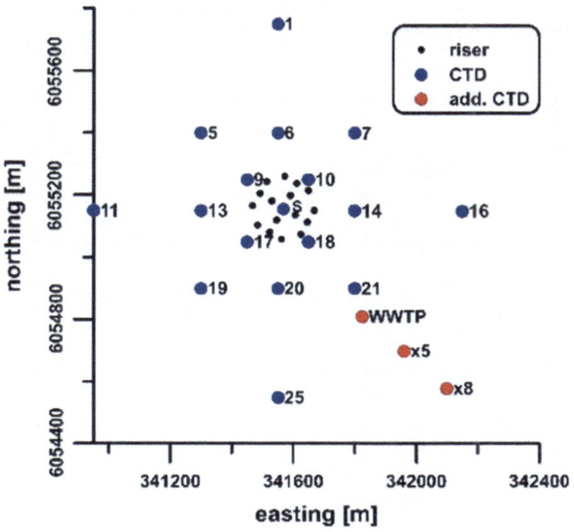

Fig. 2 Location of the monitoring points

meteorological conditions (wind and water level variation) in the area of installation. Exemplary results from two periods: April 2018—when brine saturation and discharge approached maximum permissible values, June 2008—representing conditions of low discharge and saturation, are given in Fig. 3. Registered salinity values in April 2018 show low salinity values, especially in the uppermost layer associated with inflow of fresh water originating either from the WWTP or from the Vistula river. In June 2018 brine discharge was very limited; however, explicit salinity increase in the lower part of the water column was registered. These examples confirm the complexity of salinity pattern which cannot be explained by without broader analysis of hydro-meteorological conditions in the Puck Bay. However, the measured data deliver basic information on salinity temporal variability.

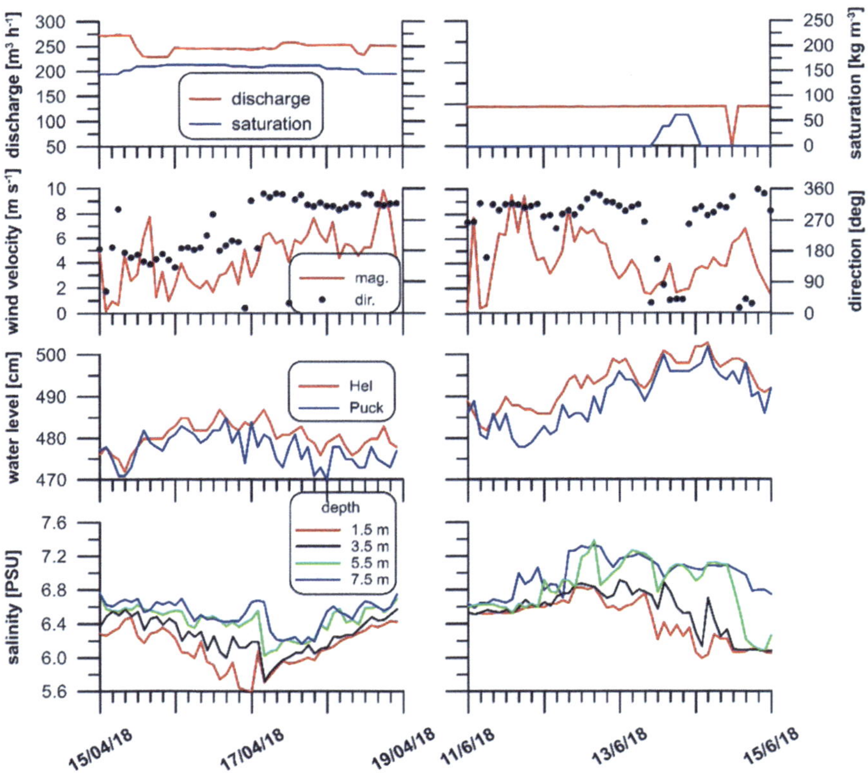

Fig. 3 Parameters of discharged brine and salinity changes in location S—exemplary results from April 2018 (left) and June 2018 (right)

2.2 *Monitoring of Brine Mixing*

Monitoring of brine mixing in the vicinity of discharge installation is on-going since autumn 2010 using salinity measurements done by CTD (*Conductivity Temperature Depth*) gauge. In the first period (X 2010–XII 2010) their frequency was almost once per month (Robakiewicz 2016), while since 2015 was reduced to twice a year. In majority, measurements were executed in 17 locations distributed symmetric with regard to the centre of installation (Fig. 2). Since 2017 the scope of measurements was extended by 3 locations in cross-section passing through centre of brine discharge installation (S) and discharge from the WWTP Dębogórze. Exemplary results from the 18th of April 2018 are given in Fig. 4. It is characteristic that salinity variation in the upper part of the water column is small (~0.1 PSU). In the bottom layer, the salinity pattern is more complex. In the northern part we observe a path of water

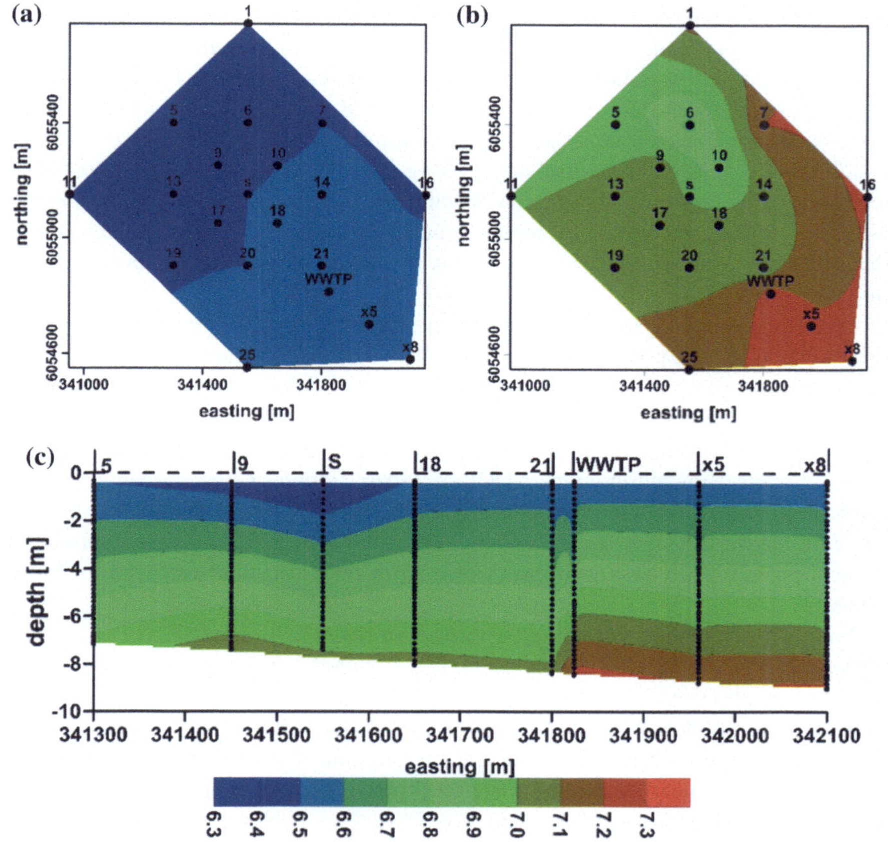

Fig. 4 Salinity spatial distribution: **a** 1 m below free surface, **b** at the bottom, **c** in the cross-section 5-x8; on 18th of April 2018

with relatively low salinity (6.8–6.9 PSU; locations 6, 10), and in the same time in the south-eastern part of the analysed area salinity reaches 7.3 PSU (Fig. 4a, b). Analysis in the cross-section 5-x8 (Fig. 4c) indicates influence of fresh water from the treatment plant on observed pattern. Knowledge of local wind conditions is insufficient to reconstruct the observed salinity patterns. Analysis of time varying water currents can be recommended. However, the presented results support general conclusion on good mixing condition in the vicinity of installation.

2.3 Support of Salinity Monitoring by ADCP Measurements

The experience gained from salinity measurements carried out within the monitoring program deliver very important data on salinity variation in the area of interest. Generally, the data are sufficient for assessing the excess salinity for operational purposes. However, there are situations when registered salinity suddenly changes without direct association with conditions and parameters of brine discharge. Especially alarming are situations of sudden salinity increase at the bottom, or salinity value approaching the permissible limit of 9.2 PSU. Quite often alarming conditions occur when execution of additional monitoring is impossible due to weather conditions. As long as brine discharge is on-going in accordance with the approved technology, the most probable reason of salinity increase is inflow of water from the Gdańsk Deep, characterized by higher salinity and lower temperature than that in the Gulf of Gdańsk. Water currents measured in situ can be used to support or reject this hypothesis.

Space and time varying water currents in the Puck Bay are predominantly wind driven. Their precise prediction using numerical approach requires detailed data set. As an alternative, it was proposed to investigate possibilities to support assessment of excess salinity by in situ measurements of currents in the vicinity of installation. The on-going field experiment started on the 17th of October 2018, and covers velocity measurements using autonomous ADCP (*Acoustic Doppler Current Profiler*) gauge located on the bottom, in the area between arms D1 and D2 of installation (see Fig. 1). The first measuring layer starts 2 m above the bottom; registrations are done in the vertical profile in division into layers of 0.25 m thickness, and gathered in 2 min intervals.

Some chosen results from the first 10 days of the registration (17.10.18 12 GMT through 27.10.18 12 GMT) are presented in Fig. 5. Salinity variations on the lowest level (7.5 m) show two phenomena of interest: A—sudden salinity increase, and B—extensive salinity decrease (see Fig. 5c). Analysis of discharge parameters (Fig. 5a) does not show any variations giving an explanation of the observed phenomena. Wind conditions measured in the location S (Fig. 5b) in connection with velocity measurements (Fig. 5c; positive u—eastward direction; positive v—northward direction) can be helpful. The sudden salinity increase (case A) occurred in a moment of a very gentle wind of changeable direction preceding the observed phenomenon. In that case, values of velocity components u and v are similar. The observed salinity

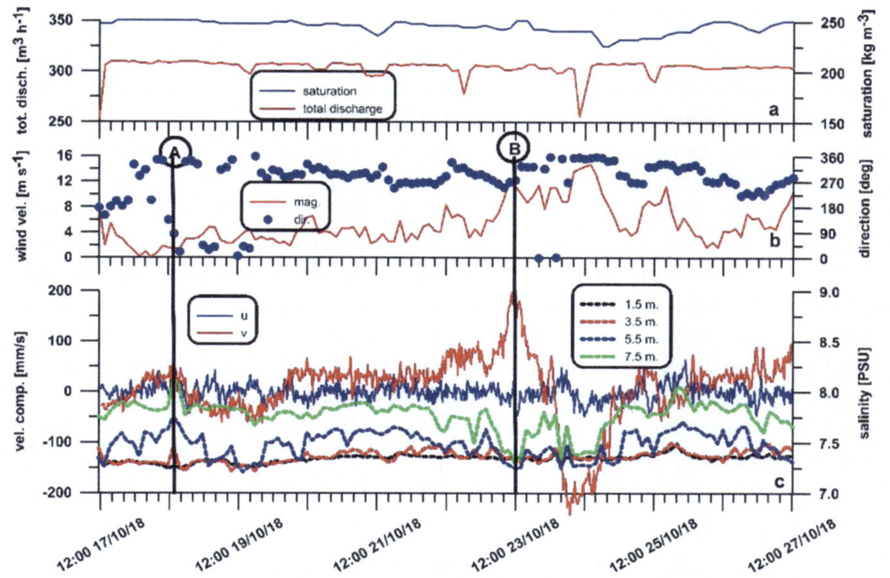

Fig. 5 Changes in the period 17–27.10.2018: **a** total discharge and saturation; **b** magnitude and direction of wind, **c** velocity components (u, v) in the layer 2–2.5 m above the bottom and salinity measured in the centre of installation (S) on four levels (1.5, 3.5, 5.5, 7.5 m below MSL)

decrease (case B) was preceded by a period of the north-eastern wind. Very strong v velocity component was associated with this phenomenon. It should be noticed that in the second case salinity measured by all gauges was very similar, suggesting intensive vertical mixing.

The presented discussion of two alternative phenomena indicates potential applicability of water currents to support operational monitoring.

3 Assessment of Excess Salinity

Functioning of the discharge installation requires continuous control of excess salinity. It can be done based on analysis of in situ salinity measurements or discharge conditions.

3.1 Excess Salinity Based on Salinity Monitoring

The water permit issued for brine discharge system obliged the operator of installation to control excess salinity using data from salinity monitoring program.

Assessment of excess salinity based on in situ data requires division of measured salinity into two components: (1) natural salinity in the ambient (i.e. salinity in case of no brine discharge) and (2) anthropogenic (excess due to brine discharge). There is no direct method to divide the measured salinity into two components.

Continuously registered data are used to assess salinity excess operationally. The proposed approach assumes that natural salinity is equal to salinity measured in the upper part of the water column (no natural stratification), while excess salinity is equal to difference between salinity measured at the bottom and in the surface layer. The method delivers the "upper limit" of excess salinity. In majority of time this method leads to a conclusion that excess salinity does not exceed 0.5 PSU. To verify the assumption taken a detailed analysis of 25 CTD data sets of in situ measurements in the vicinity of discharge installation was carried out. It has shown that the proposed assumption holds true for over 70% of situations. Higher values delivered by a simple approach require analysis of hydrodynamic conditions. Proposed measurements of currents can be an option.

3.2 *Excess Salinity Based on Discharge Parameters*

Simultaneously to excess salinity based on in situ measurement, it can be proposed an approach to estimate the efficiency of a rosette-type diffuser depending on riser spacing, the number of nozzles in each riser, and flow conditions in the receiving water body based on results of laboratory experiments (Abessi and Roberts 2016).

The dimensionless spacing parameter s_r/dF (s_r—distance between risers, d—nozzle diameter [m], F—densimetric Froude number: $F = \frac{u_0}{\sqrt{g_0' d}}$; $g_0' = \frac{(\rho_0 - \rho_a)g}{\rho_a}$, g—acceleration due to gravity [m s^{-2}], ρ_0—effluent density [kg m^{-3}], ρ_a—ambient density [kg m^{-3}], u_0—exit flow velocity [m s^{-1}]) has two asymptotic solutions: $s_r/dF \gg 1$—jets do not interact with each other, $s_r/dF \ll 1$—jets can interact with those from the neighbouring risers. In the analysed system, risers are spaced every 45 meters. Given the variability of the densimetric Froude number, the dimensionless spacing parameter varies in the range of 16.7–27.7 suggesting no interaction between jets from the neighbouring risers. Based on formulas derived by Abessi and Roberts (2016), the estimated dilution for the real discharge conditions in the Puck Bay exceeds a value of 360 determined in the pre-investment study as sufficient to achieve excess salinity lower than 0.5 PSU.

4 Conclusions

The monitoring program carried out since autumn of 2010 delivers data on salinity variations in the vicinity of installation, and based on them enables simple operational assessment of excess salinity in majority of the year. In complex hydro-meteorological conditions, theoretical analysis of brine discharge based on laboratory experiments or analysis of currents measured in situ can be proposed. Analysis of excess salinity using all available data confirms that it does not exceed 0.5 PSU.

Acknowledgements This study has been financially supported by Gas Storage Poland sp. z o.o. and Institute of Hydro-Engineering of the Polish Academy of Sciences.

References

Abessi O, Roberts PJW (2016) Rosette diffusers for dense effluents. J Hydraulic Eng. https://doi.org/10.1061/(ASCE)HY.1943-7900.0001268

Nowacki J, Jarosz E (1998) The hydrological and hydrochemical division of the surface waters in the Gulf of Gdańsk. Oceanologia 40(3):261–272

Robakiewicz M (2016) Mixing of brine waste in the Puck Bay (south Baltic Sea) in light of in-situ measurements. Oceanol Hydrobiol Stud 45(A):42–54

Robakiewicz M, Robakiewicz W (2008) Spreading of brine from proposed discharge into the Puck Bay (in Polish). Project Report, IBW PAN

Turbulence in Wall-Wake Flow Downstream of an Isolated Dune

Sankar Sarkar and Subhasish Dey

Abstract The study deals with the variations of third-order moments of velocity fluctuations, turbulent bursting and turbulent kinetic energy (TKE) budget downstream of an isolated dune mounted on a rough bed in an open channel. The experimental data demonstrates that in the wall-wake flow, the third-order moments change their signs below the crest level of the dune, whereas their signs remain unchanged above the crest. The near-wake flow is characterized by sweep events and the far-wake flow is governed by the ejection. In the near-wake flow, the mean duration of bursting events is shorter than their mean interval of occurrence. Downstream of the dune, the turbulent production and energy dissipation start with large positive values in the vicinity of the bed and decrease with an increase in vertical distance until the lower-half of the height of dune. However, all the changes in the turbulence characteristics are significant at the near-wake zones that fade away after a certain distance in the downstream.

Keywords Turbulent flow · Wake flow · Bluff-body · Shear flow · Open-channel hydraulics

1 Introduction

Flow over and downstream of an isolated dune is a subject of immense importance in several fields of environmental and engineering sciences due to its practical implications in river management. It also allows a considerable theoretical insight in wake flows. However, in spite of tremendous efforts over last few decades, the prediction of flow characteristics over a dune remains still challenging (ASCE Task Force 2002).

S. Sarkar (✉)
Physics and Applied Mathematics Unit, Indian Statistical Institute, Kolkata, West Bengal 700108, India
e-mail: sankar_s@isical.ac.in

S. Dey
Department of Civil Engineering, Indian Institute of Technology Kharagpur, Kharagpur, West Bengal 721302, India

M. B. Kalinowska et al. (eds.), *Recent Trends in Environmental Hydraulics*, GeoPlanet: Earth and Planetary Sciences,
https://doi.org/10.1007/978-3-030-37105-0_21

However, the present state-of-the-art suggests that significant progresses have so far been made in understanding the flow characteristics over dunes based on the experimental and numerical studies. McLean and Smith (1986) and Nelson and Smith (1989) studied the turbulent flow field over dunes in the context of the separation zone, reattachment point, wake region, internal boundary layer and interior, etc. Maddux et al. (2003a, b) studied the turbulent flow over a series of two-dimensional (2-D) and three-dimensional (3-D) dunes based on the experimental data. They observed that the turbulence generated by 3-D dunes is much weaker than that generated by their 2-D counterparts. Sukhodolov et al. (2006) studied the flow characteristics over natural and laboratory experimental dunes. They found that the experimental dunes showed different flow characteristics from the natural dunes. Based on the laboratory experiments, Best (2005) showed that the ejections dominate the instantaneous flow field over the crest level of the dunes and the highest instantaneous Reynolds stresses are associated with ejections and sweeps. However, the earlier studies were principally based on the beds with continuous or series of dunes. Advances in understanding the flow over and downstream of an isolated dune are very less in comparison to that of flow downstream of a series of dunes (Anta et al. 2011; Sinha et al. 2017). Furthermore, due to the geometrical features, a dune can be considered as a bluff-body in hydraulics. Earlier studies showed that the flow downstream of a bluff-body is complex, which is characterized by the modifications in the velocity, Reynolds shear stress (RSS), turbulent bursting, turbulent kinetic energy (TKE) budget, etc. Literature reveals that the flow downstream of a bluff-body has been studied extensively in last few decades (Tachie and Balachandar 2001; Kahraman et al. 2002; Akilli and Rockwell 2002; Sadeque et al. 2009; Dey et al. 2011; Sarkar and Dey 2015; Sarkar et al. 2016; Dey et al. 2018a, b, c). However, considering 2-D fixed dune as a bluff-body and studying the flow characteristics downstream of the dune in the context of bursting and the TKE budget remains absent till date. The paper, as an extension of Dey and Sarkar (2019), is therefore devoted to the higher-order moments, turbulent bursting, and TKE budget in the wall-wake flow downstream of a fixed, rough 2-D dune. Importantly, the definition sketch, experimental setup, time-averaged velocity, RSS, average turbulence intensity, turbulent length scales and Reynolds stress anisotropy have been described explicitly in Dey and Sarkar (2019).

2 Experimental Equipment and Procedure

Experiments were carried out in the Fluvial Mechanics Laboratory of Indian Statistical Institute, Kolkata, India, using a rectangular flume 20 (m) long, 0.5 (m) wide, and 0.5 (m) deep. Two rough dunes of heights of $D = 0.09$ (m) and 0.03 (m); and lengths of $l = 0.4$ (m) and 0.3 (m) were used in the experiments for Run 1 and 2, respectively, attaching one at a time to the flume bottom covering the full width (spanwise) of the flume. To be specific, an isolated dune was placed at a distance of 7 m from the inlet of the flume for each experimental run. A rough-bed was created by gluing a uniformly graded sediment of a median diameter $d_{50} = 2.49$ (mm) and the same sediment was used to roughen the surface of the dune. The experimental runs were performed under a uniform approach flow depth of $h = 30$ (cm) and a streamwise bed slope of 0.03%. During the experiments, the depth-averaged approach velocity $\overline{U}$ was 0.44 (m s^{-1}) for both the runs with a variation of approximately 1.5% due to the experimental rearrangements the runs. The shear-particle Reynolds numbers R_* ($= d_{50}u_*/\upsilon$; where u_* is the approach shear velocity and υ is the kinematic viscosity of water) were above 70, ensuring the turbulent-rough flows. A 5-cm downlooking *Vecrtino velocimeter* was used to measure the velocity with a sampling rate of 100 (Hz). Along the centerline of the flume, the data were collected at the streamwise distances of $x/l = -0.5, -0.25, 0, 0.1, 0.2, 0.3, 0.4, 05, 0.6, 0.7, 0.8, 0.9, 1, 1.1, 1.3, 1.7, 2.1, 2.5$, and 3.3. The captured velocity data were processed using the *acceleration thresholding method* for despiking (Goring and Nikora 2002). The streamwise (x-direction), spanwise (y-direction), and vertical (z-direction) time-averaged velocity components are designated as u, v, and w and their fluctuations from the time-averaged values are u', v', and w', respectively. Details of the experimental equipment and procedure are available in Dey and Sarkar (2019).

3 Third-Order Moments of Velocity Fluctuations

Third-order moments of velocity fluctuations are important parameters in characterising the turbulent flows. They present the essential stochastic information on the temporal behaviour of the velocity fluctuations, u' and w', in the form of the flux and the advection of Reynolds normal stresses. In addition, they provide a clear indication of the dominance of turbulent bursting events due to their sign conventions (Gad-el-Hak and Bandyopadhyay 1994). The general form of third-order moments is $m_{jk} = \overline{\tilde{u}^j\tilde{w}^k}$ with $j + k = 3$, where $\tilde{u}$ is $u'/(\overline{u'u'})^{0.5}$ and $\tilde{w}$ is $w'/(\overline{w'w'})^{0.5}$ (Raupach 1981). The values of j and k range from 0 to 3. Depending upon the values of j and k, the third-order moments are m_{30} [$= \overline{\tilde{u}^3} = \overline{u'u'u'}/(\overline{u'u'})^{1.5}$], m_{03} [$= \overline{\tilde{w}^3} = \overline{w'w'w'}/(\overline{w'w'})^{1.5}$], m_{21} {$= \overline{\tilde{u}^2\tilde{w}} = \overline{u'u'w'}/[(\overline{u'u'}) \times (\overline{w'w'})^{0.5}]$} and m_{12} {$= \overline{\tilde{u}\tilde{w}^2} = \overline{u'w'w'}/[(\overline{u'u'})^{0.5} \times \overline{w'w'})]$}. The higher-order moment m_{30} represents the skewness of u' referring to the streamwise flux of the streamwise Reynolds normal stress $\overline{u'u'}$, the m_{03} represents the skewness of w' signifying the vertical flux of the vertical Reynolds normal stress $\overline{w'w'}$, the m_{21} defines the advection of $\overline{u'u'}$ in the vertical direction and m_{12} defines the advection of $\overline{w'w'}$ in the streamwise direction.

Figure 1 presents the vertical profiles of m_{30} and m_{03} at the upstream and different downstream distances of the isolated dune. In Fig. 1, upstream of the dune ($x/l = -0.5$), the m_{30} and m_{03} start with a small negative and positive values, respectively, in the vicinity of the bed and increase with an increase in vertical distance without changing their signs. Downstream of the dune ($x/l = 1$–2.1), the m_{30} and m_{03} profiles possess interesting characteristics. They start with a small positive and a negative value, respectively, near the bed and increase gradually with an increase in vertical distance until they attain their peak positive and negative values at $z/D \approx 0.75$ and

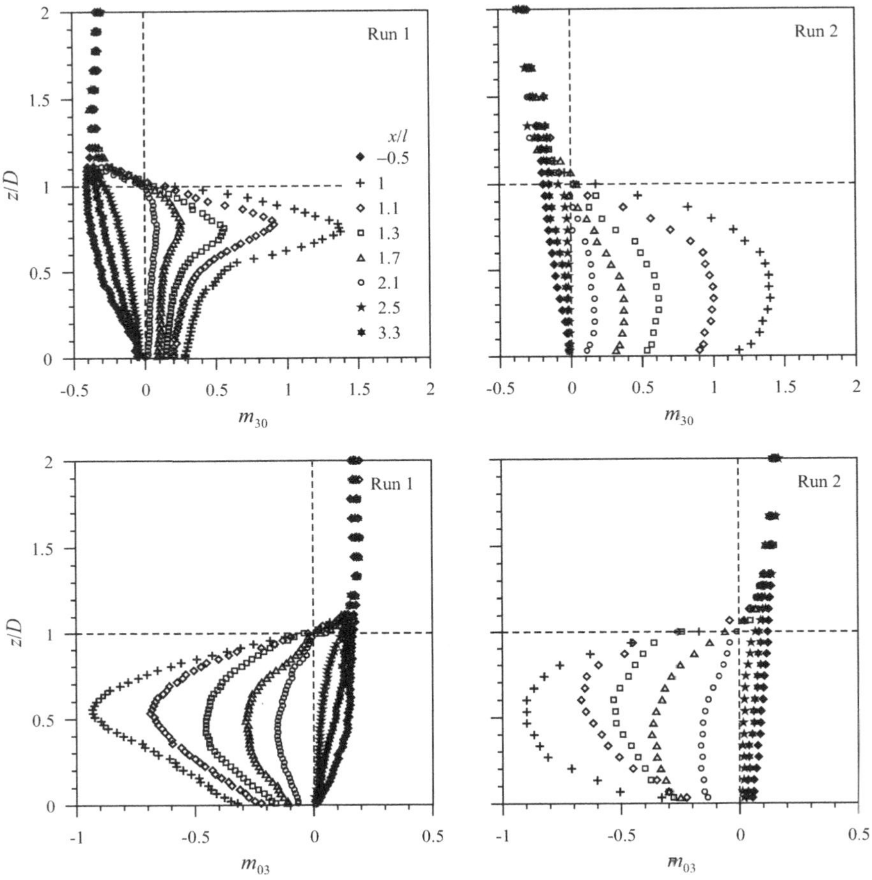

Fig. 1 Vertical profiles of third-order moments (skewness or flux of Reynolds normal stresses) m_{30} and m_{03} at different streamwise distances for Runs 1 and 2

0.5, respectively. With a further increase in z/D, the m_{30} and m_{03} decrease rapidly and change their signs at $z/D = 1$, and above that they become almost invariant of z/D. These features gradually disappear with an increase in streamwise distance x/l. However, the m_{30} and m_{03} profiles are almost similar to those in the upstream at $x/l = 3.3$.

On the other hand, the advections of the Reynolds normal stresses are also observed from the variations of m_{21} and m_{12}, as shown in Fig. 2. In general, the m_{21} and m_{12} profiles are similar to the m_{03} and m_{30} profiles, respectively, with differences in values. Upstream of the dune (($x/l = -0.5$), the m_{21} and m_{12} have small positive and negative values, respectively, near the bed and increase with an increase in vertical distance up to a certain depth. Thereafter, they decrease becoming almost invariant of vertical distance for $z/D > 1.1$. Downstream of the dune (x/l = 1–2.1),

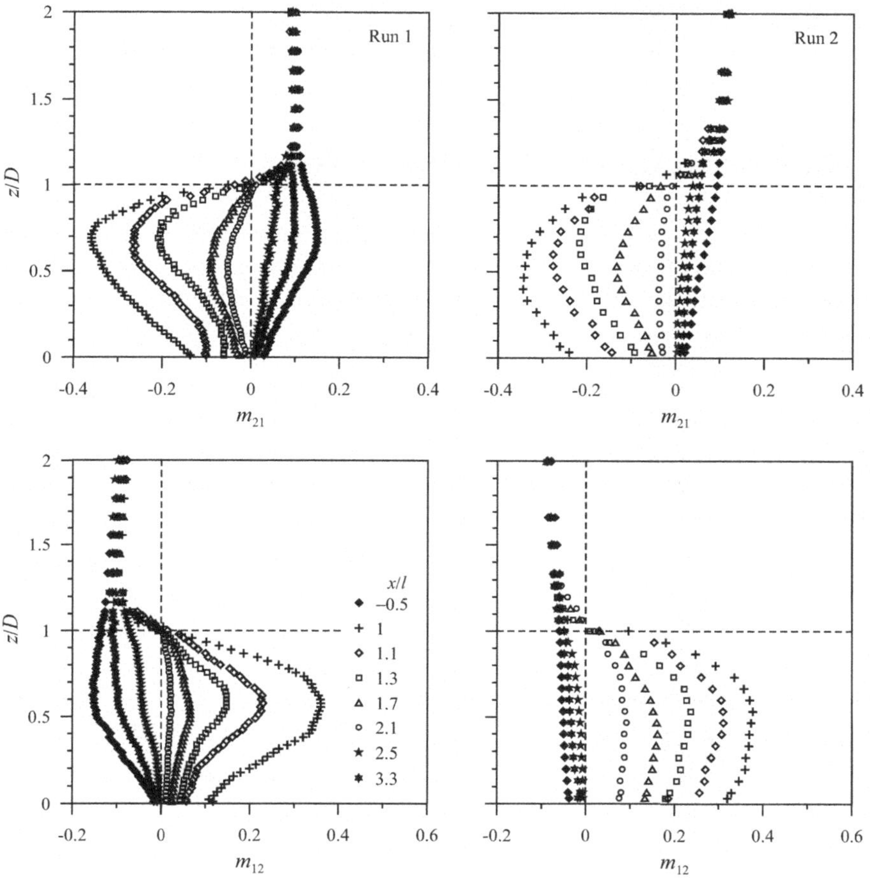

Fig. 2 Vertical profiles of third-order moments (advection of Reynolds normal stresses) m_{21} and m_{12} at different streamwise distances for Runs 1 and 2

the m_{21} and m_{12} start with small negative and positive values, respectively, near the bed and increase with an increase in vertical distance, attaining their peak values. Then, they sharply reduce and change signs at the crest ($z/D = 1$). Thereafter, the m_{21} and m_{12} rapidly attain their values of the upstream profiles. However, as one moves further downstream, the signs of the m_{21} and m_{12} start changing until their signs become same as that of upstream at $x/l > 2.1$. Importantly, the m_{21} and m_{12} follow the similar profiles to those of the upstream at $x/l = 3.3$. Thus, downstream of the dune, there exists an advection of $\overline{u'u'}$ in the upward direction and that of $\overline{w'w'}$ in the upstream direction below the crest. The analysis provides a clear indication that below the crest, a streamwise acceleration is prevalent and is associated with a downward flux causing sweeps with an advection of $\overline{u'u'}$ in the downward direction, whereas above the crest, a streamwise deceleration occurs and is coupled with an upward flux causing ejections with an advection of $\overline{u'u'}$ in the upward direction.

4 Quadrant Analysis for Turbulent Bursting

Turbulent bursting is a quasi-cyclic process that determines the conditional statistics of the fluctuations of streamwise and vertical velocity components, $u'$and w', respectively. A conventional way to estimate the bursting events is the quadrant analysis where the fluctuations u' and w' are plotted onto a $u'w'$-plane (Lu and Willmarth 1973). The turbulent bursting comprises of four events, namely (i) *outward interactions or* $Q1$ *events* ($i = 1$; $u' > 0$, $w' > 0$), (ii) *ejections or* $Q2$ *events* ($i = 2$; $u' < 0$, $w' > 0$), (iii) *inward interactions or* $Q3$ *events* ($i = 3$; $u' < 0$, $w' < 0$); and (iv) *sweeps or* $Q4$ *events* ($i = 4$; $u' > 0$, $w' < 0$), where i represents the quadrant. Using a detection function $\phi_{i,H}(t)$, the contribution of $\langle u'w' \rangle_{i,H}$ to $-\overline{u'w'}$ from the quadrant i outside the *hole-size* H is estimated as follows:

$$\overline{u'w'}^{i,H} = \lim_{T \to \infty} \frac{1}{T} \int_0^T u'(t)w'(t)\phi_{i,H}[u'(t), w'(t)]dt \tag{1}$$

where T is the time of sampling. The value of $\phi_{i,H}(t) = 1$, if (u', w') is in quadrant i and if $|u'w'| \geq H(\overline{u'u'})^{0.5}(\overline{w'w'})^{0.5}$, and $\phi_{i,H}(t) = 0$, otherwise. The relative fractional contributions toward the RSS production is given by

$$S_{i,H} = \frac{\overline{u'w'}^{i,H}}{\overline{u'w'}} \tag{2}$$

The algebraic summation of the contributions from the bursting events for $H = 0$ is unity, that is, $S_{1,0} + S_{2,0} + S_{3,0} + S_{4,0} = 1$.

According to Fig. 3, upstream of the dune ($x/l = -0.5$), the $Q2$ and $Q4$ events are the most and the second-most contributing events, respectively, to the production of RSS, whereas the $Q1$ and $Q3$ events remain insignificant across the flow depth. On the other hand, downstream of the dune ($1 \leq x/l \leq 2.1$), all the events have a large augmentation below the crest with predominating $Q4$ events. At $x/l = 2.5$, the contributions from the $Q2$ and $Q4$ events become almost equal below the crest. At further downstream ($x/l = 3.3$), the $Q2$ events dominate $Q4$ events. However, above the crest ($z/D > 1$), the $Q2$ is the most contributing events irrespective of the values of x/l.

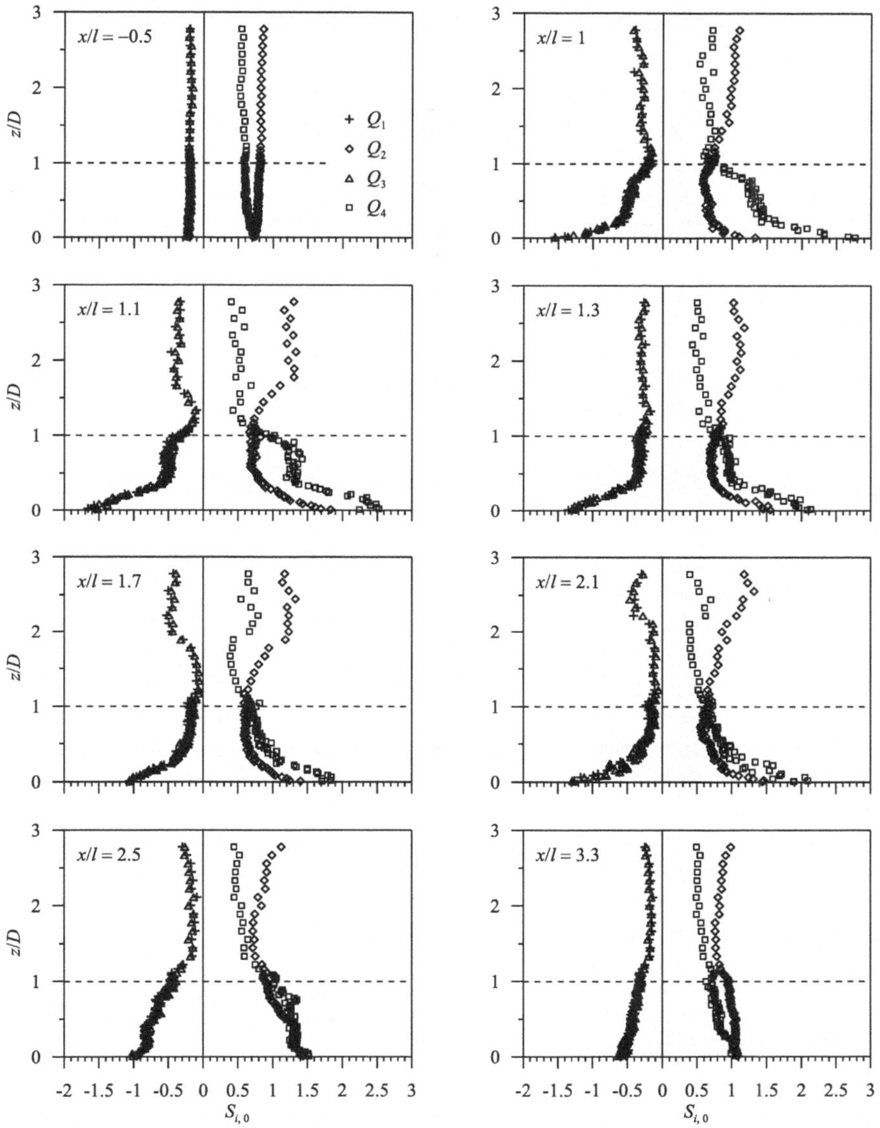

Fig. 3 Variations of $S_{i,0}$ with z/D at different streamwise distances for Run 1

5 Durations and Time Intervals of Turbulent Bursting Events

It is important to analyze the mean time-durations and mean time-intervals of ejection and sweep events, because it is evident from Dey et al. (2018b) that downstream of a bluff-body, their values change significantly. The durations and intervals of the $Q2$ and $Q4$ events are estimated from the number of their occurrences in a sample for a given hole-size H.

Figure 4 shows the nondimensional mean time-durations and mean time-intervals between two consecutive ejections t_E and sweeps t_S; and mean interval of ejections i_E and sweeps i_S for $H = 0$. The nondimensional mean time-durations of $Q2$ and $Q4$ events are expressed as T_{dE} $(= t_E u_*/D)$ and T_{dS} $(= t_S u_*/D)$, respectively, and the nondimensional mean time-intervals $Q2$ and $Q4$ events are T_{iE} $(= i_E u_*/D)$ and T_{iS} $(= i_S u_*/D)$, respectively.

It is evident from Fig. 4 that the T_{dE} and T_{dS} are almost identical across the flow depth, irrespective of the streamwise distances. Upstream of the dune ($x/l = -0.5$), the T_{iE} remains larger than T_{iS} up to the crest. At the crest ($z/D = 1$), the T_{iE} equals T_{iS}, and above that, the T_{iS} becomes greater than the T_{iE}. Interestingly, the T_{iE} and T_{iS} are always greater than T_{dE} and T_{dS} across the flow depth for a given streamwise distance, which is in confirmation with the observations of Dey et al. (2018b). On the other hand, downstream of the dune ($1 \leq x/l \leq 1.3$), the values of T_{iE} are less than those of T_{iS} across the flow depth. For $x/l \geq 1.7$, the T_{iE} remains greater than T_{iS} below the crest. Finally at $x/l = 3.3$, the values of T_{iE} and T_{iS} become almost identical to those of the undisturbed upstream ($x/l = -0.5$). However, above the crest, the overall characteristics of T_{dE}, T_{dS}, T_{iE} and T_{iS} remain unchanged with respect to vertical and streamwise distances.

It can therefore be concluded in general that the mean time-durations of the bursting events are not influenced by the dune, whereas the mean time-intervals of ejections decrease and the mean time-intervals of sweeps increase owing to the presence of the dune.

6 Turbulent Kinetic Energy Budget

In a turbulent flow, the TKE budget gives a quantitative measurement of the contributions from various TKE budget components based on a balancing relation among the TKE production rate t_P, TKE dissipation rate ε, TKE diffusion rate t_D, pressure energy diffusion rate p_D and viscous diffusion rate v_D. The TKE budget equation for a free surface flow is given as follows (Nezu and Nakagawa 1993):

$$\underbrace{\overline{u'w'}\frac{\partial u}{\partial z}}_{t_P} = \varepsilon + \underbrace{\frac{\partial f_{kw}}{\partial z}}_{t_D} + \underbrace{\frac{1}{\rho} \cdot \frac{\partial}{\partial z}(\overline{p'w'})}_{p_D} - \underbrace{\upsilon \frac{\partial^2 k}{\partial z^2}}_{\upsilon_D} \quad (3)$$

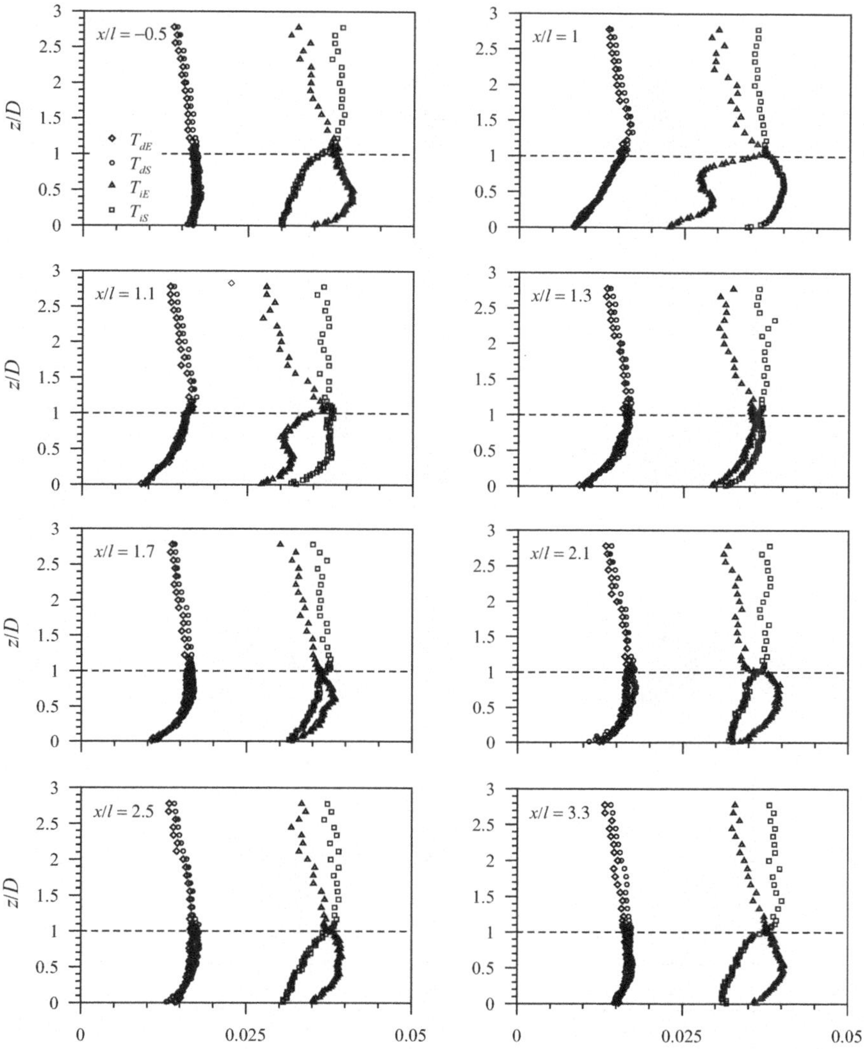

Fig. 4 Vertical profiles of mean time-durations (T_{dE} and T_{dS}) and time-intervals (T_{iE} and T_{iS}) for *Q*2 and *Q*4 events as a function of *z*/*D* at different streamwise distances for Run 1

where p' = fluctuations of hydrostatic pressure (relative to ρ), f_{kw} = vertical flux of TKE; and k = TKE given by $0.5(\overline{u'u'} + \overline{v'v'} + \overline{w'w'})$. In the present study, the values of ε were estimated from the Kolmogorov second hypothesis by using the velocity power spectra. The details of the estimation of ε are available in Dey and Das (2012). The values of t_P, ε and t_D were estimated from the measured data, whereas the p_D was estimated from $p_D = t_P - \varepsilon - t_D$. The nondimensional values of t_P, ε, t_D and p_D are given by $T_P, E_D, T_D, P_D = (t_P, \varepsilon, t_D, p_D) \times (D/u_*^3)$.

It is evident from Fig. 5 that upstream of the dune ($x/l = -0.5$), all the TKE budget components start with positive values with a sequence of magnitude $T_P > E_D > P_D > T_D$ near the bed and decrease with an increase in z/D. Above the crest ($z/D > 1$), they become very small. On the other hand, downstream of the dune ($x/l = 1$–2.1), the peak values of T_P, E_D, P_D and T_D are observed at the crest. The T_P and E_D start

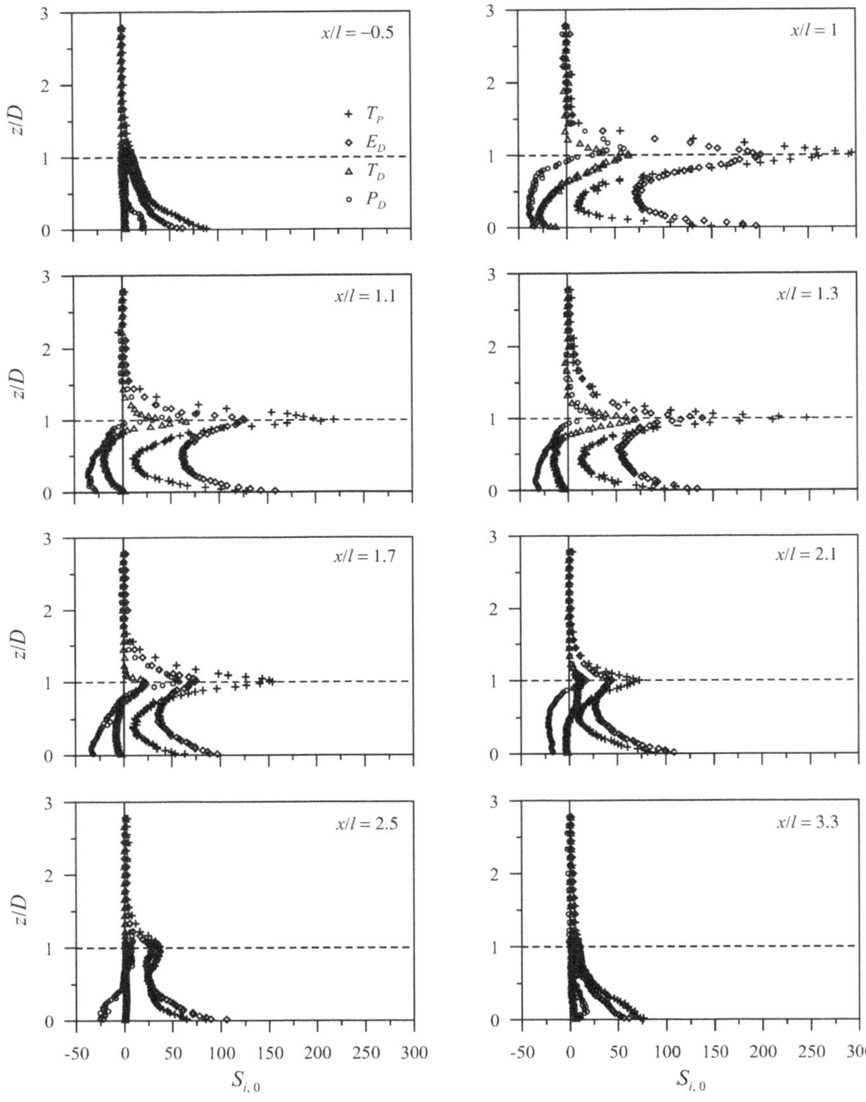

Fig. 5 TKE budget components T_P, E_D, T_D and P_D, as a function of z/D at different streamwise distances for Run 1

with positive values near the bed, whereas P_D and T_D start with negative values for $x/l = 1$–2.1. Downstream of the dune, the amplification of the absolute values T_P, E_D, P_D and T_D are observed in the vicinity of the bed, and at the crest they diminish with an increase in x/l. At $x/l = 3.3$, the T_P, E_D, P_D and T_D profiles become almost similar to those of the undisturbed upstream ($x/l = -0.5$).

7 Conclusions

Analysis of the experimental data reveals the characteristics of third-order moments and bursting events in wall-wake flow downstream of an isolated dune mounted on a rough bed. The third-order moment of velocity fluctuations suggest that downstream of a dune, a streamwise acceleration with a downward flux is prevalent below the crest of the dune, whereas above the crest a strong streamwise retardation prevails with an upward flux. The quadrant analysis for turbulent bursting shows that below the crest, the sweeps are the main contributing event; whereas above the crest, the ejections are the governing events. The mean time-durations of the bursting events are not influenced by the dunes across the flow depth but the mean time-intervals of ejections decrease and the mean time-intervals of sweeps increase below the crest. The turbulent kinetic energy budget components show an amplification of the values of all the turbulent processes that become maximum at the crest. However, all the characteristics in wall-wake flow disappear after a certain distance, signifying the recovery of the turbulence characteristics.

References

Akilli H, Rockwell D (2002) Vortex formation from a cylinder in shallow water. Phys Fluids 14(9):2957–2967

Anta J, Mera I, Pena E, Louro A (2011) Two-layer exchange flows over a dune: effect of large-scale bottom roughness. J Vis 14(2):99–101

ASCE Task Force (2002) Flow and transport over dunes. J Hydraul Eng 128(8):726–728

Best JL (2005) The kinematics, topology and significance of dune-related macroturbulence: some observations from the laboratory and field. In: Blum MD, Marriott SB, Leclair S (eds) Fluvial sedimentology VII. Special Publication of International Association of Sedimentologists, vol 35, pp 41–60

Dey S, Das R (2012) Gravel-bed hydrodynamics: a double-averaging approach. J Hydraul Eng 138(8):707–725

Dey S, Sarkar S (2019) Turbulent length scales and Reynolds stress anisotropy in wall-wake flow downstream of an isolated dunal bedform. In: Kalinowska MB, Rowinski PM (eds) Recent trends in environmental hydraulics. Springer, Berlin (in press)

Dey S, Sarkar S, Bose SK, Tait S, Castro-Orgaz O (2011) Wall-wake flows downstream of a sphere placed on a plane rough wall. J Hydraul Eng 137(10):1173–1189

Dey S, Swargiary D, Sarkar S, Fang H, Gaudio R (2018a) Self-similarity in turbulent wall-wake flow downstream of a wall-mounted vertical cylinder. J Hydraul Eng 144(6):04018023

Dey S, Swargiary D, Sarkar S, Fang H, Gaudio R (2018b) Turbulence features in a wall-wake flow downstream of a wall-mounted vertical cylinder. Eur J Mech B/Fluids 69:46–61
Dey S, Lodh R, Sarkar S (2018c) Turbulence characteristics in wall-wake flows downstream of wall-mounted and near-wall horizontal cylinders. Environ Fluid Mech 18(4):891–921
Gad-El-Hak M, Bandyopadhyay PR (1994) Reynolds number effects in wall-bounded turbulent flows. Appl Mech Rev 47(8):307–365
Goring DG, Nikora VI (2002) Despiking acoustic Doppler velocimeter data. J Hydraul Eng 128(1):117–126
Kahraman A, Sahin B, Rockwell D (2002) Control of vortex formation from a vertical cylinder in shallow water: effect of localized roughness elements. Exp Fluids 33(1):54–65
Lu SS, Willmarth WW (1973) Measurements of the structures of the Reynolds stress in a turbulent boundary layer. J Fluid Mech 60(3):481–511
Maddux TB, Nelson JM, McLean SR (2003a) Turbulent flow over three-dimensional dunes: 1. Free surface and flow response. J Geophys Res 108(F1):6009
Maddux TB, Nelson JM, McLean SR (2003b) Turbulent flow over three-dimensional dunes: 2. Fluid and bed stresses. J Geophys Res 108(F1):6010
McLean SR, Smith JD (1986) A model for flow over two-dimensional bed forms. J Hydraul Eng 112(4):300–317
Nelson JM, Smith JD (1989) Mechanics of flow over ripples and dunes. J Geophys Res 94(C6):8146–8162
Nezu I, Nakagawa H (1993) Turbulence in open-channel flows. IAHR monograph. Balkema, Rotterdam, The Netherlands, pp 77–79
Raupach MR (1981) Conditional statistics of Reynolds stress in rough-wall and smooth-wall turbulent boundary layers. J Fluid Mech 108(July):363–382
Sadeque MAF, Rajaratnam N, Loewen MR (2009) Shallow turbulent wakes behind bed-mounted cylinders in open channels. J Hydraul Res 47(6):727–743
Sarkar S, Dey S (2015) Turbulent length scales and anisotropy downstream of a wall mounted sphere. J Hydraul Res 53(5):649–658
Sarkar S, Papanicolaou AN, Dey S (2016) Turbulence in a gravel-bed stream with an array of large gravel obstacles. J Hydraul Eng 142(11):04016052
Sinha S, Hardy RJ, Blois G, Best JL, Sambrook Smith GH (2017) A numerical investigation into the importance of bed permeability on determining flow structures over river dunes. Water Resour Res 53(4):3067–3086
Sukhodolov AN, Fedele JJ, Rhoads BL (2006) Structure of flow over alluvial bedforms: an experiment on linking field and laboratory methods. Earth Surf Process Landf 31(10):1292–1310
Tachie MF, Balachandar R (2001) Shallow wakes generated on smooth and rough surfaces. Exp Fluids 30(4):467–474

Automatic Calibration of a 3D Morphodynamic Numerical Model for Simulating Bed Changes in a 180° Channel Bend

Vahid Shoarinezhad, Silke Wieprecht and Stefan Haun

Abstract Sediment transport in meanders is a complex phenomenon due to its three-dimensional behavior (i.e., the presence of secondary currents). As a result, in river bends, sediments are eroded from the outer bank, resulting in depositions at the inner bank. In order to describe the complexity of flow characteristics and sediment transport mechanisms in river bends, the use of three-dimensional numerical models is essential. Nevertheless, these models are highly parameterized; therefore, the model calibration process becomes demanding. Using optimization methods for model calibration can diminish the manual effort and avoid users' subjectivity. Besides, it helps to decrease errors caused by wrong assumptions and unrealistic combinations of involving parameters. In this study, the 3D numerical model SSIIM 2 is used to simulate the flow field as well as morphological bed changes in a U-shaped bend. The automatic calibration tool PEST is used to calibrate the morphodynamic model by altering different input parameters. The result of this study reveals that using automatic calibration can noticeably reduce the calibration time and the user-intervention in the calibration process. Comparisons between the simulated bed topography changes and the experimental data regarding the deposition and scour regions, as well as their magnitudes, demonstrate a high degree of accuracy for the automatic calibration process.

Keywords Automatic calibration · Morphodynamic models · Sediment transport · Channel bend

1 Introduction

Numerical models are nowadays extensively used to simulate the behavior of environmental systems. Many of these models, such as morphodynamic models, are characterized by a large number of input parameters. Some of these parameters can

V. Shoarinezhad (✉) · S. Wieprecht · S. Haun
Institute for Modelling Hydraulic and Environmental Systems, University of Stuttgart, Pfaffenwaldring 61, 70569 Stuttgart, Germany
e-mail: vahid.shoarinezhad@iws.uni-stuttgart.de

M. B. Kalinowska et al. (eds.), *Recent Trends in Environmental Hydraulics*, GeoPlanet: Earth and Planetary Sciences,
https://doi.org/10.1007/978-3-030-37105-0_22

only be quantified at certain locations over a limited period, and some are physically not feasible to be measured (Cunge 2003). Hence, the model accuracy and its prediction reliability strongly depend on the calibration procedure by adjustment of the uncertain input parameters. However, even by calibration, there may be a considerable amount of uncertainty in results due to the lack of error-free measurements (Muleta and Nicklow 2005). The model calibration presents in addition to a reliable prediction, also an insight into the inaccuracies of the model, and the consequences of deficiencies on the applicability of the model to reproduce physical phenomena. Model calibration has been traditionally accomplished manually by trial-and-error adjustment of parameters until reaching a satisfactory agreement between simulated and measured values. This method can become time- and cost-consuming owing to the high number of model runs, and it is highly dependent on user's knowledge of the model structure and the properties of the environmental systems (Boyle et al. 2000; Duan et al. 1993; Madsen 2000). The time-consuming practice of manual calibration induced the progression of the complex inverse modeling technique based on optimization methods (Madsen et al. 2002; Rode et al. 2007). Principally, these methods consist of three elements; an objective function to assess the differences between the model outputs and the observations, an optimization algorithm, and a convergence criterion (Vidal et al. 2007). The algorithms can be categorized into two classes: global methods based on sampling the proposed values of parameters over the entire space by multitudinous model runs, and local methods based on the point-estimation by finding the optimum point where no further progress can be achieved in the adjacent space of the parameter (gradient-based approach). Regarding the local methods, which are computationally efficient and need far fewer model runs, the initial values should be chosen wisely as the calibration process proceeds from the initial value of a parameter towards the gradient descent of the objective function (Deslauriers and Mahdi 2018). Hence, it can be simply trapped in local minima points instead of finding the global minimum (Abbaspour et al. 2001).

Due to the fully three-dimensional nature of flow in river bends, caused by the interaction of affecting forces, i.e., the transverse pressure gradient and the centrifugal force, which results in a helical flow, using a 3-D numerical model becomes indispensable. Many researchers have accomplished investigations concerning the numerical modeling of flow field and morphological process in meanders (e.g., Duc et al. 2004; Khosronejad et al. 2007; Rüther and Olsen 2005; Shams et al. 2002; Wu et al. 2000a; Ye and McCorquodale 1998). In the present study, a 3-D morphodynamic model of a 180° channel bend is developed by SSIIM 2 (Olsen 2014). The model is calibrated by using the Parameter ESTimation (PEST) package (Doherty 2016) against the experimental data obtained by Yen and Lee (1995). PEST has been successfully used for calibration and sensitivity analysis of environmental models (groundwater, hydrological, ecological, hydrodynamic, etc.) in the past (e.g., by Fabio et al. 2010; Maneta et al. 2007; Rigosi et al. 2011; Usman et al. 2018). However, to the knowledge of the authors, PEST has not been applied for morphodynamic models so far.

2 Methodology

2.1 Experimental Data

The experimental data obtained by Yen and Lee (1995) is used in this study to test the numerical model in combination with an automatic calibration tool. As an outline of their work, a U-shaped channel with 4 m central radius and 1 m width was built. Attached to the bend, an up- and a downstream straight reach (both 11.5 m long) were constructed to ensure the uniform and fully turbulent flow. A layer of 20 cm non-uniform sand with a median diameter of $d_{50} = 1$ mm and the geometric standard deviation of $\sigma = 2.5$ was placed on the bed, having a final slope of 2‰. Further details regarding the sediment characteristics can be seen in Table 1. In the experiments, the base flow rate was adjusted to $Q_0 = 0.02\ \mathrm{m^3\ s^{-1}}$ with a corresponding flow depth of $h_0 = 5.44$ cm. During the experiments, the discharge was linearly increased up to a maximum value and then lowered to the base flow (triangular-shaped hydrograph). In total, five different experiments were conducted with different durations and peak discharges.

The evolution of the bed topography and the sediment sorting subject to the unsteady flow were investigated along the bend. Among the five experiments conducted by Yen and Lee (1995), two of them (Run #1 and Run #3) are used in this study. Table 2 shows the characteristics of the applied hydrographs.

2.2 Numerical Method

For simulating the hydraulics and the morphological bed changes, the fully three-dimensional numerical model SSIIM 2, which solves the Reynolds-averaged Navier-Stokes (RANS) equations on an adaptive three-dimensional non-orthogonal grid, is employed. The model uses a finite-volume approach for the spatial discretization and

Table 1 Sediment characteristics of the experiments, applied in the numerical model

Sediment size classes								
Sediment size (mm)	0.25	0.42	0.84	1.19	2.00	3.36	4.76	8.52
Sediment fraction (%)	6.55	10.56	25.36	15.06	20.11	13.02	4.88	4.46

Table 2 Characteristics of the employed hydrographs in the simulations

Run #	Peak flow discharge ($\mathrm{m^3\ s^{-1}}$)	Peak flow depth (cm)	Duration (min)
1	0.0750	12.9	180
3	0.0613	11.3	240

an implicit scheme for the temporal discretization. Concerning the RANS equations, the Reynolds stress term is computed by the standard k–ε turbulence model. The semi-implicit method for pressure-linked equations (SIMPLE) is used to evaluate the pressure term (Patankar 1980), and the convective term is modeled by using the second-order upwind (SOU) scheme (Olsen 2014). A Dirichlet boundary condition is defined for the inflow, whereas regarding the outflow, a zero gradient boundary condition is applied. Wall laws are used for the sides and the bed (Schlichting 1979). The free water surface is calculated according to the computed pressure field using the Bernoulli equation (Olsen and Haun 2010). The grid size is adapted from a similar simulation by Fischer-Antze et al. (2009), having 254 streamwise cells, 20 cells in the lateral direction and 5 vertical cells. However, the number of cells can change during the simulation as a result of the adaptive grid and the implemented wetting and drying algorithm in SSIIM.

The computation of the sediment transport is carried out by four different formulae; Einstein (1950), Engelund-Hansen (1967), van Rijn (1984) and Wu et al. (2000b). The effect of hiding and exposure behavior of non-uniform sediment is taken into account by Wu's correction factor for the critical bed shear stresses, based on the stochastic relevance of hiding-exposure probability to the size and gradation of bed materials (Wu et al. 2000b). This method is applied for van Rijn and Engelund-Hansen formulae, which were not developed for the fractional transport rate of non-uniform sediments.

2.3 Calibration Procedure

The model calibration presented in this study is performed using the model-independent Parameter ESTimator (PEST), which can be used for calibration as well as for sensitivity and uncertainty analysis of numerical models in an automatic manner (Doherty 2016). PEST applies the gradient-based Gauss-Marquardt-Levenberg (GML) algorithm for iterative progress refinement by minimizing the sum of the squared errors between the model-generated values and the corresponding measured data. This algorithm combines the gradient descent and Gauss-Newton methods, resulting in a faster convergence to the minimum of the objective function. PEST communicates with the model by means of three user-defined files containing the instructions for the calibration process; a duplicate copy of the model input file in which the adjustable parameters can be specified; and an instruction file to interpret the model output file. In the first iteration, PEST invokes the model by the initial values specified by the user. Running the model several times according to the number of adjustable parameters, PEST fills the Jacobian matrix containing the partial derivatives of the model results with respect to the adjustable parameters. Subsequently, the upgrade vector is calculated, and parameters are altered within a user-defined upper- and lower bound. This procedure is repeated during each iteration step using the updated values from the previous iteration until finding the minimum of the objective function. For a detailed description of the theory behind PEST see Doherty (2016).

The investigated parameters for calibration in this study are as follows:

(a) Roughness height at the bed (k_s); which is usually considered to be proportional to the characteristic grain size d_n (the diameter where n percent of the bed material has a finer size) for fixed beds (i.e., Nikoradse's equivalent grain roughness). Regarding movable beds, the roughness due to the bed forms should also be added to the grain roughness. A collection of related equations is given in Garcia (2008). In this study, although the roughness coefficient had to be assumed as a dynamic parameter due to the presence of dunes, it is treated as a fixed value all over the bed with an initial value of d_{90} and a range between d_{50} and $10d_{90}$.
(b) Active layer thickness (*ALT*); which is also a dynamic value that depends on the flow conditions and the bed sediment properties. It can be defined as a function of the representative grain diameter and the bed form height. Here, the maximum grain size (d_{max}) is assumed to be the initial value for the *ALT* with a range between d_{50} and $5d_{max}$.
(c) The volume fraction of compacted sediments in the bed deposit in comparison to the water content (*VFS*); considered to be 50% with a ±10% deviation.

3 Results and Discussion

Within this study, two different inflow discharge rates (Run #1 and Run #3) were used together with four available transport formulae in SSIIM (i.e., eight different scenarios). PEST calibrated the numerical model concerning the three adjustable parameters: k_s, *ALT*, and *VFS*. The process was carried out on a desktop PC (Intel® Core™ i7, 3.60 GHz) using 8 cores with 25–35 model calls and 12–18 h of calibration time for each of the eight scenarios. Table 3 shows the calibration results. It can be seen that the thickness of the active layer and the roughness height rise by the increase of the flow discharge. This effect is mainly related to the formation and development of the dunes along the bed.

Table 3 Results of the automatic calibration for eight scenarios and three adjustable parameters

Calibrated parameter	Sediment transport formula							
	van Rijn		Engelund-Hansen		Einstein		Wu	
	Run #1	Run #3	Run #1	Run #3	Run #1	Run #3	Run #1	Run #3
k_s (cm)	0.63	0.61	0.48	0.31	2.09	1.71	1.52	1.34
ALT (cm)	2.20	1.94	1.31	1.12	3.43	2.42	2.04	1.85
VFS (%)	60	60	52	51	51	53	49	51

The sediment packing volume in comparison to the water content remains almost constant around 50%, independent of the discharge alteration for most of the bed-load transport formulae. However, this is not the case for the simulations using van Rijn's formula in which the calibrated value is found at the upper-defined boundary for both discharge rates (i.e., 60%). It means the minimum of the objective function may be outside of the physically plausible boundaries, which shows once more the importance of an appropriate selection of the upper and lower threshold values. Figure 1 presents the final bed levels according to the calibrated models of both Runs (with different discharges) along three longitudinal-sections (i.e., 10 cm from the inner wall, the central radius, and 10 cm from the outer wall).

Regarding Run #1, the maximum deposition height in the experiment occurred near the inner bank at cross-section 75° (Fig. 1a). This location and also the magnitude of the maximum bed level is predicted well by the numerical model except by using van Rijn's formula, where the maximum level of the point bar is at the middle of the bend (90°). Along the central longitudinal-section, a similar trend of the simulated pattern compared to the experiment can be seen (Fig. 1b). However, in the experiment, there are larger fluctuations of the bed level, whereas the results of the model show a smoother pattern. The maximum erosion depth in the experiment was found near the outer bank at 165°, whereas it is at 180° for all the numerical models (Fig. 1c). Here, again, a smoother pattern can be seen in the model results compared to the experimental data.

For Run #3, the location of the maximum deposition in the experiment was at approximately 75°; however, in the model, its position shifted to the upstream direction (Fig. 1d). The predicted location of the extreme value for the scour depth is at 180°, which corresponds to the experimental data (Fig. 1f). Along the central longitudinal-section, the bed-load transport formula by Wu provides a better result, although it slightly underestimates the erosion at the inflow part of the bend (Fig. 1e).

The overall performance of the numerical models is evaluated using the coefficient of determination (R^2), and the root mean squared error (*RMSE*) and is presented in Table 4. Comparing the obtained *RMSE* values, the calibrated models using Wu's formula have the best agreement with the experimental data for both Runs (lowest *RMSE* values). It is also found that the simulations conducted by van Rijn's formula provide almost the same degree of accuracy as Wu's formula.

Figure 2 shows the normalized bed deformations (by means of the bed level changes and the initial water level $h_0 = 5.44$ cm, i.e., $\Delta Z/h_0$) for Run #1 regarding the best numerical model (i.e., Wu's formula) and the measured bed level changes.

4 Conclusion

The application of a model-independent Parameter ESTimator (PEST) for automatic calibration of a 3D morphodynamic numerical model is investigated in this study. The numerical model is calibrated against laboratory data, where bed changes in a 180° channel bend were investigated. Two different experiments with different discharge

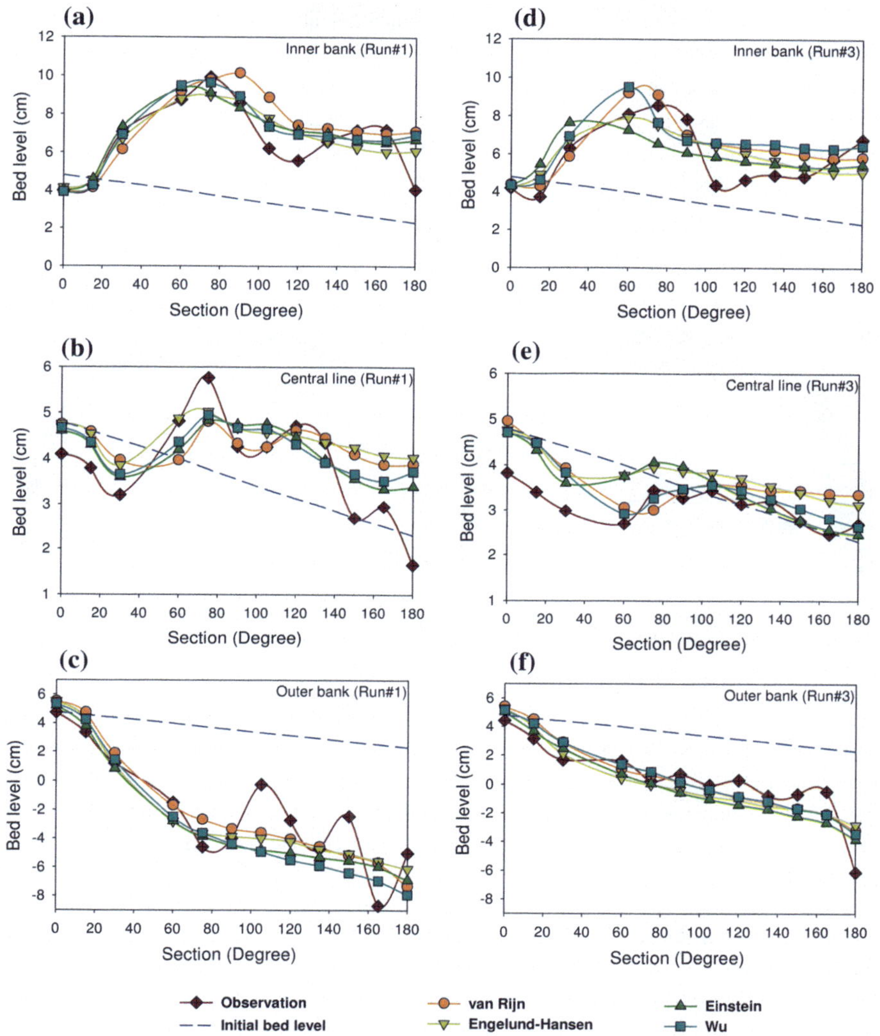

Fig. 1 Initial and final bed levels (measured and simulated), along three longitudinal sections (10 cm from the inner wall, the central radius, and 10 cm from the outer wall) for Run #1 (**a**–**c**) and Run #3 (**d**–**f**)

rates are simulated, using four various sediment transport formulae in the model. The most affecting input parameters are selected with an initial value and a reasonable range based on the literature. The results of this study reveal that the use of PEST can considerably expedite and facilitate the model calibration procedure by reducing the user-intervention. However, it should be noted that discontinuous parameters such as the selection of the sediment transport formula or some algorithms which are

Table 4 Goodness of fit for the calibrated models using different sediment transport formulae for two different discharges

Goodness of fit	Sediment transport formula							
	van Rijn		Engelund-Hansen		Einstein		Wu	
	Run #1	Run #3	Run #1	Run #3	Run #1	Run #3	Run #1	Run #3
R^2 (–)	0.90	0.89	0.88	0.83	0.89	0.84	0.90	0.89
RMSE (cm)	1.57	1.08	1.63	1.33	1.61	1.25	1.49	1.04

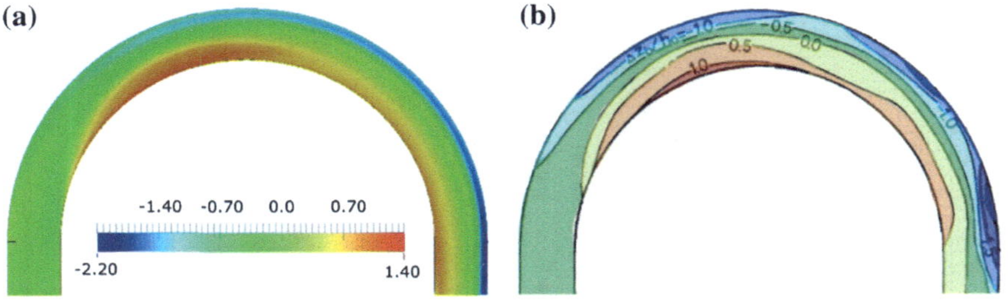

Fig. 2 Plan view of the normalized bed deformation for Run #1; **a** simulation results by using Wu's formula, **b** experimental data from the physical model test (Yen and Lee 1995)

represented as an integer value (not being a real number) within the numerical model cannot be handled by the calibration software. Another important thing which should be mentioned is that PEST is a gradient-based method; consequently, its prediction credibility is dependent on the parameter starting value. Therefore, the calibration routine can be reassessed using different parameter criteria (i.e., changing the initial value and the possible parameter range).

The performance of the calibrated model is assessed in this study by using R^2 and *RMSE*. From the results, it can be concluded that the calibrated model outputs obtained by Wu's formula have the best agreement with the experimental data. The general characteristics of the bed deformations as well as the magnitude of the deposition heights and erosion depths are well reproduced by the numerical model in most parts of the investigated domain. The formula by van Rijn using the hiding-exposure approach of Wu also provides reasonably acceptable results.

For further studies, it is suggested to compare the performance of PEST with a global based optimization method (e.g., Shuffled Complex Evolution algorithm), although the use of global methods is highly time-consuming.

Acknowledgements The first author was funded by the German Federal Ministry of Education and Research (BMBF) through the DAAD scholarship programme NaWaM.

References

Abbaspour KC, Schulin R, van Genuchten MT (2001) Estimating unsaturated soil hydraulic parameters using ant colony optimization. Adv Water Resour 24:827–841

Boyle DP, Gupta HV, Sorooshian S (2000) Toward improved calibration of hydrologic models: combining the strengths of manual and automatic methods. Water Resour Res 36:3663–3674

Cunge JA (2003) Of data and models. J Hydroinform 5:75–98

Deslauriers S, Mahdi T-F (2018) Flood modelling improvement using automatic calibration of two dimensional river software SRH-2D. Nat Hazards 91:697–715

Doherty J (2016) PEST model-independent parameter estimation user manual. Part I, 6th edn. Watermark Numerical Computing, Brisbane

Duan QY, Gupta VK, Sorooshian S (1993) Shuffled complex evolution approach for effective and efficient global minimization. J Optim Theory Appl 76:501–521

Duc BM, Wenka T, Rodi W (2004) Numerical modeling of bed deformation in laboratory channels. J Hydraul Eng 130:894–904

Einstein HA (1950) The bed-load function for sediment transportation in open channel flows. Technical Bulletin No. 1026, U.S. Department of Agriculture, Soil Conservation Service, Washington, DC, USA

Engelund F, Hansen E (1967) A monograph on sediment transport in alluvial streams. Teknisk Forlag, Copenhagen, Denmark

Fabio P, Aronica GT, Apel H (2010) Towards automatic calibration of 2-D flood propagation models. Hydrol Earth Syst Sci 14:911–924

Fischer-Antze T, Rüther N, Olsen NRB, Gutknecht D (2009) Three-dimensional (3D) modeling of non-uniform sediment transport in a channel bend with unsteady flow. J Hydraul Res 47:670–675

García MH (2008) Sedimentation engineering: processes, measurements, modeling, and practice. American Society of Civil Engineers, Reston, VA

Khosronejad A, Rennie CD, Salehi Neyshabouri SAA, Townsend RD (2007) 3D numerical modeling of flow and sediment transport in laboratory channel bends. J Hydraul Eng 133:1123–1134

Madsen H (2000) Automatic calibration of a conceptual rainfall-runoff model using multiple objectives. J Hydrol 235:276–288

Madsen H, Wilson G, Ammentorp HC (2002) Comparison of different automated strategies for calibration of rainfall-runoff models. J Hydrol 261:48–59

Maneta MP, Pasternack GB, Wallender WW, Jetten V, Schnabel S (2007) Temporal instability of parameters in an event-based distributed hydrologic model applied to a small semiarid catchment. J Hydrol 341:207–221

Muleta MK, Nicklow JW (2005) Sensitivity and uncertainty analysis coupled with automatic calibration for a distributed watershed model. J Hydrol 306:127–145

Olsen NRB (2014) A three dimensional numerical model for simulation of sediment movement in water intakes with multiblock option. Department of Hydraulic and Environmental Engineering, The Norwegian University of Science and Technology, Trondheim, Norway

Olsen NRB, Haun S (2010) Free surface algorithms for 3D numerical modelling of reservoir flushing. River Flow 2010:1105–1110

Patankar SV (1980) Numerical heat transfer and fluid flow. McGraw-Hill, New York, NY, USA

Rigosi A, Marcé R, Escot C, Rueda FJ (2011) A calibration strategy for dynamic succession models including several phytoplankton groups. Environ Model Softw 26:697–710

Rode M, Suhr U, Wriedt G (2007) Multi-objective calibration of a river water quality model—information content of calibration data. Ecol Model 204:129–142

Rüther N, Olsen NR (2005) Three-dimensional modeling of sediment transport in a narrow 90° channel bend. J Hydraul Eng 131:917–920

Schlichting H (1979) Boundary layer theory. McGraw-Hill, New York, NY, USA

Shams M, Ahmadi G, Smith DH (2002) Computational modeling of flow and sediment transport and deposition in meandering rivers. Adv Water Resour 25:689–699

Usman M, Reimann T, Liedl R, Abbas A, Conrad C, Saleem S (2018) Inverse parametrization of a regional groundwater flow model with the aid of modelling and GIS: test and application of different approaches. ISPRS Int J Geo-Inf 7:22

van Rijn LC (1984) Sediment transport, part I: bed load transport. J Hydraul Eng 110:1431–1456

Vidal J-P, Moisan S, Faure J-B, Dartus D (2007) River model calibration, from guidelines to operational support tools. Environ Model Softw 22:1628–1640

Wu W, Rodi W, Wenka T (2000a) 3D numerical modeling of flow and sediment transport in open channels. J Hydraul Eng 126:4–15

Wu W, Wang SSY, Jia Y (2000b) Nonuniform sediment transport in alluvial rivers. J Hydraul Res 38:427–434

Ye J, McCorquodale JA (1998) Simulation of curved open channel flows by 3D hydrodynamic model. J Hydraul Eng 124:687–698

Yen C, Lee KT (1995) Bed topography and sediment sorting in channel bend with unsteady flow. J Hydraul Eng 121:591–599

Sand Island Reshaping in Response to Selected Discharges: The Vistula River Returning to Its Natural State

Andrzej Strużyński, Maciej Wyrębek, Adam Nowak, Agnieszka Woś, Jacek Florek and Leszek Książek

Abstract Scientists have been researching processes leading to the formation of sand islands, but the process of their shaping when spate passes has not been accurately identified. Field measurements and numerical modelling of the Vistula river have been performed along a 1900 m section in which a sandy island with a nesting area for two species of tern exists. Numerical modelling covered a section with an island connected to the right bank of the river in 2017. As a result of observing changes in water levels and flows, two waves were selected and simplified to a uni-modal shape and run in the CCHE2D model. The following data were analysed: (i) survey measurements, discharge, and the flow velocity using an Acoustic Doppler Current Profiler; (ii) bars, islands, and water surface; (iii) the granulometric composition of bed material; and (iv) resources of the Head Office of Geodesy and Cartography in Poland (GUGiK). The purpose of the project is to determine limit discharges and related hydraulic parameters responsible for the growth or disappearance of a sand island. The limit discharge of the nesting is the moment when the whole island is submerged. The 2D modelling carried out provided data supporting the thesis that exceeding the conditions of bed material movement does not always lead to activating erosion processes. With the right load of material transported in the river bed, the transport capacity of the stream may be exhausted and, as a result, the bottom will not be eroded.

Keywords Sand river · Island formation · Fluvial processes · 2D model

1 Introduction

Natural or almost natural rivers are a dynamic environment which ensures varied sets of habitats. River bed morphology depends on many factors: geology, hydrology, land use, bed material supply, and its size. The river training works carried out in the 19th and 20th centuries have changed river bottom morphology. A regulated river in which

---

A. Strużyński (✉) · M. Wyrębek · A. Nowak · A. Woś · J. Florek · L. Książek
University of Agriculture in Kraków, Mickiewicza Ave. 24/28, 30-059 Kraków, Poland
e-mail: rmstruzy@cyf-kr.edu.pl

M. B. Kalinowska et al. (eds.), *Recent Trends in Environmental Hydraulics*, GeoPlanet: Earth and Planetary Sciences,
https://doi.org/10.1007/978-3-030-37105-0_23

intensive regulation measures are not continued will begin to revert to its natural state. Autonomous transformations of a river bed depend both on the magnitude and the duration of high water stages (Buffington 2012; Tekielak et al. 2007). Along many of the straightened and stabilized sections of the Vistula River channel, natural morphological forms, namely shallows, overdeepenings, calm ponds, sandbars, and islands, either stabilized by vegetation or uncovered, have appeared. These constitute aquatic terrestrial transition zones (Junk et al. 1989). Sand bars and islands along with their neighbouring areas provide habitat conditions for many species of water and amphibious organisms, both animals and plants stabilizing these forms (van der Nat et al. 2002; Thorp 1992; Tockner et al. 2003). Sandy islands are particularly important for conserving many bird species as their resting and nesting places (Smith and Renken 1991; Helmiö 2004; Kinzel et al. 2010; Takewka 2016). In recent years, the availability of habitats for avifauna in the form of sandy islands on the Vistula and the degree to which water flow species use these habitats has been frequently researched (Habel 2018; Pieron et al. 2013; Bukaciński et al. 2013). The occurrence, shape, and dynamics of changes to sandbars result from erosion and accumulation processes in the river channel (Buffington 2012; Michalik and Książek 2009; Strużyński et al. 2013). The reason why the authors undertook this subject are the morphodynamic processes, unfavourable from the point of view of bird fauna, which resulted in the island connecting to the right bank of the Vistula in 2017.

The purpose of this paper is to identify the conditions for the formation and disappearance of sand islands within the limited discharge (well below bankfull). There are studies which suggest that rivers with these discharge magnitudes can self-naturalize (Helmiö 2004; Kinzel et al. 2010; Takewaka 2016). The changes to the shape of the sandy island caused by the passage of two waves of various characteristics of the discharge intensity and period during which processes of bottom material erosion or accumulation start. The analyses were based on the parameters of water flow velocity, shear stresses, and changes to the island's shape.

2 Materials and Method

2.1 *Study Area*

The studied sector including the island is located in the Upper Vistula, km 281 + 500–282 + 500, approx. 2 km downstream from the mouth of the San River (Fig. 1). The catchment area at the Zawichost water gauging station is 50,655 km^2. The width of the channel is approximately 400 m, and the distance between embankments is 1 km. The measured slope of the water surface ranges from 0.00025 to 0.00329. In some cases, the San causes the discharge to increase by over 40%. Below the mouth of the San, the Vistula channel slope decreases, which coincides with the bed material accumulating in the channel (Maruszczak 1982). After the completion of river training works, the regulated Vistula channel is reverting to its natural state

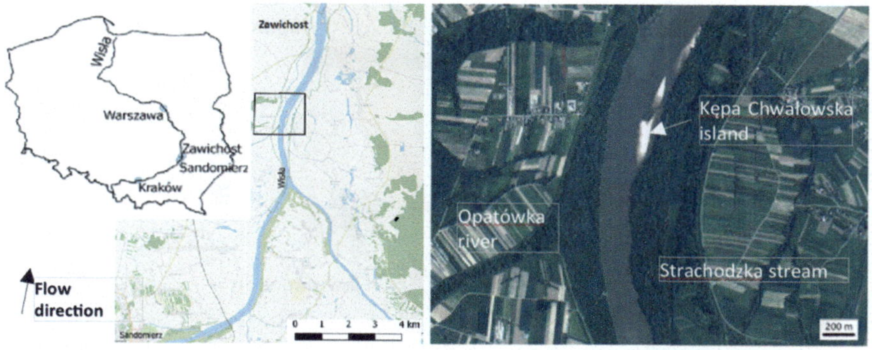

Fig. 1 Location of the island Kępa Chwałowska, km 281 + 500–282 + 500, Vistula river, 2017

and recovering habitats of valuable water and land species (Łajczak et al. 2006). A protected species is the little tern (*Sternula albifrons*), included in Annex 1 of the Bird Directive. For this species, sandbanks not connected to the bank (mid-channel bars) are a key habitat element in the nesting season. Flow regime at the Zawichost gauging station in 1981–2010 was Mean Annual Flow of 416.47 $m^3 s^{-1}$, Mean Low Flow of 139.7 $m^3 s^{-1}$, Low Low Flow of 104 $m^3 s^{-1}$, and guaranteed flow at 90% of 158 $m^3 s^{-1}$.

The island under research has formed on the convex bank. Its length is approx. 700 m, and its maximum width is 136 m for the discharge of 300 $m^3 s^{-1}$. In 2016, it was only locally overgrown with grass, while in 2018 it was seen to connect to the bank and the first shrubs appeared. This island is located on the convex bank, downstream from the mouths of small rivers, Strachodzka on the right bank and Opatówka on the left. The current creates good conditions here for the growth of the island.

2.2 *River Channel Bed Configuration Measurements*

In the material presented, fluvial processes have been researched since 2014. In 2017, the island connected to the bank. To check if the island shape modeling would lead to similar results, measurement data collected on 16 June 2016 were used (12 cross-sections, 2347 hydrometric verticals). Basic measurements of the river bed were taken from a motorboat towing a Sontek ADCP unit. Areas inaccessible by boat (of the island and sandy shallows) were measured manually using a GPS-RTK (20 cross-sections, 370 points). This measurement methodology was also used to collect data for creating the numerical water surface model (16 points of the water surface level). Measurements taken using both devices were combined, with GPS-RTK data source used as a reference. The measurements obtained using ADCP were supported by the GPS-RTK elevations (coordinate z) with spatial relation between x and y

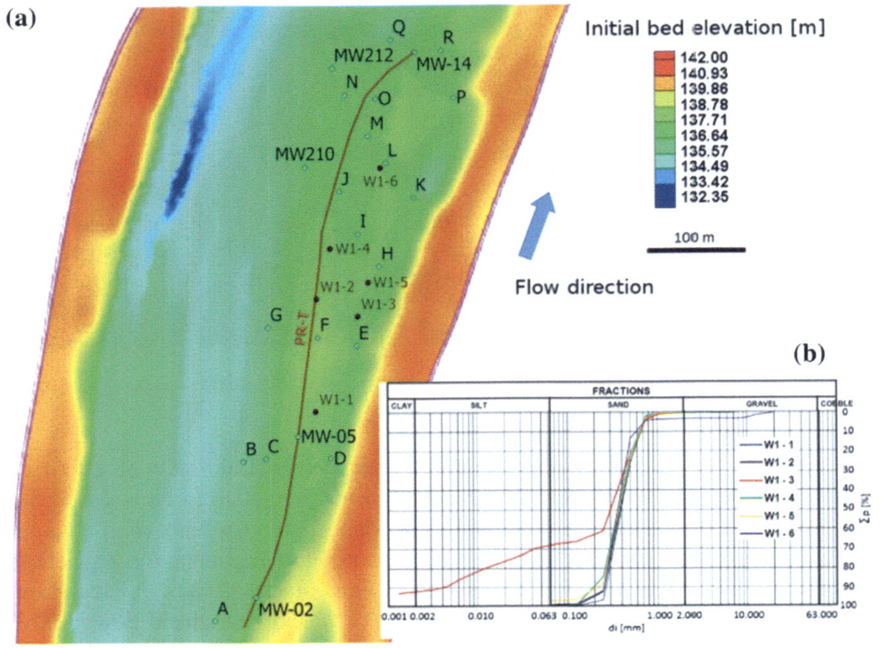

Fig. 2 Island Kępa Chwałowska, Vistula, **a** location of monitoring (blue points), sediment probes (red) and longitudinal profile PR-T, **b** grain size distribution, 2017

coordinates. As an effect, all ADCP data possess elevation accuracy of GPS-RTK (better than 0.05 m) and horizontal accuracy generally between 0.4 and 1.0 m, which is much better than the mesh spacing of the 2D model. The shape of the river valley was identified using a Digital Terrain Model with a mesh size of 0.5 m (GUGiK).

Material forming the bottom of the channel was collected in 2017: 6 samples were collected on the island from W1-1 to W1-6 (Fig. 2). Their analysis established the average conditions for the movement of bed material and the conditions in which bottom forms are created. For this purpose, the Southard and Boguchwal diagram was used (Baas et al. 2015; Strużyński and Wyrębek 2008).

2.3 Numerical Modelling of Water Flow Conditions

To model water flow conditions, a CCHE2D model was used. The model covers a section of the Vistula, 1900 m long and 540 m wide, in a rectangular 5 × 5 m mesh (Zhang 2006). Bedload transport capacity and bottom changes simulations were conducted using the Wu et al. (2000) equation (Wu 2001), which is dedicated to sand transport. The computation step was set to 5 s. The model was calibrated for steady flow conditions by correlation of the measured and calculated water surface level for discharge Q = 192 m^3 s^{-1}. At first a constant roughness of the bottom calculated using the Strickler equation was assumed. The differences between the measured and modelled levels of the water surface were reduced by correction with use of the Manning's roughness coefficient. A variance below 0.05 represents excellent accuracy, and from 0.05 to 0.10 m very good accuracy (Książek et al. 2010). The accuracy of the water surface level during the wave's culmination is below 0.05 m, which represents excellent compatibility of the numerical model. Finally, the model was classified as excellent regarding the part along the left bank and in the upper section of the island, and as very good to the downstream of the island. Initial boundary conditions for bedload transport were performed in such a way that the balance of erosion and sedimentation at the inlet was reached. The value of bedload transport rate intensity (kg m^{-1} s^{-1}) in the calibration process at the first turn was assumed based on outer software (Bartnik et al. 2005) which utilizes MPM equation (time effective when compared to the modelling process).

The main simulations were performed for unsteady flow conditions. The waves were selected after analysing the changes in the discharge at the Zawichost gauging station in 2015–2017. This data allowed us to select two waves differing in duration and culmination. To simplify the fluvial processes interpretation, the unimodal course was used. The discharge of 498 m^3 s^{-1} is slightly above the annual average (2015; Wave_1, total time 10.75 days), while 1106 m^3 s^{-1} is well beneath the bankfull discharge (2016; Wave_2, total 58.17 days). Wave_1 covered a part of the island; Wave_2 covered it completely for 17 days. The rising limb is marked as TI.0-1 (duration of 4.04 days for Wave_1 and 16.33 for Wave_2), while TI.1-2 (6.71 days for Wave_1 and 41.84 for Wave_2) indicate the recession limb.

3 Analysis of Results

Fluvial processes in distinguished wave stages were analysed along a horizontal plane (map of changes of hydraulic and morphological parameter), a vertical plane (longitudinal profile), and at selected characteristic points lying on the profiles or in their vicinity. Island bathymetry changes were analysed along 798 m of a longitudinal profile drawn across the island's top (PR-T, Fig. 2).

Modelling results as a time series were analysed at six points: MW-02, MW-05, F, J, O, and MW-14. The analysed island is not uniform in shape. On its ridge there are shallow or cross-wise rapids which become active at various water levels. For this reason, the course of waves was split into six phases shown in Table 1. Based on the measurements of the bedload, the average representative sand particle diameter in the region of the island for which the simulation of bedload transport was made was estimated at 0.00042 m. For the present sand fractions, critical stresses of the movement are 0.27 Nm^{-2}. According to the bedform phase diagram of Southard and Boguchwal (Baas et al. 2015; Strużyński and Wyrębek 2008), the medium sand which fills the Vistula channel along the analysed section is constantly moving. However, this material can be transported only in balanced conditions for the certain

Table 1 Phases of rising TI.0-1 and recession TI.1-2 limbs during wave passing

TI		Description	Time [days] Discharge [$m^3 s^{-1}$]	
			Wave_1	Wave_2
0–1	–	Initial parameters	0/252	0/278
	I	Gradual flooding of the island, causing activation of 3 streams acting on it	3.5 485	8 444
	II	Increasing flow in 3 streams (arms) leading to the total submerging of the island (nesting unsuccessful)	– [a]	5 900
	III	Flow from the total covering of the island to reaching the maximum discharge Q_{max}	1 498	3.3 1106
1–2	IV	Flow from the maximum discharge Q_{max} to the resurfacing of the island	– [a]	12.5 790
	V	Flow from the resurfacing of the island's top to the re-emergence of 3 streams (phase I)	1.75 397	14 439
	VI	The end of the simulation	3.75/259	14/307

[a]Does not occur for Wave_1

range of the acting forces (from 0.17 to 0.48 m s^{-1}—ripples forming, from 0.48 to 1.1—dunes forming) and then, after the velocity of 1.1 m s^{-1} is exceeded, increased bottom erosion and hence rapid changes in the elevation of the island will appear.

Figure 3 shows a view of changes in the shape of the island and the occurring cross-wise rapids caused by Wave_2. After the pass of Wave_2 the height of the area located downstream of point F significantly increased around points H, I and J. The PR-T profile (Fig. 4) shows changes to the shape of the middle part of the island before the inundation (TI-0—black), after maximum discharge was reached (TI-1—blue), and to the end of the modelling (TI-2—red). This profile also shows high dynamics of island bottom changes around point J. The part of the profile located at the beginning and at the top of the island subsides after the wave has passed, so this wave can be the cause for the shift of island's top towards downstream. Examples of the changes of hydraulic parameters and the location of point F (island top, Wave_2) and point MW-02 are shown in a graph (Fig. 5). At point F, the raising of the island's elevation started at the shear stress of 0.32 N m^{-2}, which is close to the calculated (mentioned at the beginning) value of critical shear stresses of 0.27 Nm^{-2}. Lowering of the island started once 0.85 N m^{-2} was exceeded and ended below the value of 2.22 N m^{-2} (see also Table 3). Hence, the island's top's altitude rose at low-water stages, which occurred at the turn of flow phases II and III (island submerged, nesting unsuccessful). Later, the island's top was eroded in phase III and in a part of phase IV, when the simulation made in the vicinity of point F once again showed that sand accumulation can occur. Finally, after the entire wave had passed, the island's top had been lowered by 0.1 m, although during the passage of the wave's peak, point F's altitude reduction reached 0.13 m. At phase TI.0-1, the altitudes of the island's top's area slightly upstream of point J were raised, forming an accumulation area

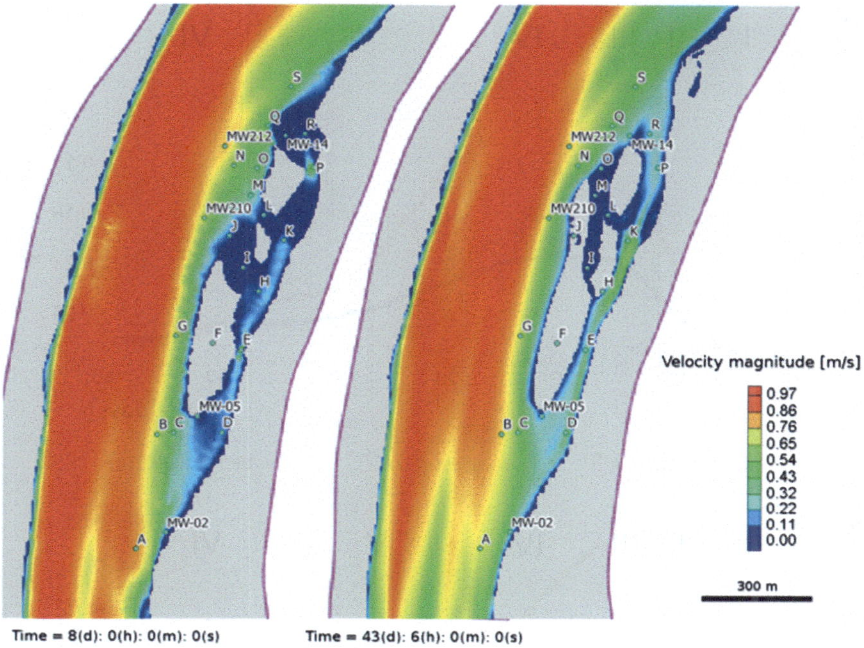

Fig. 3 Spatial velocity distribution for Wave_2 around Kępa Chwałowska Island, Vistula River, **a** rising limb TI.0-1, phase II, Q = 900 $m^3 s^{-1}$, **b** recession limb TI.1-2, phase V-VI Q = 790 $m^3 s^{-1}$

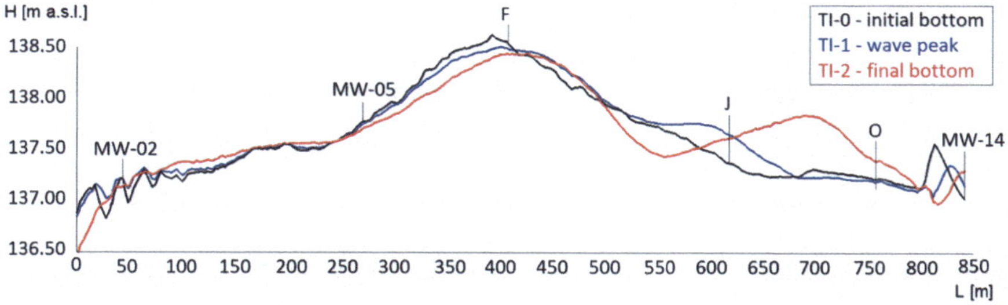

Fig. 4 Longitudinal profiles PR-T along the Kępa Chwałowska Island, Vistula; initial bottom—measured (black), blue and red—calculated bed elevation during and after passage of Wave_2, respectively

downstream of the top. The falling wave caused the local top to shift to the area between points J and O. After the wave passed (phase TI.1-2), the newly formed local top was located up to about 0.5 m higher than before the inundation.

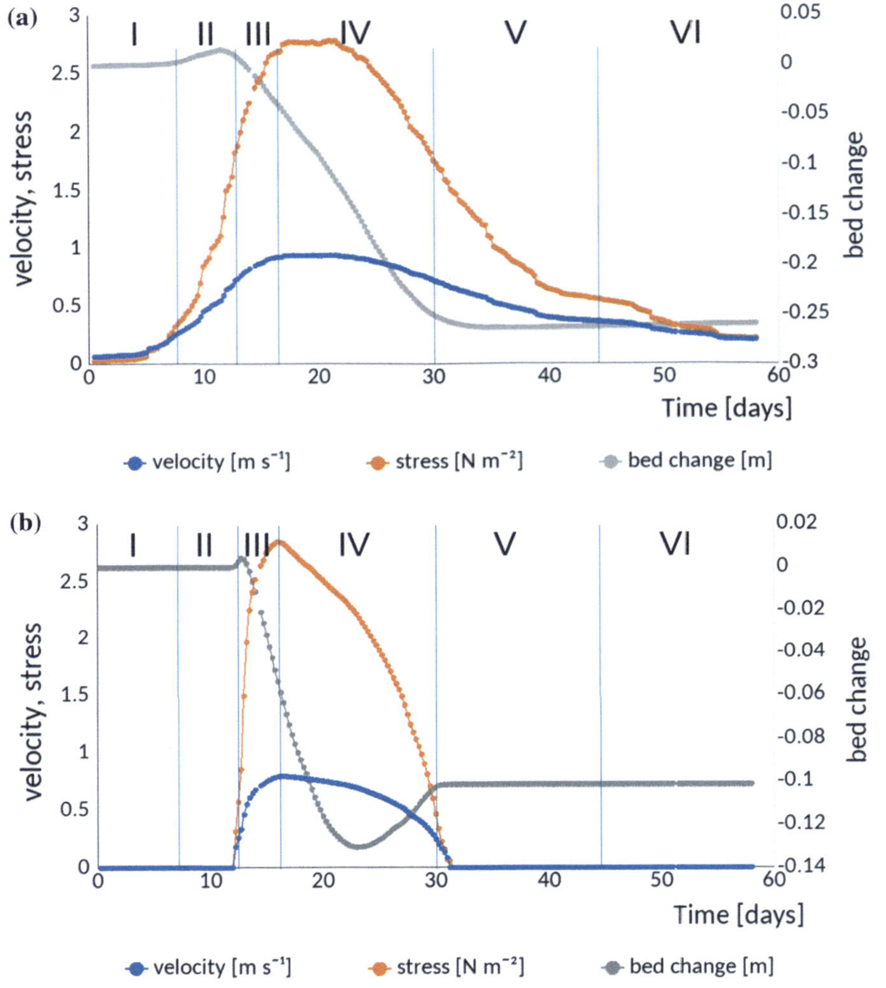

Fig. 5 Time variation of parameters for Wave_2 at monitor points: **a** MW-02—upstream part of the island, **b** F—the top of the island

The hydraulic and morphological parameters recorded in the modelled PR-T profile at the end of individual discharge phases for both Wave_1 and Wave_2 are presented in Tables 2 and 3.

Wave_1 did not cause erosion of the island. As its top was not submerged, the water stream was divided and accumulated at the upstream part of the island at point MW-05 reaching 0.03 m. The flow of water downstream from the top (point F) was slow enough to be unable to change the island's shape in this region (accumulation below 0.01 m). Already during the passage of Wave_1, a process of erosion of the

Table 2 Simulated parameter changes along profile PR-T—Wave_1

Flow phase	Point					
	MW-02	MW-05	F	J	O	MW-14
I	acc. + 0.003[a] 0.07–**0.35**[b] nm/rp[c] 0.12/0.27[d]	acc. +0.021 **0.33–0.96** rp/dn 0.32/0.51	–	no. +0.000 0.00–0.10 nm 0.03/0.13	acc. +0.000 0.00–**0.86** nm/rp 0.11/0.42	no. +0.000 0.00–0.00 nm 0.01/0.04
III	acc. +0.007 **0.40–0.50** rp 0.32/0.34	acc/er + 0.022 **1.12–1.42** dn 0.60/0.63	–	acc +0.001 0.12–0.17 nm/rp 0.17/0.18	acc. +0.002 **0.94–1.11** rp/dn 0.48/0.50	no +0.000 0.00–0.00 nm 0.02/0.03
V	acc. +0.012 **0.51–0.43** rp 0.33/0.34	er/acc + 0.023 **1.41–0.95** dn 0.57/0.63	–	acc. +0.001 0.17–0.07 rp/nm 0.15–0.18	acc. +0.003 **1.10–0.83** dn/rp 0.46/0.50	no +0.000 0.00–0.00 nm 0.02/0.03
VI	acc. +0.014 **0.42**–0.08 rp/nm 0.17/0.29	acc. +0.032 **0.92–0.33** rp 0.36/0.49	–	acc. +0.001 0.08–0.00 rp/nm 0.02/0.11	acc. +0.005 **0.75**–0.00 rp/nm 0.14/0.38	no +0.000 0.00–0.00 nm 0.01/0.03

[a]Bed process: no—no change, acc—accumulation, er—erosion; altitude change [m]
[b]Shear stresses [Nm^{-2}], bold means incipient motion exceedance
[c]Bed form description: nm—no movement, rp—ripples, dn—dunes
[d]Velocity average/maximal [ms^{-1}]

island's forehead temporarily occurred. It occurred for the discharge higher than 487 $m^3 s^{-1}$ (water surface elevation of 313 m a.s.l.). In 2015–2017, 11 inundations smaller than Wave_1—one similar, and 10 larger waves (including Wave_2)—were counted. During Wave_2, bottom material did not move, or movement conditions were restricted to ripple formation. Velocities and water depth connected to dune formation can occur only in the vicinity of points MW-05 and O during phases III and V. Wave_2 flooded the entire island, so phases II and IV were distinguished.

When water flowed above the entire island along the profile PR-T, erosion processes started at points MW-02, MW-05, F and O. In the vicinity of the island's head erosion dominated, lowering this area. However, this material could have been deposited downstream of the island's top, which happened before inundation, at point J, and located close to the island's branch. After the wave's peak had passed (Phase III or IV), this point constituted a local top of the island.

From mid-phase II to mid-phase IV at point O, intensive erosion processes can occur. Simulations show that the decreasing height of the area located around point O at phase IV may exceed 0.6 m. Bed material also accumulated in the vicinity of point MW-02. After the flood wave passed, the difference between the height of the island's top (point F) and point MW-02 decreased from approx. 1.5 m to approx. 1 m. Within almost the entire profile (except at the island's top), water velocities connected to

Table 3 Simulated parameter changes along profile PR-T—Wave_2

FP	Point					
	MW-02	MW-05	F	J	O	MW-14
I	acc + 0.004 [a] 0.03–**0.38** [b] nm/rp [c] 0.12/0.28 [d]	no 0.000 0.00–0.00 nm 0.00/0.00	no 0.000 0.00–0.00 nm 0.00/0.00	no 0.000 0.00–0.09 nm 0.01/0.11	acc +0.003 0.00–**0.55** nm/rp 0.09/0.32	no 0.000 0.00–0.01 nm 0.02/0.05
II	acc/er + 0.008 **0.40–1.89** rp/dn 0.50/0.74	acc/er -0.007 0.00–1.**45** rp/dn 0.36/0.60	acc/err + 0.004 0.00–1.**50** nm/rp 0.06/0.47	acc +0.066 0.09–**1.59** nm/rp/dn 0.35/0.65	acc/er – 0.011 **0.64–2.02** rp/dn 0.59/0.77	no/acc +0.031 0.01–**0.69** nm/rp 0.17/0.45
III	er −0.036 **2.01–2.68** dn 0.86/0.92	er −0.027 **1.52–2.07** dn 0.71/0.77	er -0.059 **1.97–2.85** dn 0.72/0.80	acc +0.226 **1.69–2.65** dn 0.79/0.87	er -0.034 **2.07–2.42** dn 0.84/0.88	acc +0.104 **0.74–1.49** rp/dn 0.59/0.69
IV	er −0.250 **2.69–1.83** dn 0.89/0.94	er −0.083 **2.09–1.65** dn 0.74/0.78	er/acc – 0.104 **2.84–0.61** dn/rp 0.64/0.80	acc/er + 0.336 **2.70–2.37** dn 0.86/0.91	er/acc -0.035 **2.42–1.21** dn 0.77/0.88	acc +0.313 **1.53–1.21** dn 0.68/0.72
V	er/acc – 0.263 **1.76–0.57** dn/rp 0.51/0.72	er/acc – 0.078 **1.62**–0.19 dn/rp 0.38/0.63	acc/no – 0.101 **0.47**–0.00 rp/nm 0.02/0.25	er +0.257 **2.33**–0.00 dn/rp/nm 0.36/0.73	acc +0.003 **1.17**–0.02 dn/rp/nm 0.24/0.57	acc/er +0.316 **1.19–0.38** dn/rp 0.35/0.56
VI	acc – 0.260 **0.57**–0.22 rp/nm 0.28/0.37	acc −0.077 0.19 – 0.00 nm 0.02/0.15	no −0.101 0.00–0.00 nm 0.00/0.00	no +0.257 0.00–0.00 nm 0.00/0.00	acc/no + 0.009 0.02–0.00 nm 0.01/0.06	er/no +0.315 **0.39**–0.00 rp/nm 0.19/0.28

[a] Bed process: no—no change, acc—accumulation, er—erosion; altitude change [m]
[b] Shear stresses [Nm^{-2}], bold means incipient motion exceedance
[c] Bed form description: nm—no movement, rp—ripples, dn—dunes
[d] Velocity average/maximal [ms^{-1}]

dune appearance prevailed (during phases II, III, IV, and V). Conditions restricted to ripple formation usually occurred along the PR-T profile during phases II and V.

The modelling results presented in this paper are consistent with field observations and measurements made along the studied section in 2014–2018. The next measurements in this reach of the Vistula were performed on 29 June 2017. These surveys in the top region of the island indicated changes similar to the simulated ones. The upstream part of the top had eroded (90 m upstream from point F, bottom change; −0.22 m) and the downstream section had accumulated (75 m downstream from point F, bottom change; +0.38 m). The observed top of the island had shifted in the direction towards the right bank. The maximal discharges of waves passing within the period from 16 June 2016 to 29 June 2017 were: 760, 630, 873, 1340 and 1060 $m^3\ s^{-1}$.

4 Conclusions

No general tendency of the island's erosion has been observed. What has been observed and modelled was only a process of island relocation and expansion in areas located within the hydraulic shadow, particularly during the receding arm of a wave with a culmination reaching the banks of the Vistula River.

The intensity of fluvial processes during the peak of the Wave_2 contributed to insignificant erosion of the island head, while the areas downstream of its top can experience both erosion and material accumulation, but accumulation processes become stronger. Even though they were very intensive during the passage of the wave, at the last phase, the moving water may, once again, have carried a part of the accumulated material from this area.

Wave_1 caused only the processes which smoothed the lower parts of the island and did not break the nesting. When flows similar to Wave_1 pass and do not cover the top of the island (point F), this may lead to gradual expansion on areas downstream of its top. Wave_2 produced a strong effect of depositing and shifting the island, although it seems that a wave that fits in the channel may cause its shape to change to more elongated and narrow (fusiform), and to relocate the island downstream. Additionally, this wave turned the breeding unsuccessful in the nesting colony.

Phases I and II, which cover a part of the island, seem not to be essential to the course of the island's formation process, although it is during phase II (i.e., from the moment water starts flowing over a significant area of the island to the moment when it is completely flooded) that fluvial processes gain in strength and become the reason for relatively large changes to the shape of the island. The passage of the wave culmination does not change the trends of processes occurrence. The clearest change in the trend of fluvial processes occurs in phases IV and V, when the local water layer covering the island disappears. Cross-flows having the nature of a sheet flow then occur in dips within the main body of the island, and these have a significant erosion capability.

An analysis of processes causing the formation or erosion of islands cannot be made locally. Neither does the analysis of isolated parameters lead to correct conclusions. Islands are irregular in shape, which reflects the local hydraulic conditions of the river section. The 2D modelling carried out provided data supporting the thesis that exceeding the conditions of bed material movement does not always lead to activation of the erosion processes. With the right load of material transported in the river bed, the transport capacity of the stream may be exhausted and, as a result, the bottom will not be eroded. For this reason, the next step in the analyses will be to establish the transport balance along different cross-sections of this island.

It should be noted that the time of the receding arm of a wave is particularly important for island formation. Waves that recede rapidly (e.g., ones caused by local rains) may contribute to island erosion; while those caused by continuous rains over a wider area are connected to the formation and growth of sandy islands.

Acknowledgements This research has been funded by the EOG PL02 2009-2012. Authors wish to acknowledge Prof. Wojciech Bartnik and Dr. Małgorzata Leja, and all who contributed during the development of the research.

References

Baas JH, Best JL, Peakall J (2015) Predicting bedforms and primary current stratification in cohesive mixtures of mud and sand. J Geol Soc 173. https://doi.org/10.1144/jgs2015-024, pp 12-45

Bartnik W, Książek L, Radecki-Pawlik A, Strużyński A (2005) On some mountain streams and rivers morphodynamical parameter characteristics using field and numerical modeling examples. In: Majewski W, Robakiewicz M (eds) Sediment transport in rivers and transitional waters. IBW, PAN, Gdańsk, 27–35

Buffington JM (2012) Changes in channel morphology over human time scales. In: Church M, Biron PM, Roy AG (eds) Gravel-bed rivers: processes, tools, environments. Wiley, Chichester, pp 433–463

Bukaciński D, Bukacińska M, Buczyński A (2013) The impact of hydrotechnical facilities on island avifauna: a case study of the middle Vistula River. Stud Ecol Bioeth UKSW 11(4):93–109

GUGiK, Head Office of Geodesy and Cartography in Poland, gugik.gov.pl/pzgik, geoportal.gov.pl

Habel M (2018) Effects of flow regulation and river channelization on sandbar bird nesting availability at the Lower Vistula River. Ecol Quest 29:43–53

Helmiö T (2004) Hydraulic geometry of cohesive lowland rivers. Boreal Environ Res 9:243–251

Junk W, Bayley PB, Sparks RE (1989) The flood pulse concept in river-floodplain systems. In: Dodge DP (ed) Proceedings of the international large river symposium (LARS). Canadian Special Publication of Fisheries and Aquatic Sciences vol 106, pp 110–127

Kinzel P, Nelso J, McDonald R, Logan B (2010) Topographic evolution of sandbars: Flume experiment and computational modeling. In: Proceedings of the 4th federal interagency hydrologic modeling conference and of the 9th federal interagency sedimentation conference, Las Vegas

Książek L, Wałęga A, Bartnik W, Krzanowski S (2010) Calibration and verification of computational model of the Wislok river by means of flood wave transformation. Infrastruct Ecol Rural Areas 8(1/2010):15–28

Łajczak A, Plit J, Soja R, Starkel L, Warowna J (2006) Changes of the Vistula River channel and floodplain in the last 200 years. Geogr Pol 79(2):65–87

Maruszczak H (1982) Wisła Lubelska. In: Piskozub A (ed) Wisła Monografia rzeki. WydawnictwoKomunikacjiiŁączności, pp 125–136

Michalik A, Książek L (2009) Dynamics of water flow on degraded sectors of polish mountain stream channels. Polish J Environ Stud 18:665–672

Pieron Ł, Hojan M, Habel M (2013) Characteristics of river sandbars on regulated reach of the Lower Vistula River. J Heal Sci 3:114–123

Smith JW, Renken RB (1991) Least tern nesting habitat in Mississipi River Valley adject to Missouri. J Field Ornithol 62(4):497–504

Strużyński A, Wyrębek M (2008) Evaluation of the Nida River main current below the perpendicular flood channel outlet. In: Hydraulic methods for catastrophes: floods, droughts, environmental disasters. Publications of The Institute of Geophysics, Monographic Volume, Polish Academy of Sciences, Ser. Hydrology E-10 (406). ISBN 978-83-88765-79-7. ISSN 0138-0133, 175-185

Strużyński A, Bartnik W, Krzysztof Kulesza K, Czoch K (2013) Hydrodynamical balance as important parameter influencing the ecological status of Carpathian rivers. Annu Set Environ Protect 15:2591–2610

Takewaka S (2016) Observation of whole flushing process of a river sand bar by a flood using x-band radar. J Mar SciEng 4:32

Tekielak T, Michalik A, Bąk Ł, Książek L (2007) Influence of floods on fluvial processes on the example of the stream Smolnik. Infrastruct Ecol Rural Areas 4:167–178
Thorp JH (1992) Linkage between islands and benthos in the Ohio River, with implications for riverine management. Can J Fish Aquat Sci 49:1873–1882
Tockner K, Ward JV, Arscott DB, Edwards PJ, Kollmann J, Gurnell AM, Petts GE, Maiolini B (2003) The Tagliamento River: a model ecosystem of European importance. Aquat Sci 65:239–253
van der Nat D, Schmidt A, Tockner K, Ward J (2002) Inundation dynamics in brainded floodplains: Tagliamento River, Northest Italy. Ecosystems 5:636–647
Wu W (2001) CCHE2D sediment transport model (version 2.1), technical report no. NCCHE-TR-2001-3. School of Engineering, The University of Mississippi, Oxford, pp 48
Wu W, Wang SSY, Jia Y (2000) Nonuniform sediment transport in alluvial rivers. J Hydraul Res IAHR 38(6):427–434
Zhang Y (2006) CCHE-GUI—Graphical users interface for NCCHE model user's manual—version 3.0, technical report no. NCCHE-TR-2006-02. School of Engineering, The University of Mississippi, Oxford, pp 158

Shallow Water Equations as a Mathematical Model of Whitewater Course Hydrodynamics

Michał Szydłowski and Patrycja Mikos-Studnicka

Abstract Predicting the positions of local hydraulic phenomena, as well as accurately estimating the depth and velocity of the water flow are necessary to correctly configure a whitewater canoeing course. Currently, a laboratory and full 3D CFD modeling are typically used in the design process to meet these needs. The article points to another possibility which can be useful at the preliminary stage of the design. The authors show that a mathematical model of depth-averaged free-surface flow can reliably predict the basic flow dynamics and location of some hydraulic local effects within a whitewater open channel. The results of a numerical simulation of the transcritical flow were compared to the flow parameters measured in a laboratory model of a constriction in an open-channel flow. A satisfactory agreement between the measurements and calculated flow parameters was observed along with the proper reconstruction of hydraulic local effects. It can be concluded that the proposed simplified approach to modeling is an efficient method to investigate the main hydraulic features of whitewater flow needed at the concept stage of the design process. This means that cost- and time-consuming laboratory and CFD modeling is necessary only at a later stage of the design, after the initial concept of the canoe course has been completed.

Keywords Whitewater hydraulics · SWE · Mathematical modeling · Laboratory modeling

1 Introduction

The canoe slalom is one of the most attractive sport disciplines in the Summer Olympic Games. The popularity of whitewater sports tends to be growing and the canoe slalom shares the top ranking of all viewed competitions in Sydney 2000, Athens 2004, Beijing 2008, London 2012 (Pollert et al. 2015) and Rio de Janerio

M. Szydłowski (✉) · P. Mikos-Studnicka
Faculty of Civil and Environmental Engineering, Gdańsk University of Technology, Narutowicza 11/12, 80-233 Gdańsk, Poland
e-mail: mszyd@pg.edu.pl

M. B. Kalinowska et al. (eds.), *Recent Trends in Environmental Hydraulics*, GeoPlanet: Earth and Planetary Sciences,
https://doi.org/10.1007/978-3-030-37105-0_24

2016. The canoe slalom was introduced for the first time as an Olympic discipline in 1972, in Munich, Germany. The canoe course was designed to copy a natural watercourse, a mostly linear channel, and set up on the Eiskanal in Augsburg, 64 km from Munich. The approximately 300-m-long venue was built of concrete simulating rocks and boulders fixed to the uneven bed and walls of the channel, as well as barn-door water deflectors. The planning and project design included a 25-m-long scale model used for flow simulations. The Augsburg Eiskanal is 10 m in width, with a drop of 4.1 m and a slope of 1.3%. The discharge of the channel fed by the River Lech equals 10 m^3 s^{-1}. The technical parameters of the Eiskanal still meet international competition and championship requirements (ORM 1972; IOC 2011).

Unfortunately, the huge costs of the artificial whitewater stadium at Eiskanal (today, over 24 million dollars) put the Olympic canoeing idea on hold for twenty years. The Canoe Slalom competition returned in 1992 during the Barcelona Olympics, becoming continuously more and more popular.

Designing courses for mountain canoeing should be the result of cooperation between hydraulic engineers, the water management sector and experienced canoeists. The parameters of whitewater courses depend on their purpose, as to whether they can be used by professionals or canoeing enthusiasts. What is also important is what kind of sport is to be performed—rafting, slalom, freestyle—and whether it is for competitive or training activities. Regardless of the purpose of the canoe course, its technical aspects should meet several criteria, such as high velocity, safety and an unchanging shape with time, and they should imitate natural mountain rivers (Lester 2012). The general design principles for whitewater channels are (Cubanova and Rumann 2009; Pollert et al. 2015):

- minimum length of 300 m—according to use in competitions like the Olympic Games and World Championships
- simple cross-section shape—trapezoidal, U-shape
- profile width not less than 8 m, optimal 10–12 m—double the size of slalom boats
- depth—minimum 0.4 m, average depth (safe for Eskimo rolls) approx. from 0.75 to 0.9 m, for freestyle canoeing approx. 1.5 m
- flow velocity—approx. 2 m s^{-1}, with some sections of greater speed
- material—concrete; roughness coefficient safe for boats
- movable obstacles, platforms, etc.—to satisfy multiple-user requirements.

Moreover, Olympic Canoe Slalom designers must consider the hydraulic features of channels in terms of the international canoe slalom rules of the ICF (International Canoe Federation), and the IOC (International Olympic Committee) requirements for the organization of Olympic contests. The maximum length of the course is 300 m and the maximum number of gates is 25 (ICF 2017). In view of the abovementioned factors, a canoe course design is not an easy task. Conceptual planning should integrate the interests of multiple-user groups in one multifunctional solution. Hosting Olympic contests or other professional competitions does not necessarily exclude the facility from commercial and recreational purposes.

2 Hydraulic Aspects of Whitewater Course Design

The growing popularity of whitewater sports requires that designers plan appropriately attractive canoe slalom courses, simultaneously ensuring the safety of their use. These features strongly depend on flow hydrodynamics, resulting largely from the geometry of the chosen venues and the method of their water supply. Since canoe courses are open channels, the flow is governed mainly by fluid weight and forces of gravity. The selection of the geometric and hydraulic parameters of the venue should be preceded by a detailed analysis of the stream moving down the channel. Predicted supercritical and subcritical flow conditions, as well as the position of some typical whitewater structures can be recognized using scale models. The geometric scale ratio should be chosen so as to ensure small dimensions of the model. Additionally, the Reynolds number should be high enough to satisfy the turbulent flow criteria even at small discharges. Similarity between the prototype and the model is usually determined by the Froude number. The Froude number has been found to be the best to characterize slalom course type and difficulty as it represents streaming in the channel as a whole. Froude number values ranging from 1.3 to 1.7 characterize a beginner whitewater course, while from 1.7 to 2.7 describe a professional venue (Goodman and Parr 1994; Pollert et al. 2015).

Based on Streeter (1964), the parameters required for a dimensional analysis can be grouped into the following categories:

- fluid properties and physical constants—density [kg m^3], dynamic viscosity [N s m^{-2}], gravitational acceleration [m s^{-2}]
- channel geometry—length [m]
- flow properties—velocity [m s^{-1}], pressure difference [Pa].

Physical modeling was used during the preparations for the London Olympics in 2012. The aim of such works was to deliver a canoe slalom course for both competition and training purposes. The design team included: architects, multi-disciplinary engineers and a whitewater rapid specialist. The models of both channels were at scale 1:10 and allowed tests to be performed for various flow features and obstacle positions. The analysis proved that the project covers all Olympic standards (Naish et al. 2011).

Before the Rio 2016 Olympics, Whitewater Parks International (WPI) in collaboration with the UK engineering firm Cundall authorized the Czech Technical University in Prague to perform hydraulic experiments on physical models (Pollert et al. 2015). As in the 2012 Olympics, the purpose of the project was to design the size and arrangement of obstacles in the channel regarding all the required hydraulic features and safety rules. Other tasks were to create different flow characteristics, stable flow and depth parameters, and proper velocities. Laboratory measurements were supported with mathematical modeling of the flow distribution.

Considering all the needs of the competitors and competition organizers, as well as the expectations of spectators, it is obvious that the key strategy for successful slalom course design is the proper recognition of local hydraulic effects combined with an exact estimation of water depths and flow velocities. From the hydraulic point of view, to create whitewater flow, the condition must be generated that causes hydraulic jumps to occur, which requires a rapidly varied flow within the course. In order to achieve such a state, whitewater courses are formed by water flowing through a channel of changing cross-sections with a steep bed, typically over submerged obstacles and through channel constrictions, all of which induce flow acceleration and transitions between subcritical and critical flows (canoeing term: 'green water'). When the flow decelerates from supercritical to subcritical, hydraulic jumps occur. These hydraulic phenomena generate air entrainment and turbulence, which canoeists call 'white water'. Hydraulic jumps of low Froude numbers are seen by canoeists as 'standing' waves, allowing surf in them. Fully developed hydraulic jumps are called 'stoppers'. 'Standing' waves are also the result of the stream flowing over an uneven bed. Such elements are known in canoeing terms as 'rollers'. The described flow features are key components of a whitewater course and are used by canoeists for performing slalom competitions. Moreover, competitors expect that all the hydraulic effects in a whitewater course will remain in the same section of the water for a period of time. Controlling these whitewater flow elements requires steady water flow conditions.

Knowing all the expectations of both the competitors and spectators, the engineers have to create an artificial mountain river in the form of a hydro-engineering structure. The whitewater course design process should consist of the physical as well as mathematical modeling of the hydraulic behavior of rapidly varied steady free-surface water flow. Both modeling methods can be used for planning the construction of the water course and the location of the obstacles in the canoe channel, so they should complement each other. The design procedure may start with simplified numerical simulations of water flow to make the first concept of the geometry of the canoe slalom course. Such simulations can provide some variants of the shape of the main channel and location of the stumbling blocks. Then the chosen variants should be studied in detail using the laboratory scale models to recognize the full structure of the hydraulic effects near the flow obstacles. Finally, the designed whitewater course elements can be verified by multidimensional numerical simulations based on computation fluid dynamic (CFD) methods. All three steps of whitewater course design (concept, design, verification) are complementary and should ensure the effective process of the creation of an attractive and safe water sport structure.

The current main approach for designing and testing whitewater courses is to use a combination of the Froude scale hydraulic model and prior experience, complemented by verification with CFD methods (Lin et al. 2008; Borman et al. 2015; Pollert et al. 2015). The problem of how complex the mathematical model and numerical methods should be to sufficiently represent the physical features of whitewater flow, depends on the aim of the numerical simulation. For example, if an analysis of the full complex structure of a hydraulic jump is desired, full three-dimensional turbulent flow modeling by CFD is undoubtedly needed. However, in order to estimate the general flow parameters and approximate position of local effects, when the inner structure of the water flow is not essential in the hydraulic analysis, two-dimensional (depth-averaged) hydrodynamic models are usually used. In our work we propose a simplified model of shallow water equations to describe the water motion through the whitewater canal. Such an approach seems to be precise enough to represent the hydraulic behavior of rapidly varied flow at the concept step of whitewater course design. The model was already successfully used to simulate rapidly varied flow due to flash floods in natural and urban areas (Szydłowski 2006), as well as a catastrophic flood caused by dam breaks (Szydłowski 2011). In this paper, the solution is verified by laboratory measurements of the water surface profile and the velocity distribution of whitewater flow in a laboratory channel. In order to carry out the experiment, a model of channel constriction was prepared allowing the transition from subcritical to supercritical flow (canoeing term: 'chute') and the formation of a hydraulic jump (canoeing term: 'stopper'). The comparison of calculations and measurements related to both characteristic whitewater course features are satisfactory.

3 Methods of Mathematical and Laboratory Modeling

As a simplified mathematical model of free-surface water flow, shallow water equations (SWE) (Toro 1997) were adopted. This two-dimensional model can be written in a conservative form as

$$\frac{\partial \mathbf{U}}{\partial t} + \frac{\partial \mathbf{E}}{\partial x} + \frac{\partial \mathbf{G}}{\partial y} + \mathbf{S} = 0 \tag{1}$$

where

$$\mathbf{U} = \begin{pmatrix} h \\ uh \\ vh \end{pmatrix}, \quad \mathbf{S} = \begin{pmatrix} 0 \\ -gh(S_{ox} - S_{fx}) \\ -gh(S_{oy} - S_{fy}) \end{pmatrix} \tag{2a,b}$$

$$\mathbf{E} = \begin{pmatrix} uh \\ u^2h + 0.5gh^2 \\ uvh \end{pmatrix}, \quad \mathbf{G} = \begin{pmatrix} vh \\ uvh \\ v^2h + 0.5gh^2 \end{pmatrix} \tag{2c,d}$$

In this system of equations h represents water depth, u and v are the depth-averaged components of velocity in x and y directions, respectively, S_{ox} and S_{oy} denote the bed slope terms, S_{fx} and S_{fy} are the bottom friction terms defined by the Manning formula and g is the acceleration due to gravity. Equation (1) can be written in another form

$$\frac{\partial \mathbf{U}}{\partial t} + div\mathbf{F} + \mathbf{S} = 0 \tag{3}$$

where vector $\mathbf{F}$ is defined as $\mathbf{F} = \mathbf{E}\mathrm{n_x} + \mathbf{G}\mathrm{n_y}$, the unit vector is $\mathrm{n} = (\mathrm{n_x}, \mathrm{n_y})^{\mathrm{T}}$.

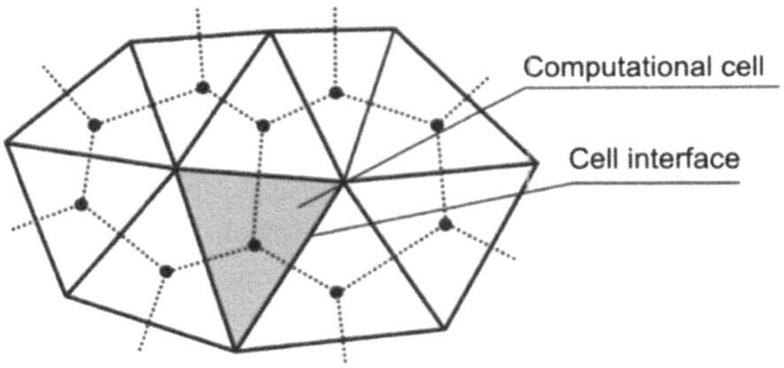

Fig. 1 Triangular computational cells for FVM

In order to integrate SWE in space, the finite volume method (FVM) was used (LeVeque 2002). Applying the FVM idea, the space domain (x, y) can be discretized, for instance, into a set of triangular cells (Fig. 1).

After the integration and substitution of integrals by the corresponding sums, Eq. (3) can be rewritten for any cell (i) as

$$\frac{\partial \mathbf{U}_i}{\partial t}\Delta A_i + \sum_{r=1}^{3}(\mathbf{F}_r \mathbf{n}_r)\Delta L_r + \sum_{r=1}^{3}\mathbf{S}_r \Delta A_r = 0 \tag{4}$$

where $\mathbf{F}_r$ is the numerical (computed at rth cell-interface) mass or momentum flux and ΔL_r represents the cell-interface length. $\mathbf{S}_r$ and ΔA_r are the components of source terms and the area of cell (i) assigned to rth cell-interface, respectively. In order to calculate the fluxes $\mathbf{F}_r$, the solution of the approximated Riemann problem was used. A description of the method was presented in detail in LeVeque (2002) and applied for rapidly varied flow modeling (Szydłowski 2011). The source term vector $\mathbf{S}$ contains two kinds of elements, depending on the bottom and friction slopes, respectively. In order to avoid the numerical integration problem, the splitting technique was applied with respect to physical processes. The computational numerical algorithm was complemented by a two-stage explicit scheme of time integration.

The experimental set-up was built at the hydraulic laboratory of Gdansk University of Technology. The test was carried out in a laboratory open channel with a horizontal concrete bed and vertical glass walls. The laboratory main channel is 10 m long and 0.38 m wide. The model of the channel constriction was located 5.0 m downstream from the main channel inflow section. The full width of the channel constriction is 0.135 m and its total length is 1.0 m. At the narrowest cross-section of the constriction the model is 0.077 m wide.

The scheme of the main laboratory channel is presented in Fig. 2. The pictures in Fig. 3 show the real view of the channel constriction installed inside the main channel. The water surface profile along the constriction was measured using a gauging needle.

The ordinates of the water surface were probed along the channel longitudinal axis (x) at 19 control points. At the same points, the vertical distribution of velocity was investigated using a hydraulic propeller to measure the velocity at points located along the depth. The scheme of the channel constriction with the distribution of control points is shown in Fig. 4.

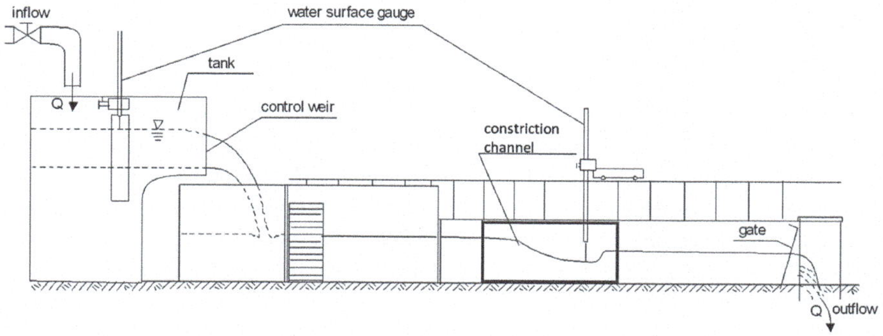

Fig. 2 Scheme of the laboratory open channel set-up

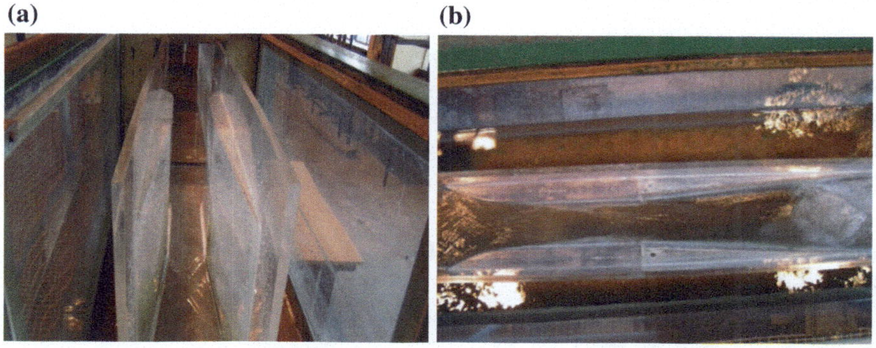

Fig. 3 General (**a**) and plane (**b**) view of the laboratory channel with a constriction

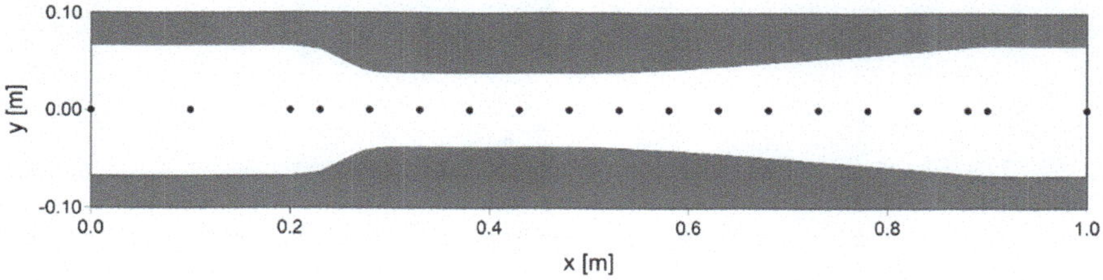

Fig. 4 Geometry of the channel constriction (dimensions in meters) with the location of probe points (dots). Flow direction from the left to the right side

The steady flow in the laboratory hydraulic system is controlled with a weir located at the inlet to the channel. For the experiment, the water discharge was equal to 0.0172 $m^3\ s^{-1}$. The bottom of the channel is flat and horizontal. The water depth is controlled with a gate located at the outlet of the channel. In order to analyze the flow in the channel constriction, an experiment was carried out on the transcritical flow. At the inflow section of the channel constriction, the subcritical flow was induced. Falling water surface and flow acceleration were observed along the channel constriction. For canoeists, the sudden decrease in water surface is the so-called ‘chute’ effect. Finally, the transcritical flow occurred inside the constriction, resulting in the supercritical movement of water along the downstream section. During the experiment, a depth of 0.16 m was forced at the outflow section of the channel constriction. Therefore, the flow near the outlet from the channel was subcritical again and a hydraulic jump was formed inside the channel constriction. The surface disturbances penetrating one another in space were present almost along the whole length of the supercritical flow section. The hydraulic jump, together with the surface wave effects and strong air penetration in the water body, are seen by canoeists as the ‘stopper’ and ‘whitewater’ effects on the canoe course, respectively. The pictures of the hydraulic phenomena observed during the experiment are shown in Figs. 3 and 5.

In order to simulate two-dimensional water flow inside the channel constriction using FVM, the real geometry of the channel had to be transformed into a numerical mesh. The 1.0 m long flow domain was covered by a mesh of triangles composed of 4296 computational cells (Fig. 6). The length of the side of a standard triangle was equal to 0.01 m.

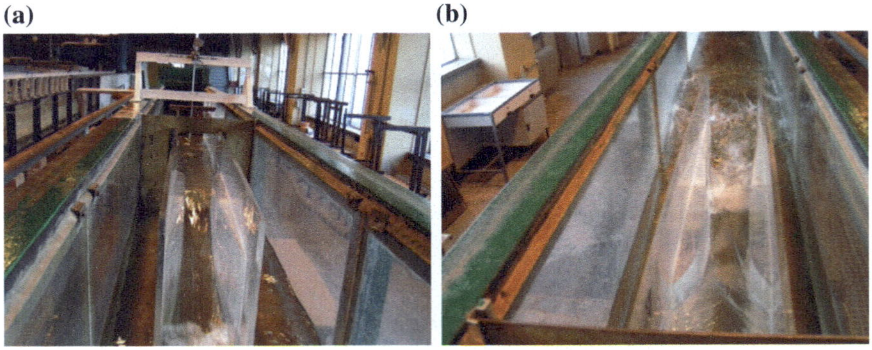

Fig. 5 Laboratory experiment in the channel constriction—downstream (**a**) and upstream (**b**) view of the transcritical flow

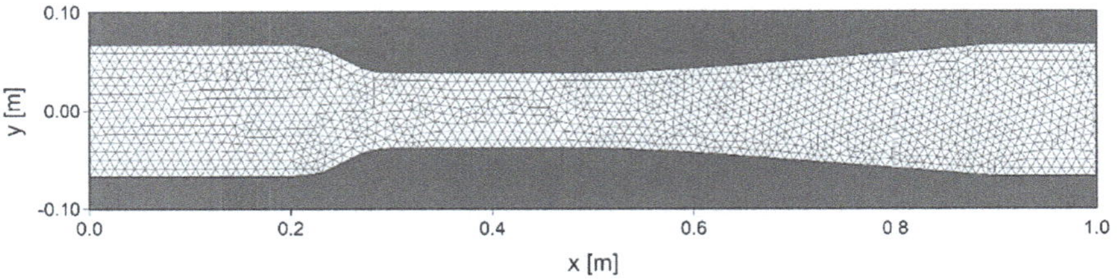

Fig. 6 Numerical mesh of the channel constriction (dimensions in meters)

Boundary conditions were imposed in accordance with the laboratory experiment. The channel walls were treated as closed boundaries and the free-slip condition was enforced. Open boundaries were located at the inflow and outflow cross-sections. At the inlet, the measured flow rate was imposed as the upstream boundary condition and at the outlet, the water depth was enforced. The laboratory experiment of steady flow through the channel constriction was simulated numerically using unsteady flow equations, starting from an initially empty channel. However, the final steady state was enforced by the boundary conditions, which were kept constant during the whole simulation. The total simulation time was equal to 10 s, after which the flow parameters did not vary in time.

4 Results of Modeling

A comparison between the measured and simulated flow parameters (water depth and mean velocity) is presented in Fig. 7. The water depth measurements were difficult to carry out due to strong water surface oscillations in the area of the transcritical flow, and intense aeration and foam in the vicinity of the hydraulic jump. Moreover, the local surface effects penetrating one another in space made the measurements difficult to carry out using simple laboratory equipment such as a gauging needle. Therefore, probes of the water level are burdened with a measurement error. Regardless of the uncertainty of depth measurements, it can be found that, in general, the computed shape of the water surface fits the observations quite well and is of acceptable quality.

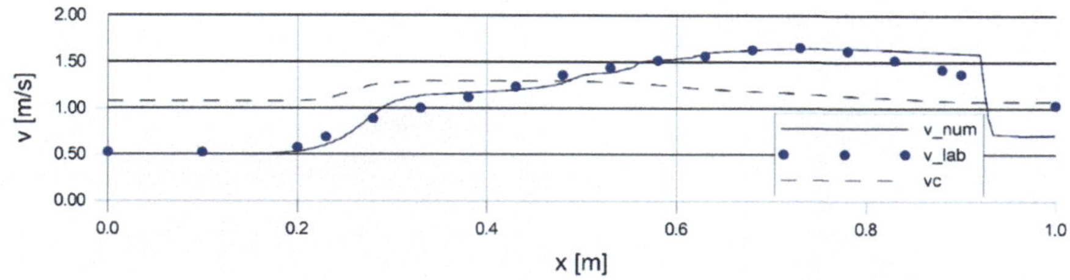

Fig. 7 Measured (lab) and calculated (num) profiles of water depth (*h*) with the critical depth line (*hc*) and velocity (*v*) with the critical velocity line (*vc*) along the channel constriction longitudinal axis

The calculated and observed water surface decreases in the constricted area. It can also be seen that the location of the crossing point of the water surface and the critical depth line is properly simulated inside the constriction, where the flow transition from subcritical to supercritical occurred.

The other flow transition between the supercritical and subcritical flow was also calculated quite precisely. The hydraulic jump, formed downstream from the channel constriction, can be seen in Fig. 7 as a sudden increase in the water depth (h) becoming greater than the critical depth. The water depth variations observed along the channel constriction are coupled with changes in the mean velocity. It can be seen again that the measured and calculated mean velocity distributions along the channel axis are similar to each other. The transcritical flow is observed in the same sections. Unfortunately, in the vicinity of the hydraulic jump, measurements of velocity with a hydraulic propeller were impossible due to the strong reverse flow and breaking surface wave. The complex vertical structure of the flow at the outflow section of the channel constriction caused a large error in the mean velocity value at the outlet from the channel. Concluding, it can be found that the location of the transcritical flow areas in the channel constriction, as well as the main flow features such as the water surface shape and the distribution of the mean velocity, match the numerical simulation quite well.

In addition to the one-dimensional analysis, the SWE numerical simulation makes the spatial, horizontal distribution of flow parameters available. In Figs. 8, 9 and 10, the spatial distributions of water depth, flow velocity magnitude and Froude number values are shown, respectively.

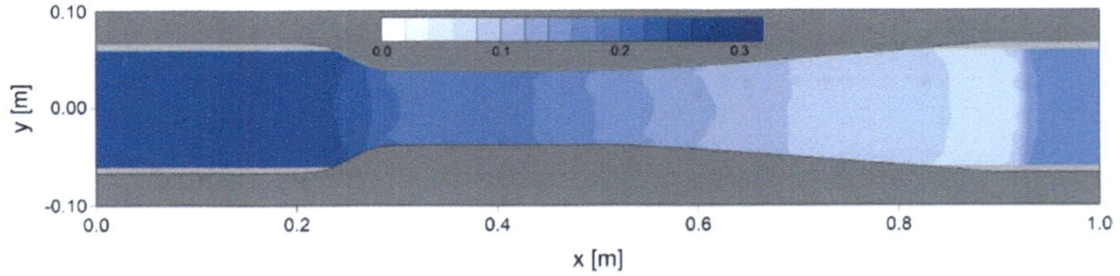

Fig. 8 Water depth [m] horizontal distribution (dimensions in meters)

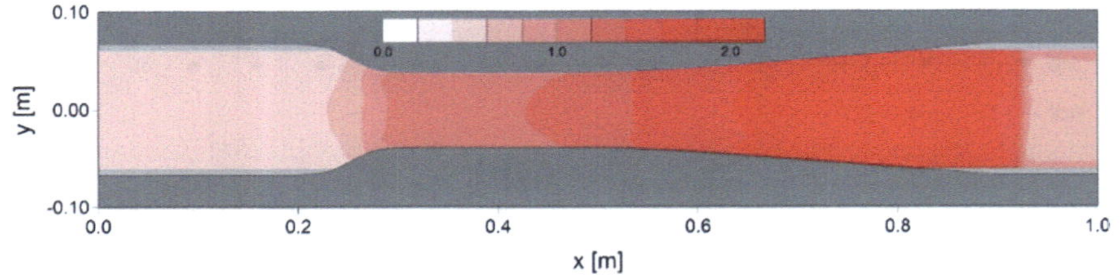

Fig. 9 Flow velocity magnitude [m s^{-1}] horizontal distribution (dimensions in meters)

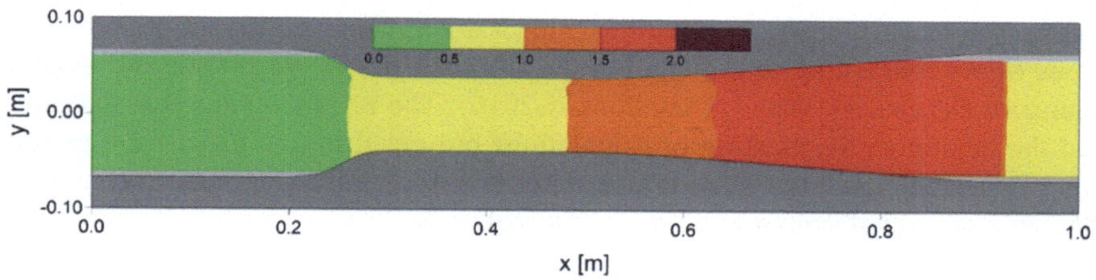

Fig. 10 Froude number [–] horizontal distribution (dimensions in meters)

It appears that the main hydraulic characteristic of water flow in the channel constriction was properly approximated. The water table in front of the constricted area, as well as the sudden decrease of the surface along the constriction can be observed in Fig. 8. The calculated shape of the water surface corresponds to the one noted during the experiment and is placed in the area of supercritical flow. Farther, at the outflow section, the sudden local water swelling of the hydraulic jump can be observed. All these flow features also occurred during laboratory tests (Figs. 3 and 5). The flow in the area of transcritical water motion can be investigated using an analysis of velocity and Froude number space distributions (Figs. 9 and 10). It can be found that the velocity magnitude field corresponds to the water depth distribution (Fig. 8). In the area of water depressions, the decrease in water depth is accompanied by the flow acceleration. In the area where the hydraulic jump occurs, the velocity is reduced. The relationship between the flow velocity and water depth can be presented as a Froude number (Fig. 10). The map of this criteria number *Fr* defines quite precisely the spatial structure of the transcritical flow, giving information about the locations of subcritical and supercritical flow zones and about the positions of hydraulic local effects accompanying the flow transitions.

5 Conclusions

The satisfactory results of the comparison of the numerical simulation and laboratory measurements carried by the authors allow one to conclude that shallow water equations can be successfully used for modelling the flow in whitewater courses. It must be underlined that the inner structure of the local whitewater flow phenomena is lost in such numerical simulations due to the vertical averaging of flow parameters in the SWE model. However, the general features of the transcritical flow are properly and sufficiently reproduced to recognize the basic characteristics and mean hydraulic parameters of the flow. This type of whitewater numerical simulation can be useful for the conceptual stage of the design process, when only rough characteristics of whitewater hydraulics are needed.

The positive laboratory verification of the numerical solution indicates the possibility of using the SWE model for hydraulic modelling of geometrically complex aquatic sports and recreational facilities. This was also confirmed by the authors' previous works concerning modelling flows in an existing body water slide and a canoe course in Drzewica (Poland) (Szydłowski 2016). The results obtained at that time and the laboratory verification of calculations presented in this paper have proved that a simplified model of free-surface water flow is an adequate numerical tool for the computer-aided design of whitewater canals.

References

Borman D, Sleigh A, Coughtrie A (2015) Three dimensional free-surface modelling with a novel validation approach. In: E-proceedings of the 36th IAHR world congress, Hague, Netherlands

Čubanová L, Rumann J (2009) Whitewater course design in Slovakia. In: Proceedings of international symposium on water management and hydraulic engineering 2009, Macedonia, Ohrid, 221–228

Goodman FR, Parr GB (1994) The design of artificial white water canoeing courses. Proc Instn Civ Engrs Mun Engr 103:191–202

International Olympic Committee (IOC) (2011) CANOE/KAYAK Slalom: participation during the history of the Olympic Games

International Canoe Federation (IFC) (2017) canoe slalom competition rules

Lester B (2012) Whitewater park design principles: an integrated approach for multiple user groups. A thesis presented to the University of Guelph, Guelph, Ontario, Canada

LeVeque RJ (2002) Finite volume method for hyperbolic problems. Cambridge University Press, New York

Lin F, Shepherd D, Slack C, Shipley S, Nilson A (2008) Use of CFD modeling for creating recreational opportunities at the Calgary Bow River Weir. In: World environmental and water resources congress, Ahupua'a

Naish C, Dungworth D, Doyle T (2011) Delivering London 2012: the Lee Valley White Water Centre. Civ Eng 165(1):19–26

Olympic Report Munich (ORM) 1972, Volume 2, Part 2 The Constructions

Pollert J jr, Pollert J, Procházka J, Chmátal P, Campbell B, Felton J, Dungworth D (2015) Physical and mathematical modelling for canoe slalom whitewater and the 2016 olympic games in Rio de Janeiro. In: Proceedings of international symposium on water management and hydraulic engineering 2016, Czech Republic, Brno, 65–75

Streeter VL (1964) Handbook of fluid dynamics. First Edition Mc Graw-Hill Book Company

Szydłowski M (2006) Mathematical modelling of flash floods in natural and urban areas. In: Transboundary floods: reducing risks through flood management. NATO Science Series: IV: Earth and Environmental Sciences, vol 72, Springer, Nederland, 143–153

Szydłowski M (2011) Application of hydrodynamics model for a case study of the Kolbudy II reservoir embankment hypothetical failure. In: Rowiński P (ed) Experimental methods in hydraulic research, geoplanet: earth and planetary sciences. Springer, London, 299–306

Szydłowski M (2016) Numerical modeling of hydrodynamics as a tool for design of the leisure and sport water structures. Acta Sci Pol Formatio Circumiectus 15(4):353–367

Toro EF (1997) Riemann solvers and numerical methods for fluid dynamics. Springer, Berlin

Velocity Distribution and Dip Phenomenon in a Large Amplitude Meandering Channel

Donatella Termini

Abstract The prediction of the velocity-dip, whereby the location of the maximum velocity occurs below the water surface, could be important for defining the flow pattern and the momentum transport processes. This study explores the dip-phenomenon in a high-curvature meandering bend. With the aid of data collected in a laboratory flume, the influence of the distance to the outer bank, the channel's curvature and the aspect ratio on the formation and position of the velocity-dip is investigated. Results show that, for the aspect ratios examined, a dip forms in the velocity profiles and its position varies along the bend as a function of the channel's curvature and the aspect ratio. Based on the measured data, an equation relating the velocity-dip to the local deflection angle, θ, and the width-to-depth ratio B/h_{av} is presented. It has been verified that the values of the velocity-dip estimated by the presented equation are in good agreement with the experimental ones.

Keywords Rivers · Flow velocity · Secondary flow · Prediction

1 Introduction

A natural meandering stream represents one of the most complex flow situations. Accurate predictions of velocity distribution along the meander wave is important for analyzing problems related to river restoration, water quality and water management. As it is known, when the flow enters a bend, the so-called helical motion, given by the combination of the transversal cross-sectional motion with the longitudinal motion (see, e.g., Yalin 1992) is produced. In bends of constant curvature of sufficient length, the cross-sectional motion does not change along the bend, creating the so-called fully developed flow conditions. In a meandering bend, which is characterized by a continuously changing curvature in the downstream direction, the cross-sectional motion varies from cross-section to cross-section (Termini and Piraino 2011). Literature studies (Blanckaert and Graf 2004; Termini 2015) have

D. Termini (✉)
Department of Engineering, University of Palermo, Viale delle Scienze, 90128 Palermo, Italy
e-mail: donatella.termini@unipa.it

M. B. Kalinowska et al. (eds.), *Recent Trends in Environmental Hydraulics*, GeoPlanet: Earth and Planetary Sciences,
https://doi.org/10.1007/978-3-030-37105-0_25

also demonstrated that the cross-sectional motion exerts an important role in the redistribution of the stream-wise velocity along the bend.

Although several works have been conducted to analyze the flow velocity pattern in meandering bends (among others Whiting and Dietrich 1993, Blanckaert and Graf 2004; da Silva et al. 2006; Termini 2009), several aspects related to the influence of the cross-sectional motion on the velocity and fluid mass distributions are still poorly understood attracting the attention of the researchers (Nikora and Roy 2012). This is important especially in high-curvature bends where the effect of channel's curvature is more accentuated than that in moderate curvature bends (e.g., Blanckaert and de Vriend 2010).

As an example, no systematic research has been conducted to analyze the vertical profile of the stream-wise flow velocity and the localization of the maximum value along a bend. Several studies conducted in straight channels highlighted the formation of the so-called velocity-dip related to the fact that the maximum velocity occurs below the free surface. Although the velocity-dip was found especially in narrow straight channels for values of the width-to-depth ratio less than 5 (Vanoni 1941; Nezu and Nakagawa 1993), some studies (e.g., Sarma et al. 1983; Tachie et al. 2003) have demonstrated that the velocity-dip occurs also for values of the width-to-depth ratio higher than 5. Yan et al. (2011) verified the existence of a central region (termed as "weak sidewall region") of the cross-section where the position of the velocity-dip depends both on the distance from the sidewalls and on the aspect ratio.

Several simple dip-modified log-wake-laws (Coles 1956; Finley et al. 1966; Kironoto and Graf 1994; Song and Graf 1996; Guo and Julien 2008) can be found in the literature but they are not universal and are unable to predict the dip location in 3D flows with high secondary currents effects. Other authors (Yang et al. 2004) proposed dip-modified log-law through an empirical function accounting for the secondary flow effect. However, these laws are formulated in straight channels and do not consider adjunctive effects due to the channel's plane-form variations.

Based on the aforementioned considerations, the present study aims to gain some insights on the evolution of the dip-phenomenon along a high-amplitude meandering bend. In particular, the main objectives of the present study are: (1) to explore the location of the maximum velocity based on available experimental data along the bend for two values of the width-to-depth ratio; (2) to investigate the relationship between the velocity-dip, the channel's curvature parameters and the aspect ratio.

2 Data and Methods

2.1 Experimental Data

The data used in the present work were collected in previous studies aimed to analyze the flow pattern along a meander wave. The meandering laboratory flume follows the sine-generated curve with a deflection angle at the inflection section $\theta_o = 110°$

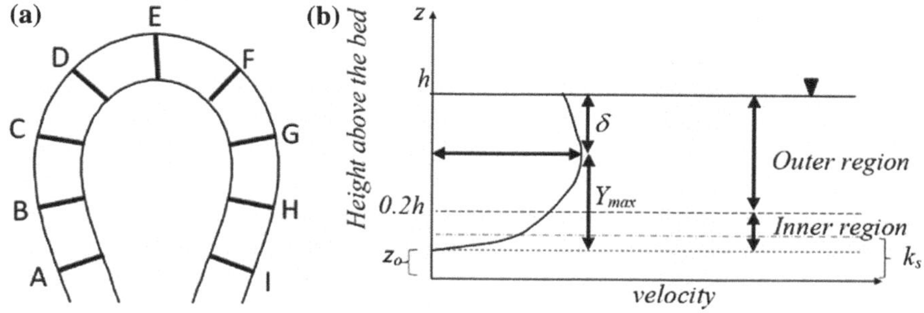

Fig. 1 **a** Meander wave and examined sections; **b** Scheme of the vertical velocity profile (z_o = reference level; k_s = representative roughness height; Y_{max} = distance of the maximum velocity from the bed; δ = velocity-dip)

and width $B = 0.50$ m. The bed is of quartz sand with medium sediment diameter of 0.65 mm. The data used in the present work were collected during runs conducted in flat bed conditions in order to isolate the effect of the channel's curvature on the kinematic characteristics of the flow (Termini 2009). In particular, run 1 conducted with a flow discharge of $Q = 0.019$ m^3 s^{-1} (channel-averaged flow depth h_{av} = 5.5 cm; aspect ratio $B/h_{av} = 9.09 < 10$) and the run 2 conducted with a flow discharge $Q = 0.007$ m^3 s^{-1} (channel-averaged flow depth $h_{av} = 3.0$ cm, $B/h_{av} = 16.67 > 10$) have been considered in the present work. Herein, in accordance with other authors (da Silva et al. 2006; Termini 2015) $B/h_{av} < 10$ indicates "small" width-to-depth ratio and $B/h_{av} > 10$ indicates "large" width-to-depth ratio. During each run, the instantaneous local longitudinal and transversal velocity components were measured along the verticals of nine transverse abscissas symmetrically to the channel axis (see details in Termini 2009) by using a 2D side-looking Acoustic Doppler Velocimeter (ADV) manufactured by SonTek Inc. Bearing in mind the specific aim of the present work, here the attention is focused only on the time-averaged longitudinal velocity component, u, and the analysis concerns the nine sections indicated in Fig. 1a.

2.2 *Theoretical Considerations*

As shown in Fig. 1b, two regions can be distinguished in the vertical velocity profile of fully turbulent flows: the outer region and the inner region. The thickness of the inner region is around 20% of the entire boundary layer thickness. Often the maximum velocity occurs below the free surface (velocity-dip δ) at a distance from the bed indicated as Y_{max}.

As it is known, the log-law can be written as follows (Yalin 1992):

$$\frac{u}{u_*} = \frac{1}{\kappa}\ln\left(\frac{z}{k_s}\right) + B_s \qquad \begin{array}{ll} B_s = \frac{1}{\kappa}\ln\frac{u*z}{\upsilon} + 5.5 & \text{for } \mathrm{Re}_* = \frac{u_* k_s}{\upsilon} < 5 \\ B_s = 8.5 & \text{for } \mathrm{Re}_* = \frac{u_* k_s}{\upsilon} > 70 \end{array} \tag{1}$$

where u_* represents the shear flow velocity, κ is the universal Von Karman's constant which is assumed equal to 0.41, ν is the water viscosity, k_s is the representative roughness height (see Fig. 1), is the roughness Reynolds number.

The shear velocity, u_*, is a parameter of Eq. (1). In straight uniform channels it is generally assumed equal to the global shear velocity $u_* = (gRS)^{0.5}$ (with g = gravitational acceleration, R = hydraulic radius, and S = energy slope). But, as previous works show (da Silva et al. 2006; Termini and Moramarco 2017), in a meandering bend the friction velocity varies from one location to another and thus it cannot be assumed constant along the bend. As literature indicates (Nezu and Rodi 1986; Nezu and Nakagawa 1993), the log-law could approximate the vertical profiles in the inner region but in the outer region it deviates from the experimental profile.

3 Results

3.1 Longitudinal Velocity Profile and Log-Law

The log-profile (1) has been used to verify how it approximates the experimental velocity profiles measured for both runs. To this aim, the friction velocity has been determined by data-fitting.

In Fig. 2, as an example, the velocity profiles predicted by log-law at the channel axis of sections A (inflection section) and E (apex section) are compared with the experimental ones, for both runs. This figure clearly shows that the log-law deviates from the experimental data and the velocity-dip forms in both cases.

In order to investigate how the position of the maximum velocity varies along the bend, its distance from the bed, Y_{max} (see Fig. 1b), has been identified for each profile. Figure 3 plots, for both runs, the values of the relative distance Y_{max}/h (where h indicates the local water depth) estimated for different relative distances from the outer bank, r/B. According to Termini and Piraino (2011), three different regions can be distinguished in Fig. 3: the outer-bank region for $r/B \leq 0.28$, the central region for $0.28 < r/B \leq 0.61$, the inner-bank region for $r/B > 0.61$. It can be observed that, apart from some exceptions, for $B/h_{av} > 10$ the ratio Y_{max}/h assumes values greater than those obtained for $B/h_{av} < 10$. This means that for $B/h_{av} > 10$ the dip δ assumes values lower than those obtained for $B/h_{av} < 10$. In the outer region, a different behavior between the two runs can be observed: for $B/h_{av} > 10$ very close to the bank (i.e. for $r/B = 0.06$) the higher peak of Y_{max}/h is found at the bend exit (i.e. between sections F and G) but for increasing distances r/B the higher peak of Y_{max}/h moves from the bend entrance (i.e. between sections C and D) at $r/B = 0.17$ to the bend apex (section E) at $r/B = 0.28$; for $B/h_{av} < 10$ the relative position Y_{max}/h assumes always the higher peak close to the apex section (section E).

The different behavior observed between $B/h_{av} < 10$ and $B/h_{av} > 10$ could be related to the fact that, according to previous findings obtained by Termini (2009), the advective behavior of flow is more accentuated in the case of $B/h_{av} > 10$ than

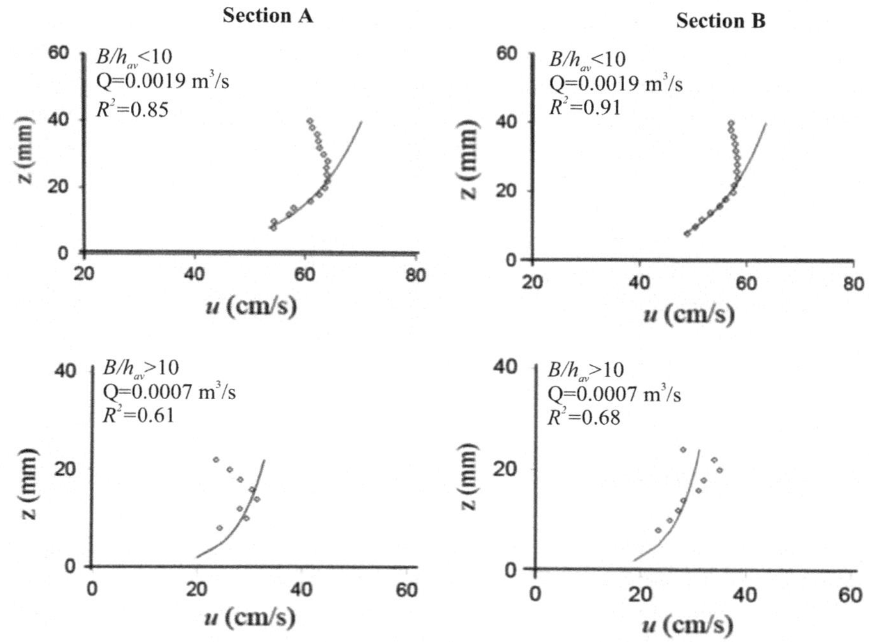

Fig. 2 Comparison between the experimental profiles and the log-law at channel axis

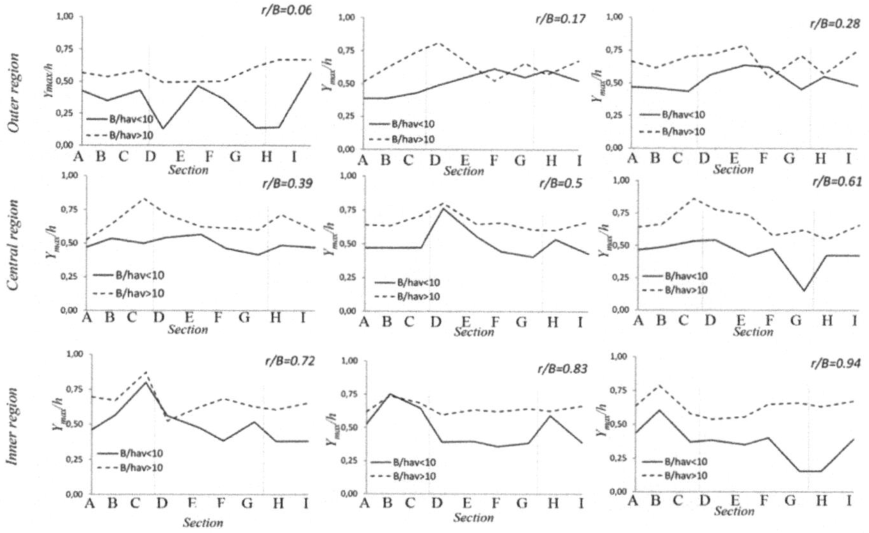

Fig. 3 Estimated Y_{max}/h along the bend for different relative distances from the outer bank

in the case of $B/h_{av} < 10$. Furthermore, as described in Termini and Piraino (2011), for $B/h_{av} < 10$ a counter-rotating circulation cell forms at the outer bank of the bend entrance maintaining the core of maximum velocity far from the bank. In the central and in the inner-bank regions, a similar behavior can be observed for the two runs. In the central region, Y_{max}/h assumes a high peak at the bend entrance and a second lower peak close to the bend exit; this behavior is especially evident for $B/h_{av} > 10$. In the inner region, the relative position Y_{max}/h assumes a behavior similar to that observed in the central region although, for decreasing distances to the inner-bank, the higher peak tends to move strictly upstream and the lower peak tends to move strictly downstream. This behavior is consistent with the trend of the downstream water elevation slope observed by Termini (2009) by using the same data as that considered in the present work.

3.2 The Velocity-Dip

Figure 3 has demonstrated that the position of the maximum velocity varies along the bend as a function of the distance from the outer bank, r/B, the channel's curvature and the width-to-depth ratio. By parameterizing the channel's curvature by the deflection angle θ of each considered section, the relation between the dip-values, δ, and the aforementioned parameters has been explored. The regression analysis of the δ-values by considering r/B and θ as independent variables has yielded the following expression:

$$\delta = cost1\left(\frac{r}{B}\right)\sin\theta + cost2\left(\frac{r}{B}\right) \tag{2}$$

where $cost1(r/B)$ and $cost2(r/B)$ are the linear functions of the distance r/B plotted in Fig. 4. It can be observed that, as r/B increases, $cost1$ decreases for $B/h_{av} < 10$ and increases for $B/h_{av} > 10$, $cost2$ increases for $B/h_{av} < 10$ and decreases for $B/h_{av} > 10$. Thus, as the distance from the outer bank increases, the coefficient of the term related to θ decreases for $B/h_{av} < 10$ and increases for $B/h_{av} > 10$.

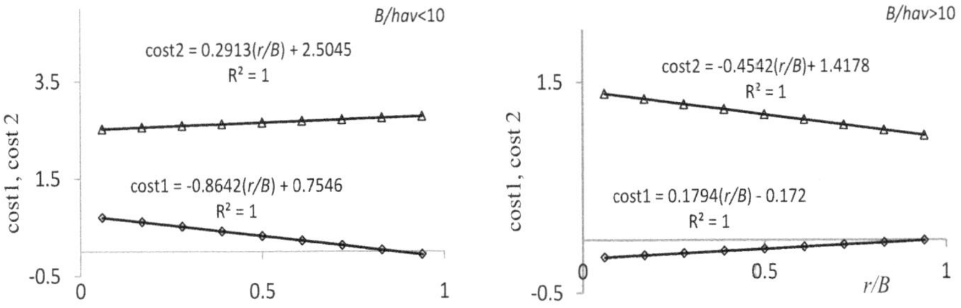

Fig. 4 Functions $cost1(r/B)$ and $cost2(r/B)$ for different relative distances r/B from the outer bank

By relating the obtained values of *cost*1 and *cost*2 to the independent variable B/h_{av}, the following equation has been obtained:

$$\delta = \left(a\frac{B}{h_{av}} + b\right)\sin\theta + \left(c\frac{B}{h_{av}} + d\right) \tag{3}$$

with a regression coefficient $R^2 = 0.8$. The coefficients *a, b, c, d* have to be determined by using experimental data; for the considered case the following values of the coefficients have been obtained: $a = -0.0536$, $b = 0.8104$, $c = -0.1925$, $d = 4.4003$. Unlike other literature expressions, Eq. (3) directly relates the velocity-dip to the local channel's curvature and the width-to-depth ratio B/h_{av}.

To verify the applicability of the aforementioned expression, in Fig. 5 both the experimental δ-values obtained in the present work and those taken from literature (see references in Table 1) are plotted against the estimated ones by using Eq. (3). Furthermore, in the same figure the δ-values estimated by applying the expression suggested by Yang et al. (2004) (see also in Termini and Moramarco 2017), which relates δ to the transversal abscissas, $\delta = 1 + 1.3e^{-r/h}$, are also compared with the experimental ones.

In Fig. 5 the error bar is defined by the root mean squared error between the estimated and the experimental values. From this figure it can be observed that, apart from some exception, the points thicken around the bisecting line, demonstrating the good agreement between the estimated and the experimental δ-values; the root mean squared error is quite low in comparison to the magnitude of the experimental

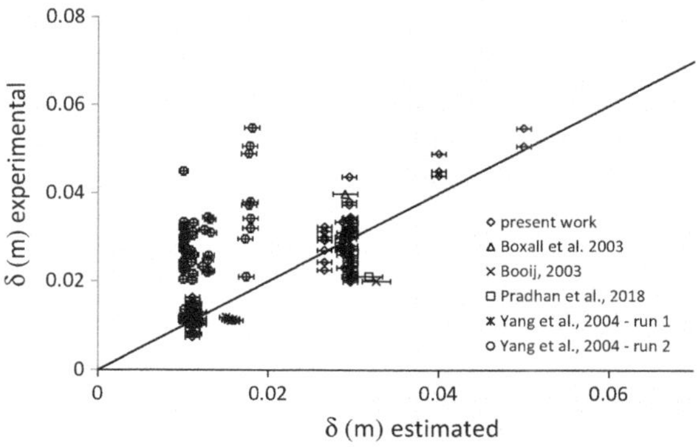

Fig. 5 Comparison between experimental and literature δ-values [m] and those estimated by Eq. (3)

Table 1 References considered for literature δ-values

Reference	Channel's planform	Section considered
Boxall et al. (2003) (see also Boxall and Guyner 2001)	Sine-generated $\theta = 60°$	apex
Booij (2003)	180° constant middle curved	Section at 135°
Pradhan et al. (2018)	Sine-generated $\theta = 110°$	apex

values. Figure 5 also demonstrates that the expression proposed by Yang et al. (2004) overestimate the values of the dip, especially in the case of run 2 ($B/h_{av} < 10$).

4 Conclusion

The results obtained in the present work have demonstrated that the distance of the maximum velocity from the bed, Y_{max}, varies along the bend as a function of the position with respect to outer bank, r/B, the deflection angle θ of the generic section, θ (which parameterizes the local channel's curvature), and the width-to-depth ratio, B/h_{av}. In particular, close to the outer-bank a different behavior between the case of $B/h_{av} > 10$ and the case of $B/h_{av} < 10$ has been observed. Such a different behavior could be due both to the fact that the convective behavior of flow is more significant for $B/h_{av} > 10$ than that for $B/h_{av} < 10$ and to the fact that a protective counter-rotating circulation cell forms close to the outer bank for $B/h_{av} < 10$, as described in Termini and Piraino (2011). In the central and in the inner-bank regions, for both the width-to-depth ratios, the relative distance Y_{max}/h assumes a high peak value at the bend entrance and a second lower peak value close to the bend exit. This behavior is especially evident for $B/h_{av} > 10$.

Based on the aforementioned results, the relation between the velocity-dip, the channel's curvature, the relative distances from the outer bank, r/B, and the width-to-depth ratio B/h_{av} has been investigated; an expression which directly relates the velocity-dip to the parameters θ and B/h_{av} has been presented. The values of velocity-dip estimated by this equation are in good agreement with the experimental dip-values both considered in the present work and taken from literature in curved channels. Despite the limited conditions investigated in this work, this equation allows us to gain insight in the variation of the velocity-dip along the bend. In order to generalize this equation, further researches should be made under further different planimetric channel's configurations and width-to-depth ratios.

References

Blanckaert K, Graf WH (2004) Momentum transport in sharp open-channel bends. J Hydraul Eng 130(3):186–198

Blanckaert K, de Vriend HJ (2010) Meander dynamics: a nonlnear model without curvature restrictions for flow in open-channel bends. J Geophys Res 115:F04011

Booij R (2003) Measurements and large eddy simulations of the flows in some curved flumes. J Turbul 4(8):1–17

Boxall JB, Guymer I (2001) Estimating transverse mixing coefficients. Proc Inst Civ Eng Water Marit Eng 148(4):263–275

Boxall JB, Guymer I, Marion A (2003) Transverse mixing in sinuous natural open channel flows. J Hydraul Res 41(2):153–165

Coles D (1956) The low of the wake in the turbulent boundary layer. J Fluid Mech 1:191–226

da Silva AMF, El-Tahawy T, Tape WD (2006) Variations of flow pattern with sinuosity in sine-generated meandering streams. J Hydr Eng 132(10):1003–1014
Finley PJ, Phoe KC, Poh CJ (1966) Velocity measurements in a thin turbulent water layer. La Houille Blanche, 6
Guo J, Julien PY (2008) Application of the modified log-wake law in open–channels. J Appl Fluid Mech 1(2):17–23
Imamoto H, Ishigaki T (1988) Measurement of secondary flow in open channel. In: Proceedings of the 6th IAHR-APD congress. Kyoto, 513–520
Kironoto BA, Graf WH (1994) Turbulence characteristic in rough uniform open-channel flow. Laboratoire de researches hydrauliques, Ecole Polytechnique Federale, Lausanne, Swizerland
Nezu I, Rodi W (1986) Open channel flow measurements with a laser Doppler anemometer. J Hydraul Eng -ASCE 112(5):335–355
Nezu I, Nakagawa H (1993) Turbulence in open-channel flows. In: Balkema AA (ed) Rotterdam, The Netherlands
Nikora V, Roy AG (2012) Secondary flows in rivers: theoretical framework, recent advances, and current challenges. In: Church M, Biron PM, Roy AG (eds)
Pradhan A, Kumar Khatua K, Sankalp S (2018) Variation of velocity distribution in rough meandering channels. Adv Civ Eng 12(Article ID 1569271):1–12
Sarma KVN, Lakshminarayana P, Lakshmana Rao NS (1983) Velocity distribution in smooth rectangular open channel. J Hydraul Eng ASCE 109(2):270–289
Song T, Graf WH (1996) Velocity and turbulence distribution in unsteady open-channel flows. J Hydraul Eng 122(3):141–154
Tachie MF, Bergstrom DJ, Balachandar R (2003) Roughness effects in low-Re open-channel turbulent boundary layers. Exp Fluids 35:338–346
Termini D (2009) Experimental observations of flow and bed processes in a large-amplitude meandering flume. J Hydraul Eng 135(7):575–587
Termini D (2015) Momentum transport and bed shear stress distribution in a meandering bend: experimental analysis in a laboratory flume. Adv Water Resour 128:128–141
Termini D, Moramarco T (2017) Application of entropic approach to estimate the mean flow velocity and manning roughness coefficient in a high-curvature flume. Hydr Res 48(3)
Termini D, Piraino M (2011) Experimental analysis of cross-sectional flow motion in a large amplitude meandering bend. Earth Surf Proc Landf 36(2):244–256
Tracy HJ, Lester CM (1961) Resistance coefficients and velocity distributions, smooth rectangular channel. U.S. Geological Survey (USGS) Water supply paper 1592-A. USGS. Washington D.C.
Vanoni VA (1941) Velocity distribution in open channels. Civ Eng ASCE 11(6):356–357
Whiting PJ, Dietrich WE (1993) Experimental studies of bed topography and flow patterns in large-amplitude meanders, 2. Mech Water Resour Res 29(11):3615–3622
Yalin MS (1992) River mechanics. Pregamon Press, London
Yan J, Tang H, Xiao Y, Li K, Tian Z (2011) Experimental study on influence of boundary on location of maximum velocity in open channel flows. Water Sci Eng 4(2):185–191
Yang SQ, Tan SK, Lim SY (2004) Velocity distribution and dip phenomenon in smooth uniform open channel flow. J Hydraul Eng ASCE 130(12):1179–1186

Numerical Modeling of Extreme Flooding for Flood Risk Assessment in the Tra Bong River Basin, Vietnam

Xuan Manh Trinh and Frank Molkenthin

Abstract Flooding and inundation are of the most concerns in Vietnam. Many extreme floods have been occurring in Vietnam frequently, especially during the period of 1999–2009. Considering the impacts of flooding plays an important role in mitigating and adapting to flood risks. In this study, a rainfall-runoff model, two hydrodynamic models, and a coupling model were successfully applied as effective tools to simulate flooding and inundation in the Tra Bong River Basin, which is located in central Vietnam and considered as an ungauged basin. These models simulated the high flows accurately by depicting all of the peak flow discharges in accordance with observed high flows. The low flows were also simulated fairly well. In general, there is a good fit between the observed and simulated hydrographs for both calibration and validation. According to the model performance, 80% of the floodplain area would be flooded during the future extreme flood event. Eventually, the model results indicate that the integration of these models is significantly suitable for the flood risk assessment in the basin.

Keywords Extreme flooding · Hydrodynamic models · Model calibration and validation · Tra Bong River Basin

1 Introduction

Van Waveren and the others (1999) have mentioned in their book that '*the development of any scientific field always needs to follow some processes such as monitoring, recording and measuring data, analysis, and explanation of natural phenomena, and*

X. M. Trinh (✉)
Hanoi University of Natural Resources and Environment, 41A Phu Dien, Hanoi 11917, Vietnam
e-mail: txmanh@hunre.edu.vn

X. M. Trinh · F. Molkenthin
Brandenburg University of Technology Cottbus-Senftenberg (BTU), Postfach 101344, 03013 Cottbus, Germany
e-mail: Frank.Molkenthin@b-tu.de

M. B. Kalinowska et al. (eds.), *Recent Trends in Environmental Hydraulics*, GeoPlanet: Earth and Planetary Sciences,
https://doi.org/10.1007/978-3-030-37105-0_26

eventually simulation and proposing solutions' Van Waveren et al. (1999). Especially, in the field of hydrology, simulations of hydrologic cycle play an essential role. There are two types of hydrological models including physical and numerical models. Physical models can be described in laboratories by aiming to have a better understanding of the processes of hydrology (Bahremend 2007). Besides that, plenty of numerical models have been increasingly developed due to their advantages such as costs, flexibility and applied technology. Many problems during flooding need to be considered regarding the prediction of flows on complicated topography (Beffa and Connell 2001).

Hydraulic and hydrological models are useful tools in order to address the real hydrological cycle in a simplified way (Long et al. 2007). Hydrodynamic models make use of several algorithms to provide a quantitative relation between the input (e.g. rainfall, meteorological data) and output data (e.g. runoff). The numerical models have been developed from the 19th century with the simplest rainfall-runoff model by Mulvaney (1851) to more complicated models such as MIKE Package developed by the Danish Hydraulic Institute (DHI); Soil and Water Assessment Tools (SWAT), HEC models developed by Hydraulic Engineering Center—USA; Flood Modeller/ISIS 1D/2D by Jacobs/Halcrow—UK, TELEMAC by EDF France and Delft 2D/3D by WL/DELFT HYDRAULICS. These models are used for understanding the hydrological processes as well as making hydrological predictions in cases where water resources management and utilization activities are implemented (Tuan 2012).

The Tra Bong River Basin is located in Central Vietnam. Annually, there are many severe floods occurring in the Tra Bong River Basin due to the effects of typhoons, which are originating from the East Sea. Particularly, in 2003, 2009, and 2013 several extreme floods happened in this river basin. Consequently, many people died and hundreds of local families had to move their households to safe places, while some even avoided the floods by climbing on the roofs of their houses in the hope of finding help from above. Therefore, considering the impacts of extreme flooding in this basin plays an essential and important role in mitigating and adapting to flood risks.

2 Overview of the Study Area

The river basin has a medium size and located in the Binh Son District, in the North of Quang Ngai Province, Vietnam. The eastern part borders the East Sea with 54 km of length of the seashore. This basin spreads widely between latitudes of 15°11–15°25 N and longitudes of 108°34–108°56 E (Fig. 1). The total river system basin area is over 700 km^2, of which more than 80% is mountainous or hilly. The river has a length of 59 km and its network density is about 0.43 km/km^2, the average altitude is 196 m and the average river slope is 10.9% (Huong 2010).

The overall topography changes rapidly from the upper part in the west to the lower part in the east of the study area. Furthermore, there is no region of transition between

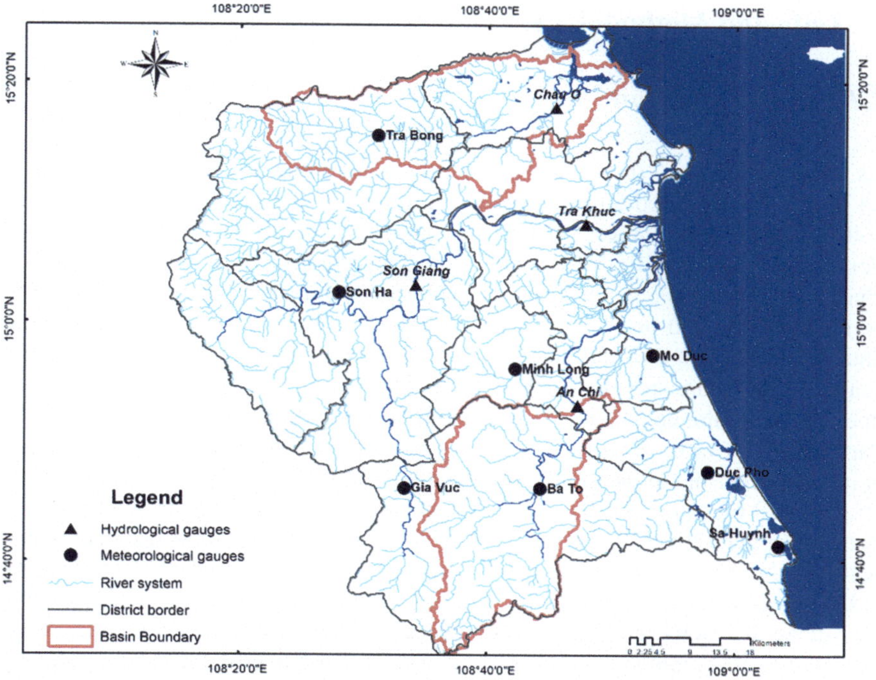

Fig. 1 Map of the study area in Quang Ngai province

the up-and downstream areas. Due to the characteristics of the basin topography, the river flow here is causing disadvantages. The flooding and inundation are annually occurring in the wet season. By contrast, the droughts are the most common natural disasters in the dry season. The monsoon tropical climate influences the entire basin, and the basin is affected significantly by the topography of the Truong Son mountain range as well as the climatic phenomena in the East Sea. The amount of rainfall in this basin is extremely high and similar to other coastal provinces in Central Vietnam. Every year, total rainfall often oscillates between 2200 and 2500 mm in the delta areas. In the highland regions, the value varies from 3000 to 3500 mm and is less than 2000 in the southern areas. The maximum rainfall happens usually in October and November and accounts for 45–55% of the total rainfall throughout the year.

3 Data and Methodology

3.1 Data Collection

As mentioned above, the basin is considered as an ungauged basin. However, a lot of data was required for this work. Accordingly, hydro-meteorological data was collected from many sources, such as the National Centre for Hydro-Meteorological Forecast of Vietnam (NCHMF), the Institute of Hydrology, Environment and Climate change (IHECC)—the Water Resources University and other experts. The data used comprises daily and one hourly discharge, rainfall, water levels, and temperature data. Geospatial data of the area and 50 cross-sections of the main river bed are available. Those are provided by the IHECC. However, cross-sections are distributed unevenly from the upstream to downstream area and do not cover the entire river. Particularly, the cross-sections are scarcely distributed at the conjunction of the river, at its tributaries and in the middle of the river.

A Digital Elevation Model with high resolution (10×10 m) and a land-use map in raster format (30×30 m resolution) are also available. The land use map with a scale of 1:50,000, which was created in 2010, was obtained from the Urban Planning Office of Quang Ngai Province. The original version of this map consisted of many classifications. However, based on the objective of this study, they were minimized by grouping them into seven main classes. The land use in this area is dominated by rural areas and developed-medium intensity areas. Furthermore, this land use map was used for assigning a set of Manning number for the two-dimensional flow model.

3.2 Methodology

The MIKE models, which have been developed by the DHI water and environment, were applied in order to investigate the flood behavior as well as to generate flooding maps in the basin. The MIKE NAM model was applied to simulate runoff from rainfall as the input for the hydrodynamic model. Then, the MIKE 11 HD was used to rout the discharge hydrograph from the upstream to the downstream of the river system. The MIKE 21 FM model was applied aiming to simulate flow on the floodplain. While the overland flow or lateral flow from the river onto floodplain was simulated by using the MIKE Flood model—a coupling model between the one-dimensional and two-dimensional models (Fig. 2).

The basin has only one meteorological station, namely Tra Bong, which is located in the upstream part of the basin (Fig. 1). One hydrological station is Chau O, which is located in the downstream part, and measures only the water level (Fig. 1). The discharge data, which is essential for model calibration and validation, however, is not available. Thus, this can be considered as an ungauged basin. Therefore, the regionalization method was significantly used in order to obtain model parameters from a similar catchment. This methodology is basically based on the assumption

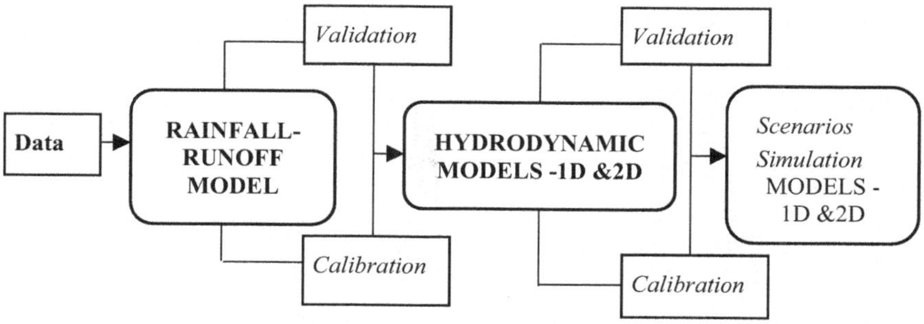

Fig. 2 Flow chart of the model's application in the Tra Bong River Basin

that if two basins have similar or identical hydrological characteristics, then their hydrological responses will be similar or identical as well.

In order to evaluate the model performance using observed data at specific control stations, a series of statistical analyses were utilized. The most important index that has been applied in many studies is thereby the Nash-Sutcliffe coefficient of efficiency (NSCE) (Nash and Sutcliffe 1970). The observed and simulated discharge values have been compared to assess the predictive accuracy of the model. NSCE numbers range from $-\infty$ to 1. When the values are from 0.9 to 1, it indicates that the model performance is perfect. When the values are from 0.8 to 0.9 and 0.6 to 0.8, the model performance is appropriate. The error coefficient is expressed with the following formula

$$\mathrm{NSCE} = 1 - \frac{\sum (Q_{cali} - Q_{obsi})^2}{\sum (Q_{obsi} - \bar{Q}_{obs})^2} \quad (1)$$

where Q_{cali}—simulated runoff at time i; Q_{bsi}—observed runoff at time i; Q_{obs}—average observed runoff at time i.

In addition to the model performance evaluation, a visual comparison and an evaluation of the peak flow error, peak time rate, volume error, and correlation coefficient were taken into account as well. Those statistical measures are very important in assessing the goodness of fit between observed and simulated values.

4 Discussion and Conclusion

4.1 Rainfall-Runoff Modeling

Due to a lack of sufficient hourly discharge data in the basin, the application of a direct calibration of the rainfall-runoff model for the basin is impossible. Accordingly, the An Chi River Basin, which is located on the Tra Khuc—Ve River Basin and nearby the study area (Fig. 1), satisfies all the assumptions corresponding to the regionalization approach.

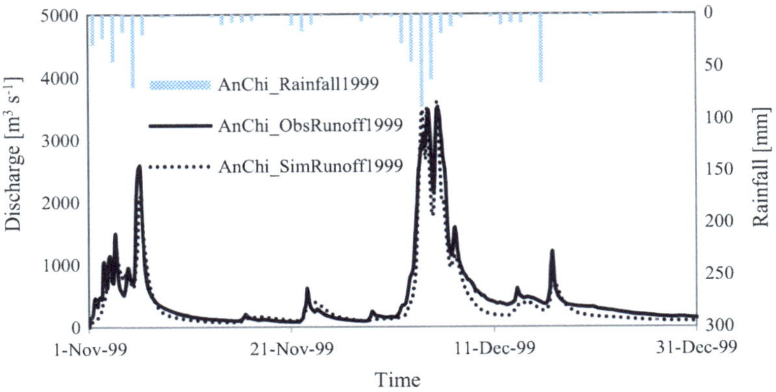

Fig. 3 The observed and simulated runoff at the An Chi station in case of the model calibration in the year of 1999

The donor basin occupies approximately 764 km^2. In order to identify optimal input values for the parameters of the rainfall-runoff model for the donor basin, three gauging stations located within and around the basin were utilized including An Chi, Ba To, and Gia Vuc stations (Fig. 1).

The hourly discharge time series and hourly rainfall time series in the year of 1999 and 2003 were used to calibrate and validate respectively due to synchronous data and typical extreme flood events in both years. The average rainfall was calculated by using the Thiessen polygon approach. The observed discharge at the An Chi station, which is the outlet of the donor basin, was applied to evaluate the model performance. The input parameters were manually obtained by using the trial and error method.

It is noticeable that there is a reasonable agreement between the model results and the observed values (Figs. 3 and 4). The shape and time of peak occurrence of the observed and simulated hydrographs are similar relatively. The high flows are rather similar to each other, while the low flows differ extensively at the beginning of the flood event because of the initial conditions. The highest values of the hydrographs are quite well simulated. The Nash-Sutcliffe coefficient of efficiency and Correlation coefficient are determined above 0.89 and 0.96 respectively. Hence, the evaluation results indicate that this model setup is reasonable and reliable in order to be used to simulate pluvial flooding in this river basin.

4.2 Hydraulic Modeling

Due to a lack of tidal data at the estuary of the river, the tidal water level, which was recorded at the Cua Dai Estuary of the Tra Khuc River, was used. Since the distance between the two estuaries is comparably short (about 30 km), it can be assumed that the tidal hydrographs are relatively homologous. The one-dimensional model was set

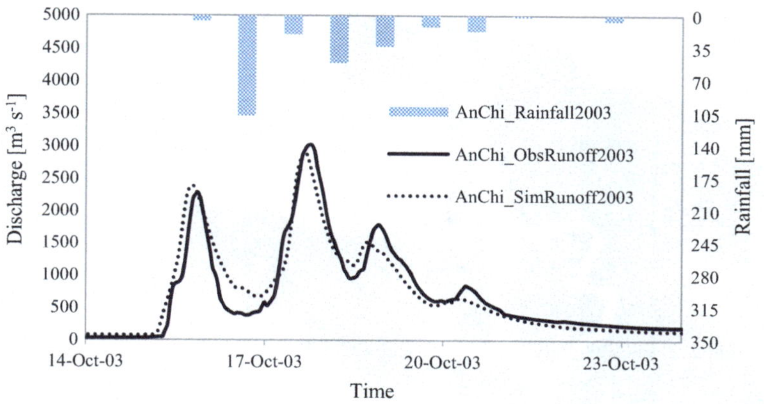

Fig. 4 The observed and simulated runoff at the An Chi station in the case of the model validation in the year of 2003

up with three reaches including one main river and two tributaries with the measured cross-sections, one upstream input point, and one downstream output point. The information for each cross-section consists of coordinates, elevation, downstream reach length, and the Manning roughness coefficient. The Manning values are varying from 0.03 to 0.12 s $m^{-1/3}$ for river banks and channel sections, respectively.

The two-dimensional model was set up with a calculated mesh deriving from the high resolution DEM. The mesh properties consist of 9224 nodes, 17,921 elements, as well as maximum area is 16,073 m^2 and the smallest angle is about 28 degrees (Fig. 5).

According to Kalyanapu (2009), the Manning's roughness coefficient is commonly used to represent surface roughness in distributed models and for large watersheds. Therefore, a manning value is assigned to each element in the mesh domain based on the suggested values for the overland surface from McCuen (1998).

The flood events from 15th October 2003 to 20th October 2003 and from 27th September 2009 to 2nd October 2009 were used for the model calibration and validation. The one-dimensional model was calibrated by changing the Manning coefficient in the main river and its tributaries, and the two-dimensional model was calibrated by changing the distributed Manning coefficient in the floodplain in order to reduce the error between the observed and simulated water levels. For the validation, a set of parameters obtained from the calibration step were applied to evaluate the accuracy of the model setup. The best results were obtained by comparing the simulated and observed hydrographs both graphically and statistically.

It is noticeable that the model performance is relatively satisfactory for both calibration and validation (Table 1). As for the calibration period, the Nash-Sutcliffe efficiency is about 0.63, the peak error is 0.013, the volume error is 10%, and the correlation coefficient is 81%. As for the verification period, the model simulated high accuracy of water levels, whereby the NSCE is 0.70 and the correlation coefficient

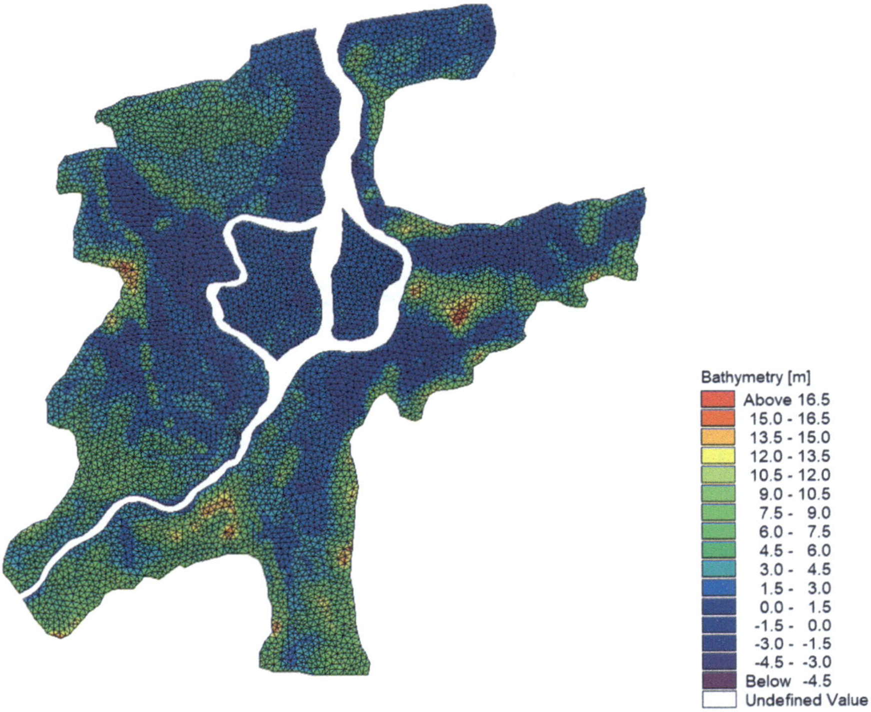

Fig. 5 The domain mesh of the floodplain

Table 1 Model performance of the MIKE flood for the calibration and validation periods

Gauging station	Period	Peak error	Correlation coefficient (%)	NSCE
Chau O	15/10/2003–20/10/2003	0.013	81	0.63
Chau O	27/9/2009–2/10/2009	0.012	92	0.70

is up to 92%. The results above indicate that this model is suitable and reliable for flooding simulation in this area. The design flood plays an important role in terms of flood defense and flood prediction. In order to build a scenario for simulating extreme flooding, a flood event with a 100 year return period as a simulation scenario was selected. For this reason, the return interval or return period places the magnitude of floods in terms of the frequency giving a probability of a particular flood event. The coupled model setup for the 100 year return period flood event is the same as the model for the validation process. The model result of flooding, in this case, is shown in Fig. 6.

During the design flood with a 100 year return period, the area of 54.6 km^2 would probably be affected by the flooding with a flood depth varying from 0.25 up to 6.2

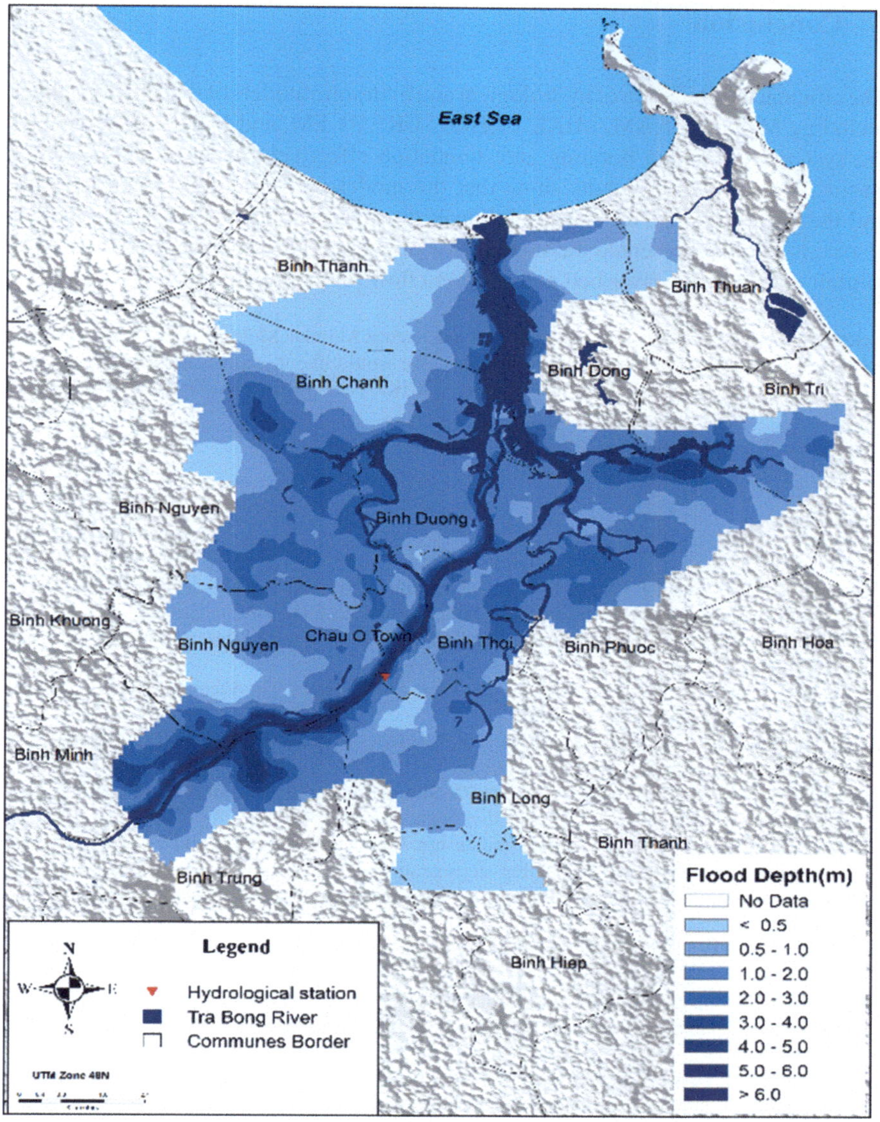

Fig. 6 Visualization of flooding in the Tra Bong River floodplain in case of the 1% design flood

meters. In this case, Binh Trung, Binh Chanh, Binh Duong, and Chau O Town, which are the districts in the downstream area of the river basin, would be the most affected areas in the floodplain.

5 Conclusion

The efficient combination of hydrological and hydraulic models of the MIKE package including MIKE 11 NAM, MIKE 11 HD, MIKE 21 FM, and MIKE Flood enables the system to simulate flooding and inundation effectively in the Tra Bong River Basin. The simulation results show that the model performances are fairly good, and the Nash-Sutcliffe efficiency and peak flows are captured well. Those models also completely represented the surface hydrological progress of the study area from rainfall to flow in the river bed and overland flow on the floodplain.

Acknowledgements We would like to thanks to Erasmus Mundus Mobility with Asian (EMMA) for their financial support to undertake this study. We would like to express the special thanks to DHI-Wasy GmbH in Germany for their support of MIKE Package license to make this possible.

References

Bahremand A, De Smedt F (2007) Distributed hydrological modeling and sensitivity analysis in Torysa Watershed. Slovakia Water Resour Manag 2008(22):393–408
Beffa C, Connell RJ (2001) Two-dimensional floodplain flow. I: model description. J Hydrol Eng 6(5):397–405
Huong NTT (2010) Flood protection in the Tra Khuc and Ve rivers. The Vietnam Institute of Water Resources Press, Quang Ngai province
Kalyanapu AJ, Burian SJ, McPherson TN (2009) Effect of land use-based surface roughness on hydrologic model output. J Spat Hydrol 9(2):51–71
Long NL, Madsen Henrik, Rosbjerg Dan (2007) Simulation and optimization modeling approach for operation of the Hoa Binh reservoir, Viet Nam. J Hydrol 336:269–281
McCuen RH (1998) Hydrologic analysis and design, Prentice-Hall, Englewood Cliffs, New Jersey
Mulvaney TJ (1851) On the use of self - registering rain and flood gauges in making observations of the relations of rainfall and flood discharges in a given catchment. Trans Inst Civl Engrs Ireland IV (II), 19–33
Nash IE, Sutcliffe IV (1970) River flow forecasting through conceptual models I. J Hydrol 10:282–290
Tuan LQ (2012) Impacts on flow regime caused by dam construction for hydropower generation in the upper Se San river basin, using SWAT and WAFLEX models. Msc thesis, UNESCO-IHE, Netherlands, p 19
Van Waveren RH, Groot S, Scholten H, van Geer FC, Wösten JHM, Koeze RD, Noort JJ (1999) Good modeling practice handbook. Dutch Dept. of Public Works, Institute for Inland Water Management and Waste Water Treatment

Application of the STIR Model to a Small River at Different River Flow Rates

Steve G. Wallis and Eleonora Dallan

Abstract The parameters of the STIR model, used with only one storage zone, were estimated by optimizing the model to a set of tracer data collected in the Murray Burn in Edinburgh. The results for cross-sectional area and dispersion coefficient were consistent with those from an independent analysis of the same data. The dispersion and transient storage parameter results suggest that the upper part of study site has different longitudinal mixing characteristics to other parts of the study site, which reflects differences in the geomorphology along the channel.

Keywords Solute transport · Rivers · Dispersion · Transient storage · STIR model

1 Introduction

The need to protect the natural environment from the effects of pollution incidents has driven the development of many mathematical models based on the principles of conservative or non-conservative solute transport in open channels (Fischer et al. 1979; Rutherford 1994). Over time, as observations revealed the weaknesses of early modelling efforts, such as the advection-dispersion equation (ADE), more processes were included in the models. Thus, the major transport mechanisms of advection and dispersion were supplemented with the effects of transient storage, creating the transient storage model (TSM), in the majority of published work since the turn of the century. It is interesting to reflect, however, that the realisation that the classical ADE model was sometimes inadequate (Nordin and Sabol 1974; Day 1975), the likely cause being temporary trapping of solute in the so-called dead zones (Valentine

S. G. Wallis (✉)
School of Energy, Geoscience, Infrastructure and Society, Heriot-Watt University, Riccarton, Edinburgh EH14 4AS, UK
e-mail: s.g.wallis@hw.ac.uk

E. Dallan
Department of Industrial Engineering, University of Padova, Via F. Marzolo 9, 35131 Padua, Italy
e-mail: eleonora.dallan@phd.unipd.it

M. B. Kalinowska et al. (eds.), *Recent Trends in Environmental Hydraulics*, GeoPlanet: Earth and Planetary Sciences,
https://doi.org/10.1007/978-3-030-37105-0_27

and Wood 1979; Beer and Young 1983) and the proposal to portray transient storage using a simple exchange mechanism between a river's main channel and its peripheral dead zones (Thackston and Krenkel 1967; Bencala and Walters 1983), all have long histories.

The promise of the TSM was augmented by the realisation that it was capable of catering both for the effects of dead zones and for the effects of the hyporheic zone (Elliott and Brooks 1997). Furthermore, by developing a variation on the standard model, which included more than one solute exchange time-scale (Choi et al. 2000; Briggs et al. 2009), the model became potentially more widely applicable to real channels. A further innovation characterised the solute exchange between the main channel and the storage zones using a general residence time distribution approach for single (Gooseff et al. 2003) or multiple storage zones (Marion et al. 2008; Bottacin-Busolin et al. 2011).

In this paper we report on the first stage of the application of one of these later models, namely the STIR (Solute Transport in Rivers) model, to a set of tracer experiments undertaken over a range of river flow rates in a small river. Whilst the larger potential of the STIR model lies in its multiple storage zone capability, here we focus on its single storage zone form. The aims of the work are to examine the variation of the model's parameters with river flow rate and to compare the results with previously published work.

2 Modelling Background

The STIR model describes conservative solute transport along a river as the combination of advection and dispersion in the main channel and retention of solute in storage zones. Considering N storage zones, solute mass balance in the main channel is expressed as:

$$\frac{\partial C(x,t)}{\partial t} + U\frac{\partial C(x,t)}{\partial x} = D\frac{\partial^2 C(x,t)}{\partial x^2} - \sum_{i=1}^{N}\left\{\alpha_i C(x,t) - \int_0^t \alpha_i C(x,\tau)\varphi_i(t-\tau)d\tau\right\} \quad (1)$$

where C is the solute concentration in the main channel, U is the velocity in the main channel, D is the dispersion coefficient in the main channel, α_i is the solute mass exchange rate for the ith storage zone, φ_i is the residence time function for the ith storage zone, τ is a dummy time variable, x is distance along the river, t is time and N is the number of storage zones (Marion et al. 2008).

In this study we consider the special case for $N = 1$ and assume an exponential residence time distribution for the storage zone. Under these circumstances the model is equivalent to the TSM (Marion et al. 2008). However, the way in which the STIR model equation is solved is rather different to the usual way in which the TSM model equations are solved. In the latter, a numerical solution approach is used to approximate spatial and temporal gradients (Wallis et al. 2013) whereas STIR employs a purely time-domain convolution method (Marion et al. 2008).

3 Tracer Data

The data used in this work came from tracer experiments undertaken in the Murray Burn. This small river runs through the Riccarton Campus of Heriot-Watt University near Edinburgh, and has previously been the focus of a travel time study (Burke 2002; Wallis 2005) and a longitudinal mixing study (Heron 2015; Wallis and Heron 2018). Herein, we use tracer data from the latter work, which used one of the reaches studied in the former work. This reach extends between its upstream boundary, designated as Site 3, and its downstream boundary, designated as Site 4. Eleven gulp-release tracer experiments were undertaken between November 2009 and November 2010 covering a river flow rate range of about 15–400 L/s. The tracer (Rhodamine WT) was injected 236 m upstream of Site 3 and water samples were collected in glass bottles at Site 3 and Site 4. The sampling interval varied between 2 min at low river flow rates to 20 s at high river flow rates. The samples were analysed under temperature controlled laboratory conditions using a calibrated Turner Designs Model 10 instrument (Heron 2015). The first two experiments were trials, which were used to fine tune the experimental work in the field. They did not yield complete and well-resolved temporal concentration profiles. Therefore, nine sets of tracer data were available for use.

In order to prepare the data for modelling purposes, three processing steps were undertaken for each experiment. Firstly, the background signal was removed; secondly, the calibration factor of the fluorometer was used to convert the data into concentrations; thirdly, the concentration data was scaled so that the area under the upstream and downstream profiles was the same. The final step is important because it removes any real or apparent non-conservative behavior of the tracer. The latter is often found in tracer data due to the presence of distributed or point inflows from land drainage or tributaries, respectively.

4 Application of the STIR Model

The STIR model, comprising a single storage zone and an exponential residence time distribution, was optimized to the processed tracer data in three different applications. Firstly, the model was optimized to the processed tracer data at Site 3 using the gulp-release at the tracer injection site as the upstream boundary condition; secondly, the model was optimized to the processed tracer data at Site 4 using the gulp-release at the tracer injection site as the upstream boundary condition; thirdly, the model was

Table 1 Physical characteristics of the reaches modelled

Reach	Length (m)	Mean width (m)	Mean slope [–]	Description
I3	236	3.5	0.021	Upper 100 m: natural channel, meandering, boulders Lower 136 m: modified channel, straight, cobbles
I4	420	3.0	0.016	Upper 100 m: natural channel, meandering, boulders Lower 320 m: modified channel, straight, cobbles
34	184	2.4	0.009	Throughout: modified channel, straight, cobbles

optimized to the processed tracer data at Site 4 using the processed tracer data at Site 3 as the upstream boundary condition. The reaches corresponding to these applications are denoted as I3, I4 and 34, respectively. Their characteristics are summarized in Table 1.

In all cases, four reach-average parameters were estimated: the cross-sectional area of the main channel, the dispersion coefficient in the main channel, the exchange rate between the main channel and the storage zones and the mean residence time in the storage zones. The ratio of storage area to main channel area was calculated from the latter two parameters (Marion et al. 2008).

In order to reduce the possibility that a too coarse temporal resolution of the tracer data might affect the results (Wallis and Manson 2018), the concentration data were interpolated to a time step of 2.5 s. Furthermore, in order to prevent longer time-scale transient storage from contaminating the results, the concentration profiles were truncated at a time equal to $4tp$, where tp is the time delay between the first rise of a profile above background and the profile peak. This procedure was aimed at ensuring that all the concentration profiles had similar lengths to their tails: it also cut off some rather poorly defined ends to some of the tails caused by spikes in the data. Optimization was achieved using a mixed approach in which a linear fitting was applied around the peak of the observed downstream profile and a logarithmic fitting was applied on the tail of the observed downstream profile. This method tends to produce better all-round fits to the whole profile than can be achieved using either a completely linear fitting or a completely logarithmic fitting. Examples of the fits for one experiment are shown in Fig. 1 for reaches I3 (upper panels), I4 (middle panels) and 34 (lower panels) using both linear (left-hand panels) and logarithmic (right-hand panels) concentration scales.

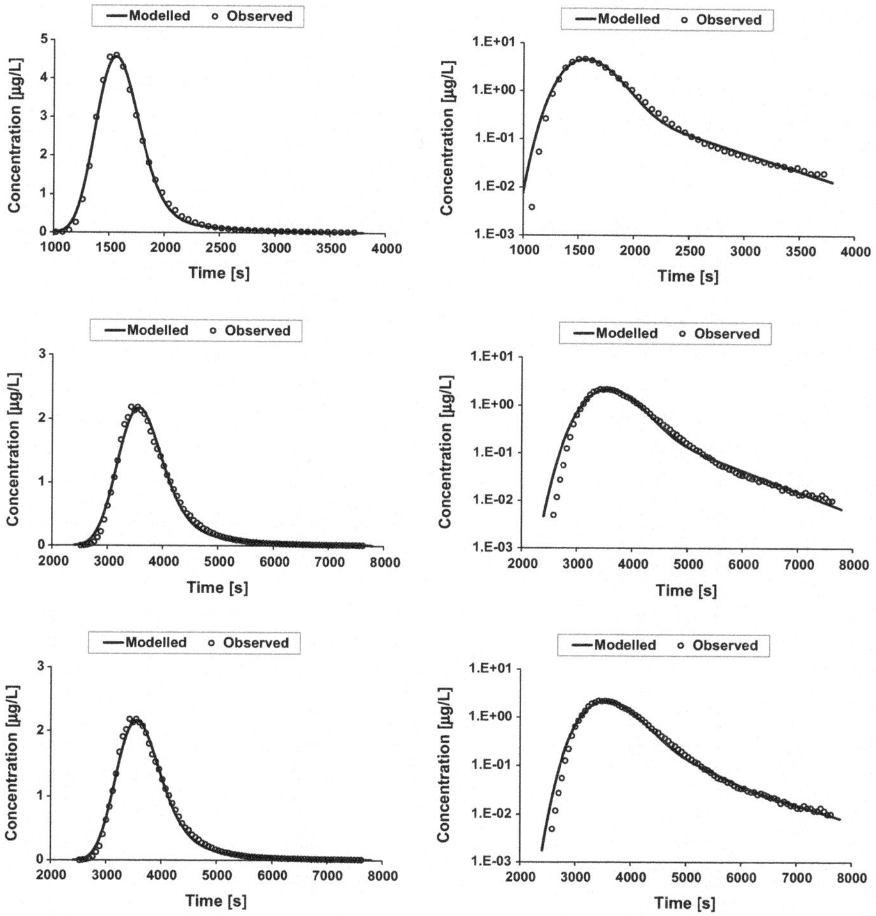

Fig. 1 Model fits for experiment 7 shown using linear (left-hand panels) and logarithmic (right-hand panels) concentration scales: upper panels, reach I3; middle panels, reach I4; lower panels, reach 34

5 Results and Discussion

Results were obtained for all three application cases for the nine experiments. Figure 2 shows optimized main channel cross-sectional areas and main channel dispersion coefficients plotted against river flow rate for the three application cases. In general, both parameters increase with river flow rate, as would be expected from the principles of open channel flow and from theoretical dispersion studies (Rutherford 1994). Furthermore, in four out of five previously published studies where the TSM was optimized to field-scale tracer data observed in a single reach at several river flow rates, a positive correlation between main channel dispersion coefficient and

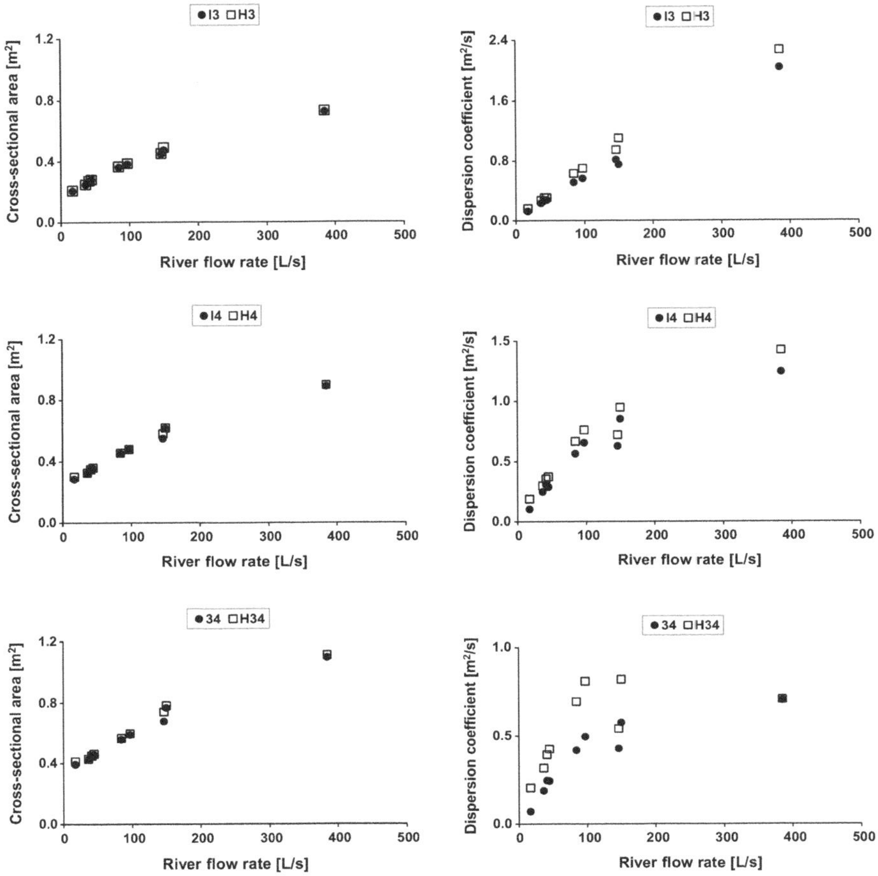

Fig. 2 Variation of optimized main channel cross-sectional area and main channel dispersion coefficient with river flow rate for three application cases: STIR results indicated by I3, I4 & 34; Heron (2015)'s results indicated by H3, H4 & H34, respectively

river flow rate was found (Hart et al. 1999; Gooseff et al. 2003; Jin and Ward 2005; Camacho and Gonzalez 2008). In the fifth study (Manson and Wallis 2018), a weak decreasing trend was found.

Support for the main channel cross-sectional areas is provided by the results of Heron (2015). His results were obtained by optimizing the analytical solution to the ADE for a gulp-release of tracer, thus obtaining direct results for cross-sectional mean velocities for cases I3 and I4. Cross-sectional areas were obtained by dividing the river flow rate by the cross-sectional mean velocities. His results are plotted on the relevant parts of Fig. 2 (left-hand side, upper two panels). In general, they are slightly larger than those from the present study (typically by 1% for reach I3 and by 2% for reach I4), because they reflect the total cross-sectional area, i.e. main channel plus storage zones. Furthermore, Heron (2015) estimated the cross-sectional mean velocity for the case 34 from a travel time based weighted average of the cross-sectional mean velocities from cases I3 and I4. The corresponding cross-sectional areas are plotted on the lower panel of Fig. 2 and are also slightly larger than the results from STIR (typically by 3%). Heron (2015) also found that his weighted average cross-sectional area results for reach 34 were consistent with estimates obtained from other methods applied directly to reach 34, such as Fischer (1968)'s routing procedure (the former being typically 2% greater than the latter).

As would be expected, the cross-sectional areas are inversely correlated with the channel slopes given in Table 1. Taking into account the different channel widths, this implies that the flow in reach 34 is generally slower and deeper than the flow in reach I3 at the same river flow rate. This is consistent with visual observations of the channel hydraulics.

Heron (2015)'s optimization of the analytical solution to the ADE for a gulp-release of tracer also provides direct results for dispersion coefficients for cases I3 and I4. These are shown in Fig. 2 (right-hand side, upper two panels), and are generally larger than the results from STIR (typically by 22% for reach I3 and by 25% for reach I4). This is expected because the results from STIR only reflect the main channel dispersion, whilst Heron's results reflect the total dispersion occurring in the reach (main channel dispersion plus the effects of transient storage). Heron (2015) also estimated the dispersion coefficient for reach 34 using the same travel time based weighted average approach referred to above. These are plotted on the lower panel of Fig. 2 and, as before, are generally larger (typically by 66%) than the results from STIR. In contrast to the close agreement between cross-sectional area estimates, however, Heron (2015)'s weighted average dispersion coefficients were about 32% greater than those obtained from Fischer's routing procedure, so the former may be overestimates. It is worth noting that the magnitudes of the dispersion coefficients in all three reaches are consistent with those found in other small rivers (Heron 2015), being of the order of 1 m^2/s.

Regarding the STIR model results, the dispersion coefficients in reach 34 are smaller than in reach I3. This is a reflection of the different physical nature of the reaches, particularly the upper part of reach I3 (see Table 1) which is a natural, non-uniform channel containing relatively large roughness elements, whilst the remaining part of reach I3 and reach 34 have been modified by landscape management to create a much more uniform and straighter channel and which contain smaller roughness elements. The generally wider and shallower flow regime in reach I3 compared to reach 34, mentioned above, also has an influence. The dispersion coefficients in reach I4 lie between those in the other two reaches, reflecting that reach I4 consists of reach I3 in series with reach 34.

Figure 3 shows optimized exchange rates, optimized mean residence times, calculated area ratio and a normalized measure of fit between the observed and modelled downstream concentration profiles. The results are plotted against river flow rate for the three application cases. The exchange rates for I4 and 34 are similar to each other, whilst those for I3 are generally larger: all three exchange rates are approximately constant over the range of river flow rate covered. This invariance contrasts with an

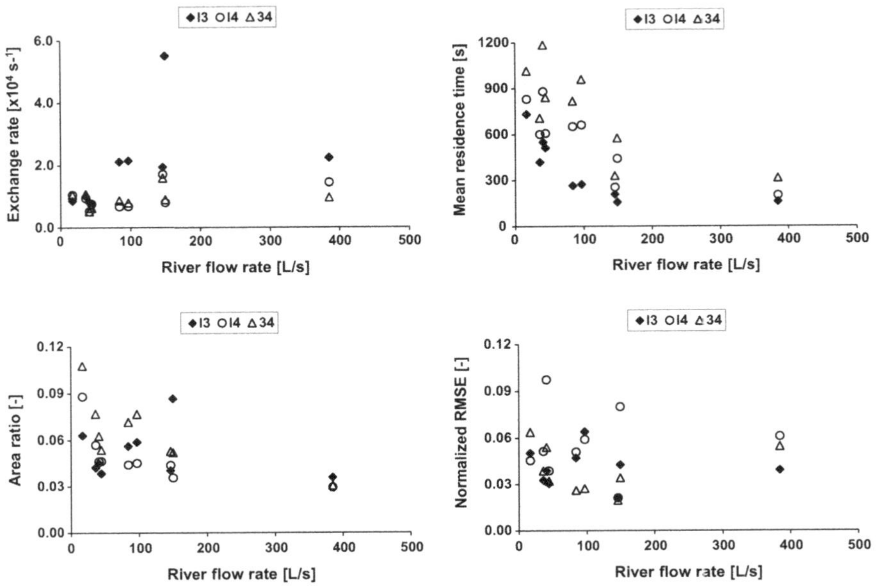

Fig. 3 Variation of exchange rate, residence time, area ratio and normalized model fit with river flow rate for three application cases I3, I4 & 34

increasing trend reported by Hart et al. (1999), Gooseff et al. (2003), Camacho and Gonzalez (2008) and Manson and Wallis (2018): there is no clear pattern at all in the results of Jin and Ward (2005).

The mean residence time for reach 34 is always larger than the mean residence time for reach I4, whilst reach I3 has the smallest residence time in all the experiments. The residence times for all three cases reduce with increasing river flow rate. The differing transient storage behavior of the reaches implied by the exchange rate and residence time results is probably a reflection of the different geomorphology of the reaches referred to above, but in the absence of more detailed information on the flow structures in the reaches it is difficult to propose a mechanism to justify this. There are no consistent differences between the area ratios for reaches I3 and I4, but the area ratio in reach 34 is the largest in the majority of the experiments. Interestingly, all three of the area ratios reduce as river flow rate increases, which is consistent with Gooseff et al. (2003) and Camacho and Gonzalez (2008). In contrast, Hart et al. (1999), Jin and Ward (2005) and Manson and Wallis (2018) found no trend. This different behavior is probably caused by the different geomorphologies of the rivers used in these studies. For example, rivers with storage zones being predominantly located in the bed may see a reduction in area ratio with increasing river flow rate because no additional storage zones become available as water levels increase. However, rivers with storage zones being predominantly located in the banks may see little change in area ratio with increasing river flow rate because new storage zones become available as water levels rise. The magnitudes of the area ratio found in the current study are towards the lower end of the ranges reported in the other studies referred to above. The model fit results reveal little substantial information. For example, model fits for reach I4 are poorer than for both reaches I3 and 34 in six of the nine experiments, whilst fits for reach 34 are poorer than for reach I3 in five of the nine experiments. There is no discernible trend between model fit and river flow rate in any of the reaches.

6 Conclusions

The parameters of the STIR model, comprising a single storage zone and an exponential residence time distribution, were found by optimizing the model to a set of tracer data for three reaches of the Murray Burn in Edinburgh. Both the cross-sectional area of the main channel and the dispersion coefficient in the main channel were found to increase with increasing river flow rate as would be expected from previous work. These results were generally consistent with an independent analysis of the tracer data by Heron (2015). In particular the STIR model dispersion coefficients were smaller than Heron's, which reflects that Heron's results quantify both the main channel dispersion and the effect of transient storage. Of the remaining STIR model parameters, the residence time and the area ratio tended to reduce with increasing river flow rate whilst the exchange rate was approximately constant. Overall, the dispersion and transient storage parameter results suggest that the upper part of reach I3 has different longitudinal mixing characteristics to other parts of the study site, which reflects differences in the geomorphology along the channel.

References

Beer T, Young PC (1983) Longitudinal dispersion in natural streams. J Env Eng Am Soc Civ Eng 109:1049–1067

Bencala KE, Walters RA (1983) Simulation of solute transport in a mountain pool-and-rifle stream: a transient storage model. Water Resour Res 19:718–724

Bottacin-Busolin A, Marion A, Musner T, Tregnaghi M, Zaramella M (2011) Evidence of distinct contaminant transport patterns in rivers using tracer tests and a multiple domain retention model. Adv Water Resour 34:737–746

Briggs MA, Gooseff MN, Arp CD, Baker MA (2009) A method for estimating surface transient storage parameters for streams with concurrent hyporheic storage. Water Resour Res 45:W00D27. https://doi.org/10.1029/2008wr006959

Burke NA (2002) Travel time and flow characteristics of a small stream system. PhD Thesis, Heriot-Watt University, United Kingdom

Camacho LA, Gonzalez RA (2008) Calibration and predictive ability of longitudinal solute transport models in mountain streams. Environ Fluid Mech 8:597–604

Choi J, Harvey JW, Conklin MH (2000) Characterizing multiple timescales of stream and storage zone interaction that affect solute fate and transport in streams. Water Resour Res 36:1511–1518

Day TJ (1975) Longitudinal dispersion in natural channels. Water Resour Res 11:909–918

Elliott AH, Brooks NH (1997) Transfer of nonsorbing solutes to a streambed with bedforms: theory. Water Resour Res 33:123–136
Fischer HB (1968) Dispersion predictions in natural streams. J San Eng Div Am Soc Civ Eng 94:927–944
Fischer HB, List EJ, Koh RCY, Imberger J, Brooks NH (1979) Mixing in inland and coastal waters. Academic Press
Gooseff MN, McGlynn BL, McGlynn RS (2003) Transient storage processes and stream discharge recession in a headwater stream, Maimai, New Zealand. In: Proceedings of North American benthological society annual meeting, May 2003 (Poster)
Hart DR, Mulholland DJ, Marzolf ER, DeAngelis DL, Hendricks SP (1999) Relationships between hydraulic parameters in a small stream under varying flow and seasonal conditions. Hydrol Process 13:1497–1510
Heron AJ (2015) Pollutant transport in rivers: estimating dispersion coefficients. MPhil Thesis, Heriot-Watt University, United Kingdom
Jin H-S, Ward GM (2005) Hydraulic characteristics of a small coastal plain stream of the southeastern United States: effects of hydrology and season. Hydrol Process 19:4147–4160
Marion A, Zaramella M, Bottacin-Busolin A (2008) Solute transport in rivers with multiple storage zones: the STIR model. Water Resour Res 44:W10406. https://doi.org/10.1029/2008WR007037
Nordin CF, Sabol GV (1974) Empirical data on longitudinal dispersion in rivers. USGS, Water Resources Investigations, pp 74–20
Rutherford JC (1994) River mixing. Wiley
Thackston EL, Krenkel PA (1967) Longitudinal mixing in natural streams. J Sanit Eng Div Proc Am Soc Civ Eng 93:67–90
Valentine EM, Wood IR (1979) Experiments in longitudinal dispersion with dead zones. J Hyd Div Am Soc Civ Eng 105:999–1016
Wallis SG (2005) Experimental study of travel times in a small stream. In: Czernuszenko W, Rowinski PM (eds) Water quality hazards and dispersion of pollutants. Springer, pp 109–120
Wallis SG, Osuch M, Manson JR, Romanowicz R, Demars BOL (2013) On the estimation of solute transport parameters for rivers. In: Rowinski P (ed) Experimental and computational solutions of hydraulic problems. Springer, pp 415–425
Wallis SG, Heron AJ (2018) Experimental determination of longitudinal dispersion in a small stream. In: Proceedings of the 5th IAHR European division conference, Trento, Italy (rpsonline.com.sg/rps2prod/iahr2018)
Wallis SG, Manson JR (2018) Parameter estimation of the transient storage model: sensitivity to discretization. In: Proceedings of the 5th IAHR Eur European division conference, Trento, Italy (rpsonline.com.sg/rps2prod/iahr2018)

The manufacturer's authorised representative in the EU is Springer Nature Customer Service Centre GmbH, Europaplatz 3, 69115 Heidelberg, Germany. If you have any concerns regarding our products, please contact ProductSafety@springernature.com

Printed and bound by CPI Group (UK) Ltd, Croydon, CR0 4YY
30/09/2026
02225443-0017